BEHAVIORAL TREATMENT OF DISEASE

NATO CONFERENCE SERIES

I Ecology
II Systems Science
III Human Factors
IV Marine Sciences
V Air-Sea Interactions
VI Materials Science

III HUMAN FACTORS

BEHAVIORAL TREATMENT OF DISEASE

Edited by

Richard S. Surwit and Redford B. Williams, Jr.
Duke University Medical Center
Durham, North Carolina

Andrew Steptoe
St. George's Hospital Medical School
University of London
London, England

and

Robert Biersner
Naval Medical Research and Development Command
Bethesda, Maryland

Published in cooperation with NATO Scientific Affairs Division

PLENUM PRESS · NEW YORK AND LONDON

Library of Congress Cataloging in Publication Data

NATO Symposium on Behavioral Medicine (1981: Karas Cape, Greece)
    Behavioral treatment of Disease.

    (NATO conference series. III, Human factors; v. 19)
    "Proceedings of a NATO Symposium on Behavioral Medicine, held June 30–July 3,
1981, at Porto Carras, Chalkidiki, Greece" — T.p. verso.
    "Published in cooperation with NATO Scientific Affairs Division."
    Includes bibliographical references and index.
    1. Medicine and psychology — Congresses. 2. Behavior therapy — Congresses. 3.
Medicine, Psychosomatic — Congresses. I. Surwit, Richard S. II. North Atlantic Treaty
Organization. Division of Scientific Affairs. III. Title. IV. Series: [DNLM: 1. Behavior
therapy. 2. Behavior. W3 N138 v. 19 / WM 425 N279b 1981]
R726.5.N37 1981   616'.001'9   82-12363

ISBN-13: 978-1-4613-3550-4   e-ISBN-13: 978-1-4613-3548-1
DOI: 10.1007/978-1-4613-3548-1

Proceedings of a Nato Symposium on Behavioral Medicine, held
June 30–July 3, 1981, at Porto Carras, Chalkidiki, Greece

© 1982 Plenum Press, New York
Softcover reprint of the hardcover 1st edition 1982

A Division of Plenum Publishing Corporation
233 Spring Street, New York, N.Y. 10013

DEDICATION

This book is dedicated to the memory of Philip Handler, former James B. Duke Professor of Biochemistry and President of the National Academy of Sciences. We are indebted to him for his valuable assistance in the organization of this Symposium.

PREFACE

Behavioral Treatment of Disease: A NATO Symposium on Behavioral Medicine was held on June 30th through July 3rd, 1981 at Porto Carras, Neo Marmaras, Greece. It was a multi-disciplinary meeting which provided an opportunity for North American and European scientists from ten different NATO member countries to share the emerging principles and technology of behavioral treatment of disease. In addition, it served as a forum whereby continued high level research in the area was stimulated.

Financial support for the symposium was initially provided by the Scientific Affairs Division of the North Atlantic Treaty Organization as part of their continuing series of scientific symposia. Funds made available by a grant from the United States Office of Naval Reseach permitted widespread international participation in the symposium. We would like to thank each of these organizations for their support in making the symposium possible. In particular, we would like to thank Dr. B. A. Bayraktar of the Scientific Affairs Division of NATO and Dr. Donald Woodward of the U.S. Office of Naval Research. Though both of these men were unable to attend the meeting personally, they provided invaluable assistance in its planning.

The organization of this symposium was carried out in consultation with a special organizing committee which consisted of John Basmajian, Niels Birbaumer, John Boulougouris, Charles Mertens, and Dieter Vaitl. We would like to express our particular appreciation to Barbara Jackson and Cynthia Mongeon who prepared these precedings for publication and who were responsible for coordinating various aspects of the symposium from its inception.

Finally, we would like to thank Mr. John Carras for the warm hospitality he provided to all participants which helped make the meeting a success.

Richard S. Surwit, Ph.D.
Redford B. Williams, Jr.,
Andrew Steptoe, D. Phil
Robert Biersner, Ph.D., Commander, USN

Durham, North Carolina
September, 1981

CONTENTS

SECTION VI: CHRONIC PAIN

SECTION VII: MILITARY APPLICATIONS

SECTION VIII: THEORETICAL ISSUES IN BEHAVIORAL MEDICINE

BEHAVIORAL TREATMENT OF DISEASE:

INTRODUCTION

Richard S. Surwit, Ph.D.
Department of Psychiatry
Duke University Medical Center

This volume is the result of a NATO Symposium on Behavioral Medicine focusing on behavioral treatment of disease. It was part of the NATO Human Factors Conference and Symposia Program of 1981. Eighty-four scientists from ten NATO countries participated. This was the first major international meeting specifically devoted to behavioral treatment of disease. The meeting was intended to summarize the basic scientific and technical knowledge of the area. It more than achieved its goal.

Behavioral Medicine is broadly defined to encompass the application of principles from all behavioral sciences to medical problems. It is significantly different from its related discipline, psychosomatic medicine, which concerns itself with hypothetical intrapsychic phenomena such as personality characteristics and neurotic conflict. Behavioral Medicine is an empirical discipline, whose efforts are directed at discovering and manipulating the relationship between observable behaviors, physiological function and disease. As such, it has given birth to a new technology whereby principles of behavior control are applied to the treatment, prevention and rehabilitation of various physical disorders.

The objective of this symposium was twofold. First, it was intended to be a forum for high-level interchange of information among investigators seriously engaged in research in this area in laboratories throughout Europe and North America. The reader will note that chapters contained in this volume come from investigators from laboratories based in England, Germany, Belgium, the Netherlands as well as Canada and the United

States.   The   perspective   of   this   volume   is   thus   truly
international.

The second objective of this symposium was to provide an
overview of the current state of the art of behavioral medicine.
This volume helps fulfill this goal.   Participants in the
symposium were asked to modify their presentations so that each
of the following chapters is a review of the state of behavioral
science as it relates to the understanding and treatment of
specific disease areas.   Selected data-based papers from the
wide-ranging poster session have been published as a special
issue of the Journal of Psychosomatic Research, Volume 26, 1982.

The   organization   of   this   volume   follows   that   of   the
symposium itself.   The sections of the book are devoted to
specific   topic   areas   on   coronary   heart   disease,   ischemic
disorders, hypertension, neuromuscular rehabilitation, headache,
and chronic pain.   Each section begins with a consideration of
how the pathophysiology of the relevant problem lends itself to
behavioral intervention.   This review is followed by in-depth
reviews of specific sub-areas of the general problem usually
focusing on treatment.

The first topic area to be discussed is coronary heart
disease.   Redford Williams reviews the pathophysiology of
coronary disease and emphasizes the role of the central nervous
system in both the development and the expression of the
disorder.   Chandra Patel, drawing largely from her own work in
London, describes how behavioral techniques can be used to
prevent the development of coronary heart disease in a
population at risk.   James Blumenthal then discusses the use of
behavioral techniques in preventing the recurrence of coronary
events in patients already affected.   Finally, Ethel Roskies
reviews the research on the utility of behavior modification in
altering Type A behaviors.

In a related section, Richard Surwit and Robert Freedman
talk about the treatment of the vasospasms of Raynaud's disease
and Raynaud's phenomenon, while Derek Johnston explores the use
of a variety of behavioral techniques in treating the ischemic
pain associated with angina pectoris.

The third section of the book deals with the major problem
of hypertension.   Redford Williams summarizes the of the
disorder with special emphasis on the contribution of the
sympathetic nervous system to the expression of this disorder.
The treatment of hypertension is then reviewed in separate
chapters by David Shapiro, Andrew Steptoe, Guido Godaert, and
Dieter Vaitl.   Each chapter emphasizes a different behavioral
approach and the entire section gives the reader a good idea as

to the spectrum of behavioral methodologies available in the treatment of this condition.

The next section of the book deals with the use of behavioral techniques in neuromuscular rehabilitation. John Basmajian provides an overview of the utility of EMG feedback procedures in the development of neuromuscular control. Joseph Brudny then describes how EMG feedback can be used in the rehabilitation of patients with torticollis and other torsion dystonias. Bernard Engel reviews his and other work on the behavioral treatment of fecal incontinence, a program which is so successful that it has become the treatment of choice for this disorder. Finally, Neal Miller presents a general overview of how automated devices designed to record movement and activity can aid the behavioral treatment of muscular disorders in general.

The fifth section of the book is devoted to the behavioral treatment of headache. John Graham, a pioneer in the understanding of the pathophysiology of headaches and their pharmacological treatment, reviews the mechanisms of headache and sets the stage for the application for behavioral techniques in treatment. Jackson Beatty then reviews the application of biofeedback approaches to the understanding and treatment of vascular headaches. Niels Birbaumer and Birget Kroner then separately review a variety of different behavioral techniques in the treatment of both muscle contraction and migraine headaches. As in the section on hypertension, these various chapters provide the reader with a multifaceted overview on the broad spectrum of procedures available to the behavioral as well as medical practitioner.

Chronic pain constitutes the next specific topic area addressed in the symposium. Pathophysiological mechanisms of pain are reviewed from the perspective of a neurosurgeon by Blaine Nashold. His emphasis on the role of the central nervous system in the disorder suggests the utility of behavioral procedures in analysis and treatment of this problem. Bernard Tursky then reviews his pioneering work in the psychophysical assessment of pain. This chapter is complimented by one by Wilbert Fordyce, in which he describes how chronic pain can be seen as a behavioral as well as a physiologic problem and how it can be manipulated environmentally. Finally, Francis Keefe describes how a variety of behavioral procedures can be used in treatment of pain syndromes.

In the last topic of the book, Robert Biersner reviews the application of behavioral technology to problems of health and performance in military settings. The military experience contains many of the stresses and demands encountered in civilian life in an environment which is often well suited for controlled behavioral study.

In the final section, Arthur Bachrach considers the implications of the experimental analysis of behavior for the developing field of behavioral medicine, stressing the inherent differences between behavioral and biomedical sciences. Stephen Weiss and Gary Schwartz, then discuss the interdisciplinary nature of behavioral medicine. They emphasize the need to consider the interaction of variables from both the biomedical and behavioral sciences in the etiology, pathogenesis and treatment of disease. Joseph Brady and Charles Mertens then discuss the place of behavioral medicine in conventional psychiatry.

The reader should note that although this symposium was comprehensive, it was not exhaustive. Limitations of time necessitated that we selectively review the applications of behavioral techniques to those disease areas in which the most progress has been made. In so doing, the exciting work dealing with the behavioral treatment of seizure disorders, dental problems, ophthalmologic difficulties, etc. had to be excluded.

Behavioral Treatment of Disease thus represents an authoritative and up-to-date appraisal of the impact of behavioral technology on modern medical care. Ten years ago, such an international symposium could not have taken place; the term "behavioral medicine" did not exist. In 1976, the Scientific Affairs Division of NATO sponsored a conference entitled, Biofeedback and Behavior. While this symposium restricted itself largely to biofeedback and most of its clinical relevance was by implication rather than application, it did herald the field which we review here. Behavioral Medicine is now a reality. The growth of this new field depends both on continued research, as well as the introduction of proven clinical techniques in everyday practice, and it is hoped that this volume will serve as stimulus to both.

BEHAVIORAL MECHANISMS IN THE PATHOPHYSIOLOGY OF

CORONARY HEART DISEASE

Redford B. Williams, Jr.
Department of Psychiatry
Duke University Medical Center
Durham, North Carolina

Before we can consider the application of behavioral approaches in the prevention, treatment and rehabilitation of coronary heart disease (CHD) -- the central theme of this Section -- it is necessary to consider first the various means whereby behavioral factors appear to be playing a role in the etiology and course of CHD. In this chapter I shall first review the epidemiologic evidence indicating a role of behavioral factors in the precipitation of acute clinical events, the pathogenesis of the atherosclerotic plaque and the clinical outcome of relief of anginal pain with either medical or surgical management. Following this initial review, I shall consider the various physiological and neuroendocrine mechanisms whereby the associations shown in the epidemiological research might be mediated.

Behavioral Factors in the Precipitation of Acute CHD Events

The most persuasive case for the role of any behavioral or psychosocial factor in the etiology and pathogenesis of any disease is that presented by the now voluminous body of research findings documenting the increased rate of CHD events in persons displaying the Type A behavior pattern (see Dembroski, Weiss, Shields, Haynes, & Feinleib, 1978, for a detailed review of this evidence). As detailed in the now definitive Western Collaborative Group Study (Rosenman, Brand, Jenkins, Friedman, Straus & Wurm, 1975), Type A men experience about twice the frequency of acute CHD events, including all CHD, myocardial infarction, angina pectoris, recurrent myocardial infarction and death, when compared to Type B men over an 8½-year follow-up period. This increased rate of CHD events among Type A men is not an artifact of increased levels of other risk factors in

5

Type A men remains significant even when statistical adjustment is made for all known risk factors. In a reanalysis of the audio taped interviews used to determine behavior pattern in the Western Collaborative Group Study, Matthews, Glass, Rosenman, and Bortner (1977) found potential for hostility and competitive drive to be subcomponents of the global Type A behavior pattern which accounted for most of the increased CHD risk among Type A men.

In addition to the Western Collaborative Group Study, which evaluated CHD risk in over 3000 middle-aged men, the prospective association between Type A behavior pattern and increased rates of CHD events has also been documented in the Framingham Study. Using a questionnaire designed to assess Type A behavior pattern, Haynes, Feinleib, Levine, Scotch and Kannel (1978) found Type A behavior as assessed by this means also to be associated with increased rates of CHD events.

In 1978 the National Heart, Lung and Blood Institute sponsored a conference at Amelia Island, Florida, at which the evidence relating Type A behavior pattern to CHD risk (as contained in Dembroski et al., 1978) was subjected to critical review by a distinguished panel of behavioral and biomedical scientists. Based on this review, it was concluded that "...Type A behavior--as defined by the structured interview used in the Western Collaborative Group Study, the Jenkins Activity Survey and the Framingham Type A behavior scale -- is associated with an increased risk of clinically apparent CHD in employed, middle-aged U. S. citizens. This risk is greater than that imposed by age, elevated values of systolic blood pressure and serum cholesterol, and smoking and appears to be of the same order of magnitude as the relative risk associated with the latter three of these other factors" (Review Panel on Coronary-prone Behavior and Coronary Heart Disease, 1981). Thus, it now appears reasonable to accord to the Type A behavior pattern the status of a major risk factor for CHD, comparable to cigarette smoking, elevated blood pressure and elevated serum cholesterol levels.

This is only a first step, however, and the question still remains as to how Type A behavior leads to excess CHD events. One possible mechanism would be that in the presence of pre-existing coronary atherosclerosis, Type A behavior is associated with excess cardiovascular arousal, such that the narrowed coronary arteries are unable to meet myocardial oxygen demands with the result being the emergence of clinically apparent CHD. Another possibility -- not necessarily exclusive of the first -- is that Type A behavior is also associated with increased rates of coronary atherogenesis. If this latter mechanism is also operative, then the implications of Type A behavior pattern for pathogenesis assume even greater importance with regard to efforts at primary prevention; for if Type A behavior is found to accelerate coronary atherogenesis, a process that is felt to begin early in life long before the expression of clinical manifestations, then attempts to reduce the CHD risk associated with Type A behavior should also begin early in life, as is now the case with efforts to control hypertension, prevent the initiation of cigarette smoking, and identify and treat various forms of hyperlipidemia. To evaluate the possible role of Type A behavior in coronary atherogenesis, I shall now review the available evidence relating Type A behavior pattern to coronary atherosclerosis.

Type A Behavior Pattern and Coronary Atherosclerosis (CAD)

In the absence of safe non-invasive means of assessing the coronary vasculature, the only means of evaluating the coronary vasculature is coronary arteriography, now the standard, definitive diagnostic technique in clinical practice. Since this diagnostic procedure is not without risk, it is not possible at present to evaluate the relation of Type A behavior pattern to CAD in "normal" population samples. Thus, the generalizability of findings in patients referred for diagnostic coronary arteriography is open to question, since relationships of behavioral factors to CAD in this highly selected population who have gone through an extensive screening process before referral to tertiary medical centers may not hold in the population at large. Nevertheless, it is this patient population with suspected CHD which accounts for much of the medical costs associated with CHD in the U.S. at the present time. Over 300,000 coronary arteriograms are performed annually in the U.S. currently, and from these 300,000 patients are selected the entire population of patients -- currently over 100,000 -- who undergo coronary bypass graft (CABG) surgery at a cost in excess of $1.5 billion annually. Therefore, even though the generalizability of findings in the population undergoing coronary arteriography is open to question, it is abundantly clear that even if implications are relevant only for this selected population, it is a very important population to study. Moreover, it is by no means clear that findings in this clinical population are not applicable to the general population.

Three studies were conducted at about the same time in the mid-1970s assessing the relationship between Type A behavior pattern and arteriographically documented coronary atherosclerosis. One of these studies (Zyzanski, Jenkins, Ryan, Flessas & Everist, 1976) used the Jenkins Activity Survey (JAS) to assess Type A behavior pattern and found in a study of 94 men that those with two or more major coronary arteries with luminal narrowing of 50% or more scored higher on all four scales of the

JAS than did patients with zero or one artery with 50%
narrowing. Multivariate analysis of their data showed only the
JAS Type A scale to be associated with increased CAD
independently of degree of angina pain, age, prior experience of
myocardial infarction and various anxiety and neuroticism scales
derived from the MMPI. In addition to the JAS Type A score, low
levels of denial and high levels of anxiety and depression were
also found independently and significantly related to increased
CAD.

Frank, Heller, Kornfeld, Sporn and Weiss (1978) reported an
association between Type A behavior pattern and increased
severity (assessed in terms of the number of major coronary
arteries with a greater than 50% occlusion) of CAD among a
sample of 124 men and 23 women undergoing diagnostic coronary
arteriography. This study extended the findings of the Zyzanski
et al. (1976) study in several important respects. First, the
sample included women, and the relation between Type A behavior
and CAD was found to be equally strong in women compared to men.
Second, the traditional risk factors (age, sex, cholesterol,
smoking history and hypertension) were all evaluated with regard
to CAD severity and entered first in a multiple regression
analysis prior to entry of Type A behavior pattern. Despite
adjustment for all these physical risk factors, Type A behavior
pattern was found to account for a significant proportion of
disease, above and beyond the cumulative effects of the physical
risk factors. Third, the means of assessing Type A behavior
pattern was the structured interview (Rosenman, 1978), instead
of the JAS, which was not employed in this study.

The third study was that of Blumenthal, Williams, Kong,
Schanberg and Thompson (1978), which in many respects replicated
the Frank et al. (1978) findings, and contained some further
refinements, as well. Like the Frank et al. (1978) study, this
sample of 142 patients included both men and women, though a
greater proportion (75%) were men. The Blumenthal et al. (1978)
study also made simultaneous statistical adjustment for
traditional physical risk factors: age, sex, cigarette smoking,
blood pressure and cholesterol level.

Thus, the Blumenthal et al. (1978) study essentially
replicated all the major findings of the Frank et al. (1978)
study: Type A behavior pattern was found associated with
increased CAD severity, independently of physical risk factors
and among women as well as men (Figure 1). In addition, the
Blumenthal et al. (1978) study went beyond the Frank et al.
(1978) study in several respects. First, it employed the JAS in
addition to the structured interview as a measure of Type A
behavior. It was noteworthy that, whereas interview-determined
Type A behavior was found strongly related to CAD levels, as

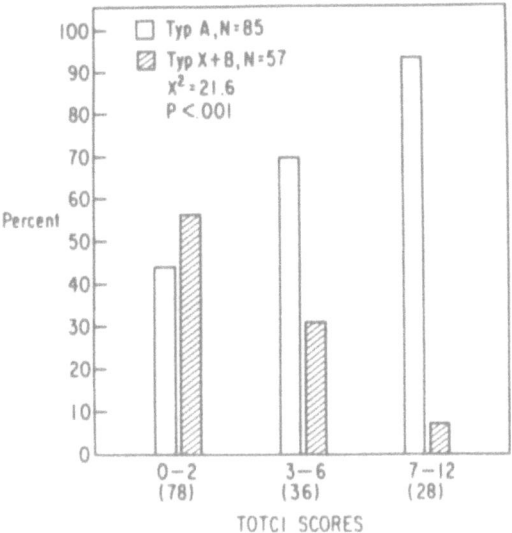

Figure 1. Relationship between severity of coronary atherosclerosis (TOTCI = Index of coronary atherosclerosis; 0-2 = mild, 3-6 = moderate, 7-12 = severe) and Type A behavior pattern. (Reproduced from Blumenthal et al., 1978).

noted above, JAS-determined Type A behavior did not relate significantly to CAD level, despite a moderate and significant correlation between JAS scores and interview results. A second aspect of the Blumenthal et al. (1978) study was the measurement of an index of genetically determined sympathetic nervous system activity -- serum levels of the norepinephrine-synthesizing enzyme dopamine-beta-hydroxylase (DBH) -- in a subset of 83 patients. No relationship was found between DBH levels and CAD severity, suggesting that genetically determined increased levels of sympathetic activity are not involved in atherogenesis. It should be noted, however, that this only pertains to genetically determined, chronic levels of sympathetic activity, not necessarily to acute behaviorally or situationally induced increases. While the reasons for the failure to replicate the JAS results of Zyzanski et al. (1976) may never be conclusively resolved, it appears likely that the presence of a substantial proportion of women and patients from rural, rather than urban backgrounds may have been factors.

In contrast to the above series of studies which found a relationship between Type A behavior and CAD severity -- one using only the structured interview to assess Type A behavor, one using only the JAS and one using both -- Dimsdale and coworkers have found no relationship between arteriographically documented CAD and Type A behavior pattern, whether the latter is assessed by the JAS (Dimsdale, Hackett, Hutter, Block & Catanzano, 1978) or the interview (Dimsdale, Hackett, Hutter, Block & Catanzano, 1979). The reasons for the failure of Dimsdale and coworkers to replicate the earlier findings are not clear at present. Among the possible explanations are population differences among the various medical centers, erroneous behavior pattern assessments using the structured interview, and simply a chance failure to replicate.

Based upon our earlier finding (Blumenthal et al., 1978) of a relationship between interview-determined Type A behavior pattern and CAD, we have been collecting in a more systematic manner a wide gamut of psychological, social and behavioral information on all patients undergoing coronary arteriography at Duke University Medical Center since 1976. The overall study design is shown in Figure 2. All the behavioral/psychosocial as well as clinical and catheterization variables are stored in the Cardiology Data Bank, thus greatly facilitating the simultaneous evaluation of relationships between behavioral/psychosocial variables and indices of CAD outcomes. We have been attempting

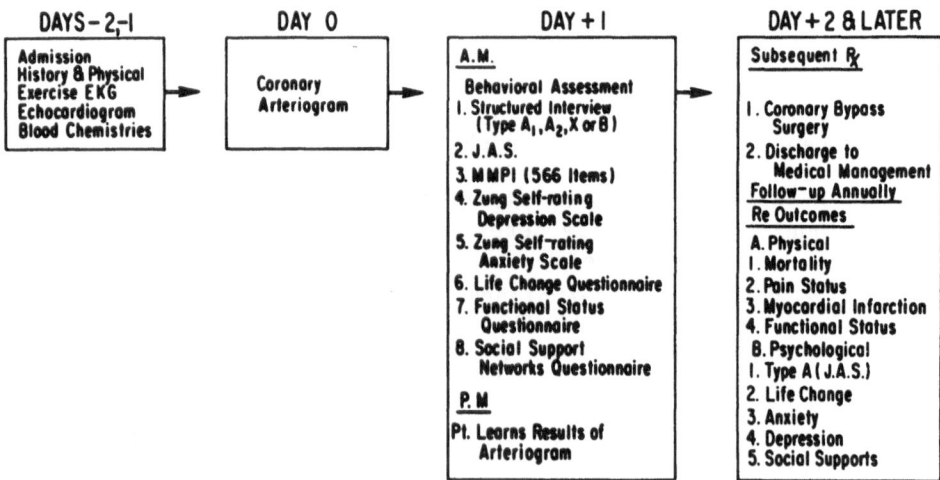

Figure 2.   Duke project to evaluate role of psychosocial factors
             in CHD outcomes: overall study design. (Reproduced
             from Williams et al., 1980b).

in our continuing studies to define with greater precision than the global Type A behavior pattern those psychological/ behavioral characteristics which relate to increased CAD levels, as well as psychosocial characteristics which predict such important CHD outcomes as relief of anginal pain with various forms of treatment.

We first addressed the question of the relationship between Type A behavior pattern as determined by the interview and CAD in a far larger sample (424) of patients than had been previously reported in any of the earlier studies. As shown in Figure 3, we again found (Williams, Haney, Lee, Kong, Blumenthal & Whalen, 1980b) a significant relationship between Type A behavior and CAD: over 70% of the 319 Type A patients had at least one artery with a clinically significant (70% or greater) occlusion, in comparison to only 56% of the 105 Type B patients with a clinically significant occlusion. Based upon earlier work by Matthews et al. (1977) showing potential for hostility to be associated with increased risk of CHD events in the Western Collaborative Group Study sample, we scored the MMPI protocols on these 424 patients for Cook and Medley's (1954) Hostility (Ho) Scale. An interesting, non-linear relationship was found (Figure 4) between Ho score and the presence of a significant occlusion: among those patients who scored in a very low range on the Ho scale --10 or less-- reflecting an essential absence of distrust for and dislike of people in general, there was observed only a 48% rate of significant CAD. In contrast, among patient groups scoring anything higher than 10 there was observed a 70% rate of significant disease. We next evaluated the joint relationship between Type A behavior pattern and Ho score with CAD (controlling for sex, since females are less often Type A and have lower Ho scores on average). As shown in Figure 5, there is a striking progression in the proportion with significant CAD, ranging from only 12.5% in Type B women who score 10 or less on the Ho scale to 44% among Type A women with Ho scores greater than 10, and progressing to 82% among Type A men with Ho scores greater than 10. Both low Ho score and Type A behavior pattern were independently related to presence of significant CAD --a result most likely of the fact that Ho scores of 10 or less were present equally in the Type A and non-Type A groups.

These findings indicate that it is possible to identify psychometrically determined individual difference measures that relate to CAD independently of the behaviorally determined (via the structured interview) Type A characteristic. We have suggested (Williams, Haney, Lee, Harrell, McKinnis, Blumenthal, Rosati, Kong & Sabiston, 1980a) that this reluctance to endorse items reflective of dislike for and distrust of others may also

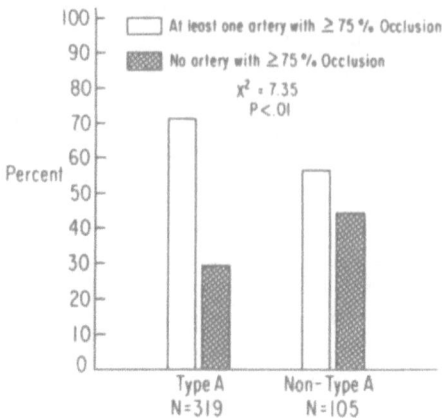

Figure 3.   Type A behavior pattern and proportion with
            significant coronary atherosclerosis.
            (Reproduced from Williams et al., 1980b).

help to explain the relative absence of CHD in Japan, where,
despite a very high prevalence of high blood pressure, the CHD
rate is the lowest of industrialized countries. In addition,
very recent and preliminary analyses of Ho scores in a
large-scale prospective study of CHD suggest that low Ho scores
are also predictive of reduced CHD mortality over long-term
follow-up -- suggesting that the apparent protective effect of
low Ho scores with regard to CAD in a clinical population may
also extend to protect free living populations with regard to
CHD mortality as well.

Psychological Factors and Relief of Anginal Aain

        Thus far in this chapter we have focused on the possible
role of behavioral factors in the pathogenesis of CHD. It is
likely that once CHD is clinically manifest, psychological
characteristics of the patient may also play a role in the
course of the disease. At the recent consensus conference on
coronary bypass surgery, a number of interesting facts emerged
concerning the current clinical management of CHD (Kolata,
1981). Of the over 100,000 patients who currently are treated
with coronary artery bypass graft(CABG) surgery, only 43% have
lesions (left main obstruction or three-vessel disease) for
which there is evidence that CABG prolongs survival; for the
remaining 57% the only documented benefit is that CABG surgery
achieves better relief of anginal pain than standard
pharmacologic therapy. While the status of the coronary
arteries and indices of left ventricular function are potent
predictors of survival, there are presently no known physical

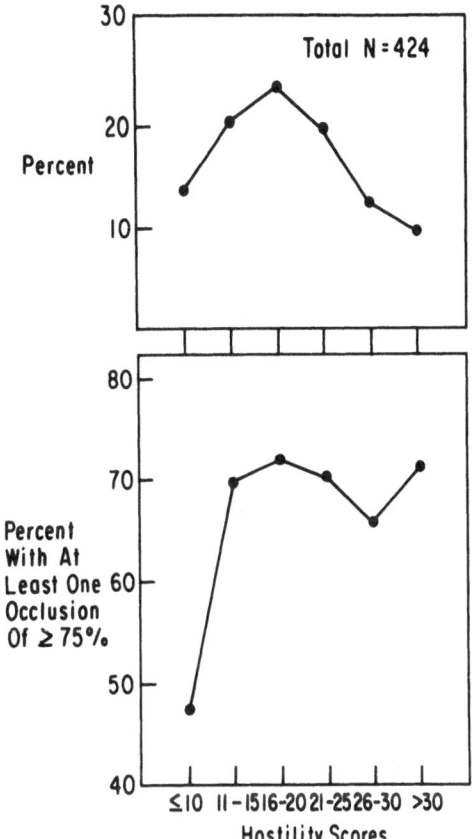

Figure 4.   Relationship between Hostility level and coronary
            atherosclerosis.   (Reproduced from Williams
            et al., 1980b).

characteristics which are predictive of anginal pain relief with
medical management.   If it were possible to identify reliably
those 57,000 patients who currently undergo CABG surgery only
for the benefit of pain relief because they did not receive
adequate pain relief with medical management prior to the start
of the eventually unsuccessful course of pharmacotherapy, then
more   intensive   attention   --   using   both   more   vigorous
pharmacologic approaches and innovative behavioral approaches --
might prove successful in a substantial proportion in achieving
adequate pain relief.   The result could be a substantial
savings, both in terms of financial savings at the rate of over
$15,000 per operation and in terms of reduced risk and suffering
occasioned by the surgical procedure itself.

      Based on earlier studies showing that patients with low
back pain and high scores on the MMPI Hysteria (Hy) and
Hypochondriasis (Hs) scales fail to obtain pain relief with back
surgery (Sternbach, 1974), we predicted several years ago that

high Hy and Hs scores would also identify patients with CHD who
are unlikely to obtain relief of anginal pain with medical
management.   We have recently reported findings (Williams, et
al., 1980a) which confirm this prediction.   Except for presence
of   right   coronary   artery   atherosclerosis,   no   physical
characteristics were found either in the clinical historical
data   or   the   cardiac   catheterization   data,   which   relibly
predicted which patients would achieve anginal pain relief with
medical treatment.   In contrast, both Hy and Hs scores on the
MMPI were strongly predictive of pain relief.   Patients scoring
high on Hy and Hs were far less likely to have obtained
clinically significant pain relief at six-months follow-up than
patients with lower Hy and Hs scores.   In addition, it was found
that certain social variables were also predictive of pain
relief with medical management.   Among those patients who were
actively employed at the time of their cardiac catheterization,
nearly twice as large a proportion had achieved clinically
significant pain relief in comparison to patients who were not
employed at the time of catheterization.   It should be noted
that this prediction of anginal pain relief by psychosocial
characteristics was not a function of underlying physical

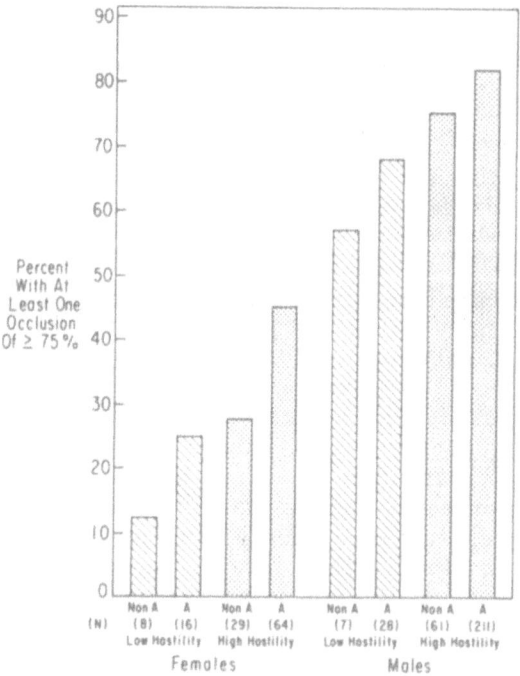

Figure 5.   Relationship of gender, Type A behavior pattern
           and Hostility level and coronary atherosclerosis.
           (Reproduced from Williams et al., 1980b).

characteristics, since the statistical approach used adjusted for any correlation between physical and psychosocial characteristics. That is, the significant prediction of pain relief by psychosocial factors was statistically independent of any underlying physical characteristics of the patients.

Based on these findings, it appears that prospective assessment of both psychological and social characteristics (along lines suggested in Figure 1) among patients being considered for medical management of their angina could serve to pinpoint those patients most unlikely to respond to medical management with adequate relief of their angina. Such patients might be targeted for special attention -- e.g. more intensive pharmacological therapy or such behavioral approaches as biofeedback or stress management training -- with the result that a greater proportion might achieve angina relief. If so, a substantial proportion of those patients who now become medical treatment failures and go on to CABG surgery might be helped to obtain relief, with substantial savings in terms of both cost and suffering associated with surgical management.

Type A Behavior and CHD: Physiologic and Neuroendocrine Mechanisms

As outlined earlier in this chapter, it is now generally accepted that Type A behavior pattern is a major risk factor for acute CHD events and, probably, for increased CAD as well. Studies documenting both cardiovascular and catecholamine hyperresponsivity among Type A persons to a variety of behavioral challenges have led to the hypothesis that such hyperresponsivity represents a mechanism for the expression of excess coronary events and CAD among Type A persons (Williams, Friedman, Glass, Herd & Schneiderman, 1978). In the remainder of this chapter I shall review some of the earlier evidence leading to this hypothesis and then conclude by describing a theoretical approach to this issue which has guided recent research of our group at Duke, as well as some of the specific findings of that research.

The currently available evidence concerning the physiologic hyperresponsivity hypothesis is based on studies which evaluate either cardiovascular or catecholamine responses to behavioral challenge among Type A and B subjects -- rarely are both cardiovascular and catecholamine responses evaluated in the same study, and the secretion of other neuroendocrine parameters to behavioral challenge among Type A and B subjects has not been assessed at all. In addition, there has been some inconsistency among the available studies regarding which response parameters are enhanced among Type A persons. For example, during work on an impossible jigsaw puzzle, Type A subjects showed excessive

norepinephrine (NE) but not epinephrine (E) secretion (Friedman,
Byers, Diament & Rosenman,1975); while during a competitive TV
"pong" game with harassment, Type A subjects showed
hyperresponsivity only with regard to E secretion (Glass,
Krakoff, Contrada, et al., 1980). With regard to cardiovascular
hyperresponsivity, some studies find only enhanced systolic.
blood pressure response among Type A subjects, while in others
only heart rate response, or both systolic blood pressure and
heart rate responses are enhanced among Type A subjects (Manuck
& Garland, 1979; Krantz, Schaeffer, Davia, Dembroski,
MacDouglas, & Shaffer, 1980; and Dembroski, MacDougall, Herd, &
Shields, 1979). These studies have employed a wide variety of
behavioral challenges. Perhaps the variability in the findings
to date stems from a differential responsivity of Type A and B
subjects to specific aspects of these varying behavioral
challenges.

In trying to understand how differing forms of challenge
might elicit different patterns of response, the work of Mason
(1968) is particularly relevant. Mason has proposed that a
given neuroendocrine parameter does not respond to a behavioral
challenge in isolation, but as one component of a broad array of
multiple, concurrent responses; in addition, different types of
challenge may elicit different patterns of neuroendocrine
response. Based on Mason's suggestions, on theorizing by the
Laceys (1974), and on our own prior finding of muscle
vasodilatation during mental work behavior and muscle
vasoconstriction during sensory intake behavior(Williams,
Bittker, Buchsbaum & Wynne, 1975), we have proposed (Williams,
1975) that these two behaviors are associated with qualitatively
distinct patterns of both cardiovascular and neuroendocrine
response, and that enhanced expression of these patterns could
provide an explanation for the increased CHD risk observed among
Type A persons.

Accordingly, in a recent study we evaluated a broad array
of both cardiovascular and neuroendocrine parameters in response
to a mental arithmetic (mental work) and a reaction time
(sensory intake) task. Subjects were 31 young male college
undergraduates: 13 were Type A and 18 were Type B according to
the student version of the JAS. Findings indicated that Type A
subjects are hyperresponsive, but only with respect to certain
parameters and only during certain tasks. As shown in Table 1,
both Type A and B subjects showed a muscle vasodilatation during
mental arithmetic performance as indexed by a significant increase in forearm blood flow (FBF) and a significant decrease in
forearm vascular resistance (FVR). The size of this vasomotor
response (muscle vasodilatation has long been known to occur
during mental arithmetic periormance, and is felt to be part of
the "defense reaction") was significantly larger, however, for

the Type A subjects. The heart rate and blood pressure responses of the Type A and B subjects did not differ. With regard to neuroendocrine responses during mental arithmetic performance, all subjects showed significant increases (Table 1) in NE, E, cortisol and prolactin. Again, Type A subjects showed larger increases in E, NE and cortisol. In contrast to the other neuroendocrine parameters, where Type A hyperresponsivity was found, the prolactin response for Type A and B subjects was virtually identical.

With the exception of no change in FBF and FVR, during the reaction time task all cardiovascular parameters showed significant increases, though smaller than during mental arithmetic. Among neuroendocrine parameters, only NE and E showed significant increases during reaction time performance, though smaller than during mental arithmetic. Cortisol and prolactin were unchanged during reaction time performance. In contrast to mental work, where Type A subjects showed selective hyperresponsivity, during reaction time performance there were no differences between Type A and B subjects with respect to either cardiovascular or neuroendocrine parameters. Katkin, Goldband and Medine(1979) have found, however, that among subjects with a positive family history of cardiovascular disease (+FH) Type A subjects do show vasomotor hyperresponsivity during a reaction time task in comparison to Type B subjects. Therefore, we evaluated the Type A/+FH interaction with respect to cardiovascular and neuroendocrine response during our reaction time task. A strongly significant interaction effect was found for cortisol response (P=.008); while not significant (P=.11), a similar tendency was found for diastolic blood pressure response. The nature of this interaction for both cortisol and diastolic blood pressure response was the same as found in the Katkin et al. (1979) study: during reaction time performance Type A subjects show hyperresponsivity relative to Type B subjects only in the presence of a genetic predisposition to cardiovascular disease. We have subsequently replicated this Type A/+FH interaction finding in two additional studies -- again, Type A subjects were found to be hyperresponsive relative to Type B subjects only in that subsample of subjects with a +FH of cardiovascular disease. Despite their cardiovascular and neuroendocrine hyper-responsivity, Type A subjects showed no better performance than Type B subjects on either the mental arithmetic or reaction time tasks.

These results support our initial hypothesis. During mental work, Type A subjects showed selective hyperresponsivity: in muscle vasodilatation but not heart rate or blood pressure; in E, NE and cortisol but not prolactin. In contrast, during sensory intake behavior, Type A subjects were not hyperresponsive as a group, but only in the presence of a genetic predisposition to cardiovascular disease, replicating the earlier finding of Katkin et al., (1979).

Our findings have several potentially important impli- cations for understanding mechanisms underlying the increased CHD risk documented for Type A persons earlier in this chapter. The combination of catecholamine and cortisol hyperresponsivity could be particularly relevant in explaining the excessive rate of pathogenesis of both CAD and the precipitation of acute clinical CHD events among Type A persons. Cortisol is known to both stimulate catecholamine synthesizing enzymes and inhibit a major catecholamine degrading enzyme -- COMT -- and increase the sensitivity of adrenergic receptors to a given level of neuro- transmitter (Kvetnansky, 1980). Hence, among Type A persons, the physiologic and metabolic effects of any given level of sympathetic activation could be potentiated by the concomitant enhanced adrenocortical arousal. That such a mechanism may indeed be playing a role in accelerated atherogenesis among Type A persons is supported by the observation among Air Force personnel that elevated serial plasma cortisol levels during an oral glucose tolerance test are associated with increased severity of arteriographically documented coronary atheros- clerosis (Troxler, Sprague, Albanese, et al., 1977).

Our finding of enhanced catecholamine response to a specific form of behavioral challenge -- mental work with no requirement ot attend to senory inputs -- is consistent with earlier reports (see above) of plasma catecholamine hyper- responsivity among Type A persons. In contrast to these earlier studies, however, where only NE or E response -- never both -- was enhanced in Type A subjects, in our study we find both catecholamines to increase more during mental arithmetic per- formance among Type A subjects. In contrast, during a sensory intake task, Type A and B subjects do not differ in their physiologic responses, except when we take into account the presence/absence of a family history of cardiovascular disease. The earlier studies may have failed to find Type A/B differences as uniform as these due to a greater expression in those studies of individual differences in response, resulting from the use of complex tasks with mixed requirements to attend to visual stimuli (sensory intake), as well as plan strategies and cope with various forms of harassment (mental work).

To the extent that the specific cardiovascular and neuro- endocrine hyperresponsivity we have found among Type A subjects during performance of mental work and sensory intake tasks in the laboratory is also occurring among Type A persons while performing these ubiquitous behaviors in the real world, our findings could help to identify basic mechanisms mediating the increased CHD risk among Type A persons. Such knowledge could be applied to the rational formulation of both pharmacologic and behavioral interventions designed to reduce CHD risk.

SUMMARY

In this chapter I have tried to provide an introduction to the subsequent chapters in this Section dealing with behavioral approaches to prevention, treatment and rehabilitation of CHD, by delineating the various means whereby behavioral and psychosocial factors play a role in the etiology and course of CHD. First, it is now generally accepted that Type A behavior pattern is a major CHD risk factor. This conclusion is based on the repeated demonstration, in large-scale prospective studies, of increased rates of CHD events among Type A persons, as well as an abundance of evidence that among persons referred for diagnostic coronary arteriography Type A persons also show increased levels of coronary atherosclerosis. Second, among persons in whom CHD is already clinically manifest, it appears that psychological and social characteristics are predictive of the important treatment outcome of relief of anginal pain. Finally, in numerous laboratory studies Type A subjects have been found to exhibit both cardiovascular and neuroendocrine hyperresponsivity when confronted with a variety of behavioral challenges. Of particular interest in this latter regard is the new observation of cortisol hyperresponsivity among all Type A subjects during mental work, as well as among Type A subjects with a positive family history of cardiovascular disease during sensory intake behavior.

REFERENCES

Blumenthal, J. A., Williams, R. B., Kong, Y., Schanberg, S. M. & Thompson, L. W. Type A behavior pattern and coronary atherosclerosis. Circulation, 1978, 58, 634-639.

Cook, W. W. & Medley, D. M. Proposed hostility and pharisaic-virtue scales for the MMPI. Journal of Applied Psychology, 1954, 38, 414-418.

Dembroski, T. M., MacDougall, J. M., Herd, J. M., & Shields, J. L. Effect of level of challenge on pressor and heart rate response in Type A and B subjects. Journal of Applied Social Psychology, 1979, 9, 209.

Dembroski, T. M., Weiss, S. M., Shields, J.L., Haynes, S. G., & Feinleib, M. (eds.) Coronary-Prone Behavior. New York: Springer-Verlag, 1978.

Dimsdale, J. E., Hackett, T. P., Hutter, A. M., Block, P. C. & Catanzano, D. M. A personality and extent of coronary atherosclerosis. American Journal of Cardiology, 1978, 42, 583-586.

Dimsdale, J. E., Hackett, T. P., Hutter, A. M., Block, P. C. & Catanzano, D. M. Type A behavior and angiographic findings. Journal of Psychosomatic Research, 1979, 23, 273-276.

Frank, K. A., Heller, S. S., Kornfeld, D. S., Sporn, A. A. & Weiss, M. D. Type A behavior pattern and coronary atherosclerosis. Journal of American Medical Association, 1978, 240, 761-763.

Friedman, M., Byers, S. O., Diament, J., & Rosenman, R. H. Plasma catecholamine response of coronary prone subjects (Type A) to a specific challenge. Metabolism, 1975, 24, 205-210.

Glass, D. C., Krakoff, L. R., Contrada, R. H., Hilton, W. F., Kehoe, K., Mannucci, E. G., Collins, C., Snow, B. & Elting, E. Effect of harassment and competition upon cardiovascular and catecholamine responses in Type A and Type B individuals. Psychophysiology, 1980, 17, 453-463.

Haynes, S. G., Feinleib, M., Levine, S., Scotch, N. & Kannel, W. The relationship of psychosocial factors to coronary heart disease in The Framingham Study. American Journal of Epidemiology, 1978, 107, 384.

Katkin, E. S., Goldband, S. & Medine, B. Cardiovascular responses of Type A and Type B subjects differing in family history of heart disease. Paper presented at Annual Meeting, Society for Psychophysiological Research, Cincinnati, 1979.

Kolata, G. B. Consensus on bypass surgery. Science, 1981, 211: 42-43.

Krantz, D. S., Schaeffer, M. A., Davia, J. E., Dembroski, T. M., MacDouglas, J. M., & Shaffer, R. T. Paper presented at the Annual Meeting of Society for Psychophysiological Research, Vancouver, B.C., October, 1980.

Kvetnansky, R. Recent progress in catecholamines under stress. In Usdin, E., Kvetnovsky, R. & Kopin, I. J. (eds.) Catecholamines and Stress: Recent advances. Amsterdam: Elsevier/North Holland, 1980.

Lacey, J. I. & Lacey, B. C. On heart rate responses and behavior: A reply to Elliot. Journal of Personality and Social Psychology, 1974, 30, 1.

Manuck, S. B. & Garland, F. N. Coronary-prone behvior and cardiovascular response. Psychophysiology, 1979, 16, 13€.

Mason, J. W. A review of psychoendocrine research on the pituitary-adrenal cortical system. Psychosomatic Medicine, 1968, 30, 791.

Matthews, K. A., Glass, D. C., Rosenman, R. H., & Bortner, R. W. Competitive drive, pattern A and coronary heart disease: A further analysis of some data from the Western Collaborative Group Study. Journal of Chronic Disease, 1977, 30, 489-498.

Review panel on coronary-prone behavior and coronary heart disease. Coronary-prone behavior and coronary heart disease: A critical review. Circulation, 1981, 63, 1199-1215.

Rosenman, R. H. The interview method of assessment of the coronary-prone behavior pattern. In Coronary-Prone Behavior. Dembroski, T. M., Weiss, S. M., Shields, J. L. (eds.) New York: Springer-Verlag, 1978.

Rosenman, R. H., Brand, R. J., Jenkins, C. D., Friedman, M., Straus, R. & Wurm, M. Coronary heart disease in the Western Collaborative Group Study: Final follow-up experience of 8½ years. Journal of American Medical Association, 1975, 233, 872.

Sternbach, R. A. Pain patients: Traits and treatment. New York: Academic Press, 1974.

Troxler, R. G., Sprague, E. A., Albanese, R. A., Fuchs, R., & Thompson, A. J. The association of elevated plasma cortisol and early atherosclerosis as demonstrated by coronary angiography. Atherosclerosis, 1977, 26, 151-162.

Williams, R. B. Physiologic mechanisms underlying the association between psychosocial factors and coronary heart disease. In Gentry, W. D. & Williams, R. B. (eds.) Psychosocial aspects of myocardial infarction and coronary care. St. Louis: C.V. Mosby Company, 1975.

Williams, R. B., Bittker, J. E., Buchsbaum, M. S. & Wynne, L. C. Cardiovascular and neurophysiologic correlates of sensory intake and rejection: I. Effect of cognitive tasks. Psychophysiology, 1975, 12, 427-434.

Williams, R. B., Friedman, M., Glass, D. C., Herd, J. A. & Schneiderman, W. Mechanisms linking behavioral and pathophysiological processes. In Dembroski, T.M., Weiss, S. M., Shields, J. E. (eds.) Coronary-Prone Behavior. New York: Springer-Verlag,1978.

Williams, R. B., Haney, T. L., Lee, K. L., Harrell, F. E., McKinnis, R. A., Blumenthal, J. A., Rosati, R. A., Kong, Y. & Sabiston, D. C. Psychosocial factors and pain relief in patients with CHD. Presented at American Heart Association, 53rd Scientific Session, November, 1980.

Williams, R. B., Haney, T. L., Lee, K. L., Kong, Y., Blumenthal, J. A. & Whalen, R. E. Type A behavior, hostility and coronary artherosclerosis. Psychosomatic Medicine, 1980, 42, 539-549.

Zyzanski, S. J., Jenkins, C. D., Ryan, T. J., Flessas, A. & Everist, M. Psychological correlates of coronary angiographic findings. Archives of Internal Medicine, 1976, 136, 1234-1237.

PRIMARY PREVENTION OF CORONARY HEART DISEASE

Chandra Patel

Department of Epidemiology
London School of Hygiene & Tropical Medicine
London, England

Coronary Heart Disease (CHD) is the most common of all causes of death in Western Countries. Approximately 40% of the deaths in men aged 45-64 years are attributed to CHD. Although coronary care units and coronary by-pass surgery may have contributed to the quality of life in individual cases, it has generally been recognized that prevention only can reduce the great burden of mortality from coronary heart disease in our countries. Many studies conducted over the last 35 years or so have identified numerous risk factors; amongst them are high blood pressure, raised serum cholesterol, cigarette smoking, diabetes, obesity, sedentary life and positive family history. The first three are known as the major risk factors because of the stronger and more consistent association. However, it is difficult to reconcile with the widespread view that reduction in these risk factors will result in the prevention of the disease.

Marmot and Winklestein (1975) re-examined the data from the National Pooling Project (Intersociety Commission, 1970). Table 1 shows classification of 7352 men between 30-59 years by the risk factor status at entry and the number of cases that occurred in each group within ten years of follow-up.

In the highest CHD risk group with the presence of all three risk factors, 14% of the individuals developed CHD in the next ten years. In other words, 86% of the high-risk group did not develop CHD in ten years without any intervention. Alternatively, it can be seen that 83% of the CHD cases could not be predicted by the presence of all three risk factors as only 17% of the cases came from this high-risk group. If the

Table 1. Ten-Year Incidence of CHD in 7352 men aged 30-59

Risk factor status	10 yr. incidence % affected	Cumulative % of case
All three	14	17
Any two	9	58
Any one	5	94
None	2	100

Based on the National Co-operative Pooling Project
Adapted from Marmot and Winklestein (1975)

high-risk CHD group is enlarged by including population with
two or more risk factors, we can predict 58% of CHD cases.
However, this increase in sensitivity is gained at the expense
of specificity. Thus, of those people in this enlarged high
CHD risk group approximately 9% could be expected to get CHD.
If there were 100% successful pharmacological measures, they
would have to be applied to 100 people to save nine lives, and
if the compliance rate and degree of success expected from this
hypothetical drug or drugs is, say, only 50% each -- (and this
is not a far-fetched idea at all as we often accept drugs at
much lower efficiency) -- then the whole exercise is likely to
save 2.5 lives. Against that, one must also take into
consideration the possible hazards of drugs, as we have already
seen in the case of Practolol, and recently Atromid-S
(Committee of Principal Investigators, 1978), as well as the
total cost.

Many studies reported in the literature indicate the
weakness in predicting future CHD from the level of risk
factors. For example, Gordon, Garcia-Palmier, Kagen, Kennel,
and Schiffman (1974) compared incidence rate of CHD found in
studies which had used uniform methods in Framingham, Honolulu
and Puerto Rico. The incidence of CHD in Framingham was
two-three times higher than that in Puerto Rico and Honolulu.
When univariate controls were made each for blood pressure,
serum cholesterol and smoking, the excess in Framingham
persisted. In other words, the higher rate in Framingham could
not be accounted for by the conventional risk factors.
Smoking, for example, had no effect on the incidence rate of

CHD in Puerto Rico. Similarly, in the "Seven Countries Study" (Keys, 1970), smoking was not found to be a significant risk factor.

The literature abounds with such discrepancies for every single conventional risk factor. Thus, it is reasonable to conclude that although the ability to predict CHD from the presence of these widely accepted risk factors is very impressive, a substantial proportion of CHD must occur for reasons other than these risk factors. Maybe it is because of this that a number of intervention trials using diets and drugs have consistently shown poor results (reviewed by Mann, 1977). When a disease is responsible for the death of one in every twelve men under the age of retirement (Rose, 1976), when their families and communities need them most, the very least we can do is to mobilize all our resources and explore any new hypothesis. It is my aim to put forward a hypothesis and support it with the results of studies carried out so far. There is a lot of work that needs to be done before the hypothesis can be proven or the appropriate intervention can be shown to be successful, but at least a start has been made.

Stress Hypothesis

Psychosocial occupational stress has been considered by some to be an important risk factor, but there are more who strongly refute this. I suggest that psychosocial stress may be a causative factor (see Figure 1) which working through appropriate neuroendocrine stimulation and biochemical disturbances, leads to sudden death, probably through the electrical disturbance of the conducting system. It may also lead to myocardial infarction through gradual development of atherosclerosis, thrombosis or prolonged vasospasm of the coronary arteries. The disease can occur directly without the involvement of the conventional risk factors or it can occur indirectly through a network of known and some as yet unknown risk factors. It is well recognized that emotional stress can be one of the factors which lead to overeating, alcohol drinking or cigarette smoking. It may also lead to aggressive or Type A behavior (Friedman & Rosenman, 1974) and physical inactivity, possibly by promoting early fatigue (Nixon, 1976). Overwhelming stress can cause glucose intolerance or hyperglycemia (Hinkle & Wolf, 1952). Its contribution to hypertension (Henry & Cassel, 1969; Gutman & Benson, 1971) and hypercholesterolemia (Thomas & Murphy, 1958; Friedman, Rosenman, & Carrol, 1958) is gradually being revealed.

If the hypothesis is right, we can expect to prevent CHD by dealing with psychosocial stress and increasing the individual's coping ability. The attraction of this hypothesis

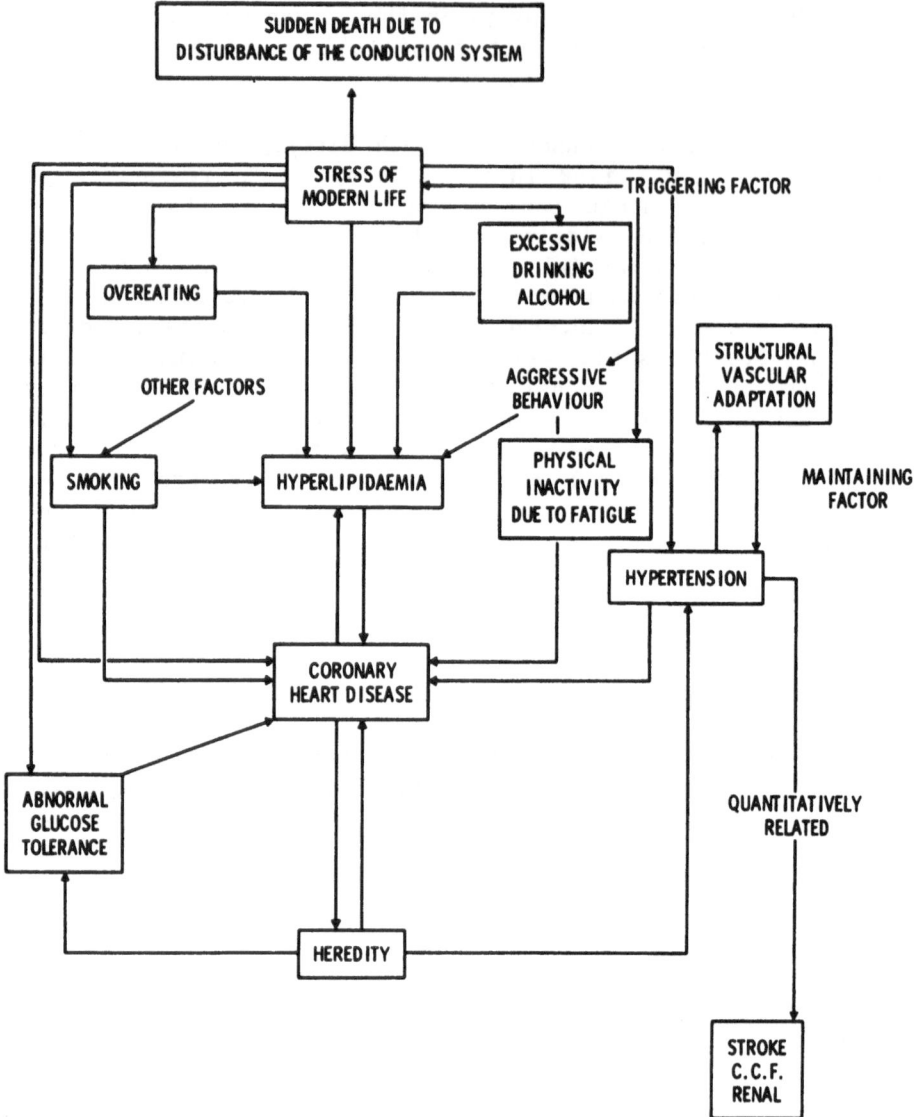

Figure 1.   Relationship between emotional stress and coronary
            heart disease: a hypothesis

is that it can explain why some patients who get myocardial
infarctions are overweight, while others have high serum
cholesterol or high blood pressure, depending upon their
genetic susceptibility. It can also explain the cases who do

not have any of the recognizable risk factors. It is a working hypothesis upon which we should gather evidence without joining any specific lobby or being overawed by the eminence of the faction who would like us to believe that there is no case. The studies which I carried out earlier showed how hypertension may be reduced by behavioral methods aimed at counteracting environmental stress. In the last few years, other colleagues have joined me and we have extended the work to study the effect on all three major risk factors. Following is a discussion of each of these three risk factors to see if by adding a new dimension, psychosocial stress, we can explore the possibility of at least strengthening our intervention program.

Essential Hypertension

When I started my work in the early 1970's, we knew that lowering blood pressure with antihypertensive drugs, at least in moderate and severe hypertension with diastolic blood pressure of 105 mmHg or more, reduces the incidence of these complications (Veterans Administration, 1967, 1970). Yet in a number in surveys, both of the United States and the United Kingdom (Miall & Chinn, 1974), it had been shown that about half of the hypertensives were not detected, and of those detected half were not treated. Of those treated only half were ideally controlled. Even when the treatment was accepted, earlier studies had shown that the incidence of uremia, heart failure and stroke had gone down, but that of myocardial infarction remained unchanged (Breckenridge & Dollery, 1970; Veterans Administration, 1967, 1970).

Recently, two studies, one Australian (Management Committee, 1980) and the Hypertension Detection and Follow-up Program (HDFP, 1979), from the U.S., have shown that lowering of blood pressure in mild hypertension is beneficial. However, if we are to treat all patients, including those with mild hypertension, which constitutes over 70% of all hypertensives (HDFP, 1979) we shall end up putting a third of the middle-aged population (Hawthorne, Greaves, & Beevers, 1974) on anti-hypertensive drugs. It is too early to say how the treatment will be accepted by the people with diastolic blood pressure of around 90 mm. and what the safety record of long-term medications will be. So all in all, there are still very good reasons to develop a non-drug approach which is safe and accepted by the patients, which is effective not only in reducing blood pressure but also in preventing myocardial infarction.

The Concept of Self-Control

The concept of autonomic control or biofeedback has created a considerable amount of interest in the last few years. In the biofeedback control of blood pressure, the biofeedback instrument gives an objective measure of blood pressure to the patient, who tries to control it by an act of volition. This knowledge of objective measure helps him to control his pressure, but we must remember that with every change in the objective measure, there is an associated subjective state which is responsible for that change.

It is generally agreed that hypertension results from an interaction between genetic predisposition and environmental factors. There is nothing we can do about inheritance. But, looking into the environmental factors, it seems that environment working through the mind has been incriminated (Pickering, 1968). In order to counteract environmental influences, one has to either remove the noxious environments or learn to respond to them more appropriately. The evidence suggests that the environments of industrialized, urbanized, competitive society are pathogenic (see Henry & Cassal, 1969 for review). However, most of us would not want to relinquish the achievements of modern civilization and revert back to primitive, rural society. So the only possible thing to do is, for a susceptible individual, to learn to modify his physiological response to his environment, so as to mitigate his genetic tendency.

Therapeutic Behavior

This is based on the understanding of the physiological response involved and consists of the following:

Appraisal. The patient is told that the intensity of the response depends upon his mental evaluation of a particular situation. The response is mobilized when a person perceives situations as threatening or over-demanding. Audio-visual methods are used to demonstrate appropriate and inappropriate responses in everyday life, as well as realistic and unrealistic fears and aggression. In other words, the patient is made more and more aware of his inappropriate responses and given an opportunity as well as the know-how to correct them.

Breathing Exercise. He is taught simple breathing exercise at first. It is known that breathing is erratic when a person is excited, yet slow and regular when he is calm and composed. By a simple, rhythmic diaphragmatic breathing exercise, a certain amount of physical calmness is induced.

This exercise can be performed anywhere and in any position without anybody even noticing it. This is followed by deep muscle relaxation.

Deep Muscle Relaxation. The person is asked to lie down, close the eyes and systematically relax each part of the body. He is told that for full benefit he should do this exercise with empty stomach and bladder. The fact that deep muscle relaxation reduces the intensity of the hypothalamic response is evident from animal experiments. For example, increase in proprioception through passive movements produced the intensity and a greater rise in blood pressure, while a decrease in proprioception through curarization decreases the intensity and causes a smaller rise in blood pressure when the hypothalamus is electrically stimulated (Gellhorn, 1964; Hodes, 1962; Hess, 1957). It is thought that the intensity of the response is directly proportional to the amount of sensory input to the brain. In this context, it is also interesting to note that an increase in isometric contraction, such as a tight hand grip or the carrying of a heavy suitcase, a considerable rise in blood pressure has been observed (Lind, Taylor, & Humphreys, 1964). Maybe we live in a world with far too much sensory stimulation. It is assumed that a reduction in sensory input due to relaxation would reduce the sympathetic responsiveness of the hypothalamus, and eventually lower blood pressure.

Meditation. After a few sessions in breathing exercise and deep muscle relaxation, a type of mental relaxation is introduced in the form of passive concentration and eventually meditation. Whether any of the patients reaches anywhere near a meditative state is highly questionable, but whatever mental calm he can get is useful. One definite advantage, however, is that it at least prevents sleep. Relaxation is very conducive to sleep because of the very mechanisms of sleep (Magoun, 1963), but if the patient is allowed to sleep then the whole concept of voluntary control is nullified. Meditation is also known to change the EEG pattern into a more synchronized one with high amplitude, slow wave pattern of the relaxed brain not passing into sleep (Wallace & Benson, 1972). It is also known to increase coherence between two hemispheres as well as between the anterior and posterior parts of each hemisphere (Banquet, 1973).

Biofeedback. The points so far discussed in the program are aimed at reducing the levels of arousal. The biofeedback principle is used here to train patients more efficiently to shift into a low arousal state. One of the two very simple instruments are used. A Galvanic Skin Resistance (GSR) machine which informs the patient about his level of skin resistance

and indirectly of the level of his arousal, or an EMG
(Electromyographic Feedback Machine) which continuously
measures and displays his level of muscular tension. The idea
behind this procedure is that the knowledge of results
reinforces the learning. In addition to this relaxation
feedback, the patient is also given an overall feedback of his
blood pressure level at the end of each session. Every success
the patient has is taken as an opportunity to raise his
self-esteem and his motivation to continue the program on a
long-term basis. Each session lasts about half an hour. In my
practice, the sessions vary from three per week for three
months to two per week for six weeks. In addition, the patient
is asked to practice twice a day on his own for 15 to 20
minutes. Recent studies have included loaning the patient an
instruction cassette tape for home practice.

Integration of Behavior. The last point in the plan is
desensitization. It is assumed that one needs to desensitize
the patient against the whole environment of industrialized,
urbanized society. A counter-conditioning method is used in
which fear or aggression, inducing stimulus is paired with
another neutral stimulus, such as relaxation, which inhibits
fear or aggression (Wolpe, 1958). For example, car driving is
one of the modern activities which raises blood pressure in
some individuals and causes aggression. What the patient is
asked to do is to take one deep breath relax and "let go" at
every red traffic light or intersection. He uses the same
method before answering a telephone, speaking in public, during
an interview, while waiting for a bus or in a dentist's
surgery, and so on. A list of stressful situations can be made
up by an individual to suit his requirement. A tiny, colored
paper disc is stuck to his wrist watch dial, so every time he
looks at the watch he is reminded to relax.

Studies of Behavior Therapy in Hypertension

Patel (1973, 1975) studied twenty hypertensive patients.
The only criterion was that they were known to be hypertensive
for at least a year. The hypertensive controls were added from
the age and sex register. The groups were comparable with
respect to mean duration of hypertension and severity of
hypertension. Most were controlled on antihypertensive
medications. At the end of three months training program of
three sessions per week, mean systolic pressure was reduced by
20.4 mmHg, while the mean diastolic pressure was reduced by
14.2 mmHg in the treatment group. In addition to this highly
significant reduction in blood pressure (P = < .001), it was
also possible to reduce drugs in 12 patients ranging from
33-100%. The control group patient also attended the same
number of sessions and had the same number of blood pressure

measurements made, but instead of being trained in behavior modification, was asked to lie down and rest. The average reduction in systolic and diastolic pressures in this group were 0.5 and 2.1 mmHg respectively. These were not significant and their drug requirement remained unchanged.

In a condition like hypertension, where the spontaneous fluctuation is so common, the value of any therapy can be assessed by its long-term effectiveness. The patients in both groups were followed monthly up to one year. Except for some minor changes in blood pressure and drug requirements in both the groups, the reduction obtained in the treatment group was maintained. In some patients the blood pressure started to go up. They turned out to be mostly those who could not continue the regular practice of relaxation. However, with restarting the relaxation practice, blood pressure came down again. The lesson learned during this follow-up study is the clear importance of motivating these patients for a life-long discipline of practicing and integrating relaxation into their daily activities.

Having seen fairly impressive results, it became important that we should have more convincing evidence about the effectiveness of this therapy. A randomized controlled study was, therefore, carried out (Patel & North, 1975). In order to avoid any subjective and objective bias which may have existed in the previous study, all the blood pressure measurements were made by an experienced nurse using a random zero sphygmo-manometer (Wright & Dore, 1970). The antihypertensive medications were kept constant throughout the study. We had 34 patients who were known to be hypertensive for at least six months with original diastolic blood pressure of 110 mmHg or more before stabilizing them on medications. They came on three separate days for baseline blood pressure measurements which were averaged to give the initial value (see Figure 2). They were then randomly allocated to either treatment group or a control group. Both the groups attended twice a week for six weeks for half-hour sessions. The treatment group patient was offered the treatment already described while the control group patient was asked to lie down and relax in his own fashion. The number of blood pressure measurements were kept the same in both the groups. After the six-week program, each patient was followed up once every two weeks for three months. The averages of all the measurements made during the follow-up are given as final values for each group (see Figure 2). The blood pressure fell by a small but significant degree in the control group. However, the fall in both systolic and diastolic blood pressures in the treated group was much greater and the differences were highly significant (P < .001). The average drops in the treated group were 26 mm systolic and 15 mm

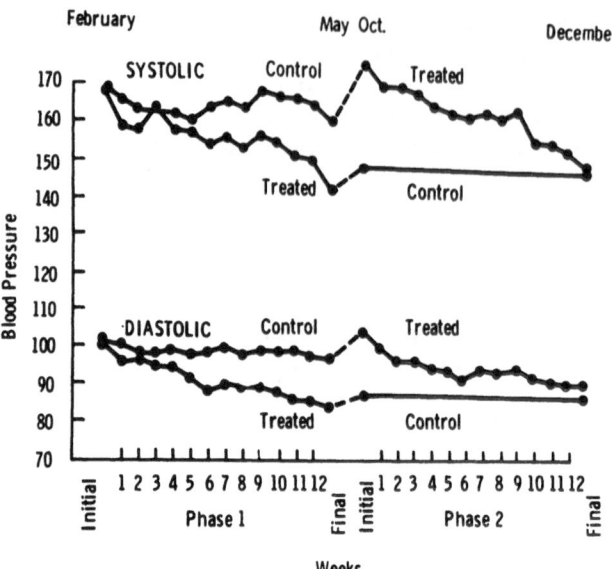

Figure 2.   Average measurements during three month
            followup as final values

diastolic pressures. The respective drops in the control group
were 8 and 4 mmHg. After two months, all the patients were
recalled. Surprisingly, the pressures in the control group had
gone back to their original levels, while most of the
reductions obtained in the treated group were maintained. The
treated group was, of course, asked to continue with the
practice of relaxation and its integration into everyday life.
Perhaps the results in the control group are typical of a
placebo group in the sense that the placebo effects persist as
long as placebo factors are operating, whatever they may be.

        By chance, the random allocation had not divided all
the independent variables equally between the groups.
Therefore, we decided to extend the study to an additional
phase during which we offered treatment to the previous control
group over a six-week period while the already treated group
now became the control group and attended once at the beginning
and again at the end of Phase 2 for reference purposes. As can
be seen in Figure 2, the blood pressure of this newly treated
group fell in a similar fashion and at the end of six weeks,
the average blood pressure of this group came down to the level
of the previously treated group, showing once more that the
treatment is effective, when carried out in the manner
described.

Relaxation in Reducing Serum Cholesterol

Friedman et al. (1958) have shown that occupational deadlines and increased pressure of work in a group of accountants prior to tax deadlines were associated with a small, but significant, increase in serum cholesterol. They have also shown in a prospective study that in people with aggressive Type A personality there is a higher level of serum cholesterol and higher mortality from CHD (Friedman & Rosenman, 1958, 1974). Other investigators have also shown a rise in various blood lipids in association with experimental and other naturally occurring psychosocial stress (Thomas & Murphy, 1958; Dreyfuss & Czaczkes, 1959). If relaxation-based behavior modification can reduce high blood pressure as well as emotional and occupational stress, it is possible that it may also reduce blood cholesterol and other lipids.

In a pilot study involving 14 hypertensive patients (Patel, 1976), the results over a six-week period showed a highly significant reduction in mean systolic pressure from 170.6 to 147.9 mmHg and diastolic pressure from 102.5 to 89 mmHg. The mean cholesterol level in this group of patients was reduced from 241.6 to 217 mg/100 ml (P = <.001). Thirteen out of 14 patients showed some decrease in serum cholesterol. However, this was an uncontrolled study.

In another pilot study (Patel & Carruthers, 1977), four groups of subjects were studied. A normotensive group of 18 subjects acted as control; another normotensive group of 18 subjects was treated by the biofeedback-relaxation-meditation-behavioral modification program; a group of 22 hypertensives and a group of 18 current smokers were similarly treated. The results showed significant reduction in blood pressures in all the treated groups with no significant change in the control group. Plasma cholesterol, triglycerides and free fatty acids were reduced significantly in some but not in all the treated groups, with no change in the control group. In smokers, the number of cigarettes smoked was reduced by 48% at the end of the first week, dropping gradually week by week until an approximate 80% reduction was achieved by the end of a six-week training period. However, not all the reduction was maintained, but at six-month follow-up the total number of cigarettes smoked was still 60% less than the original. Most of these smokers wanted to give up smoking and had themselves approached the local anti-smoking clinic for help. There are dangers in extrapolating results from a volunteer group to smoking population in general.

Multiple Risk Factor Intervention by Behavior Modification

Having seen a fairly convincing evidence that relaxation-based behavioral modification can reduce hypertension and some indication that it may also reduce serum cholesterol and cigarette smoking, a randomized controlled trial was set up to see if all the major risk factors can be reduced at once in unselected group of people engaged in full-time jobs (Patel, Marmot, & Terry, 1981). One thousand, one hundred thirty two employees of a large manufacturing industry between 35-64 years of age were screened. Those with two or more risk factors were reexamined. The risk factors were defined as an average of two measurements of blood pressure to be 140/90 or more; serum cholesterol of 6.3 mmol/1 or more and current cigarette smoking of ten or more cigarettes per day. If the person still qualified on the grounds of two or more risk factors at the second examination, he was invited to participate in the study and if he consented a further sample of blood was withdrawn for plasma renin activity (PRA) and plasma aldosterone assays. Two hundred and four or 88% of those qualified consented and were randomly allocated to treatment and control groups. The groups were comparable at entry as seen in Table 2.

The management for both groups consisted of ten minutes of individual counseling about their risk factors and distribution of health education literature on blood pressure, smoking or dietary fats. The treatment group in addition followed a behavior modification program consisting of one-hour group sessions once a week for eight weeks and a three-hour of stress management education. Training was carried out in groups and consisted of breathing exercise, deep muscle relaxation and meditation, which was enhanced by a specially developed multi-circuit GSR feedback instrument. The stress education program consisted of understanding stresses and strains; effect of psychosocial stimulation on blood pressure and coronary heart disease; the rationale of relaxation; the WHAT, WHY and the HOW of meditation and learning to integrate relaxation in everyday life. The subjects were requested to practice relaxation meditation twice a day and were loaned the instruction cassette for home practice for eight weeks. They were further assessed at eight weeks and again at eight months.

The results (Table 3) showed significantly greater reduction in systolic and diastolic pressures in the treatment group whether the analysis included the whole group or was confined to the high-risk subgroup (see Table 2) whose initial pressure was 140/90 or more (P = .001). These reductions were maintained at eight-month follow-up. Serum cholesterol was

Table 2.   Comparison between treatment and control
           groups at entry to the Study.

	Biofeedback n = 104	Control n = 97
No completed 8 weeks protocol	99	93
% male	61	62
% female	39	38
Mean syst. B.P.)$_1$	145.2	144.2
Mean diast. B.P.)^1	87.4	87.9
Mean syst. B.P.)$_2$	160.1	158.9
Mean diast. B.P.)^2	100.1	98.2
% with syst. B.P. $\geqslant 140$ mmHg	64	62
% with diast. B.P. $\geqslant 9$ mmHg	49	51
Mean cholesterol (mmol/l)	6.9	7.1
No. of cig/day in smokers	19	20
% smokers	82	70

1 = Whole group
2 = High risk group who met defined risk level criteria

significantly lower in both groups at eight weeks as well as at
eight months (P = <.001).   However, the greater drop in the
treatment group was only significant at eight weeks and was
confined to the high-risk subgroup (P = <.025).

Our dietary analysis showed that groups were similar at
entry in their intake of total calories and animal fats and
that both groups had reduced their intake of saturated fats and
increased polyunsaturated fats.   To the degree that the dietary
assessments (Heart Disease Prevention Project questionnaire,
Rose, Heller, Pedee, & Christie, 1980) were accurate and
unbiased, the control group had made slightly greater change in
the dietary fats.   Therefore, the greater reduction in serum
cholesterol evident at eight weeks in the treatment group is
likely to be due to relaxation.   Body weight was measured at
entry and at each follow-up examination.   Despite the reported

dietary changes, there was no significant change in either group. It is further evidence that blood pressure and cholesterol differences between the groups were unrelated to dietary or weight changes.

Finally, for people in the two groups who initially were smokers, 68% in the treatment group claimed to have reduced the number of cigarettes smoked per day, compared with 39% in the control group. Of these, 11% in the treatment group compared with 7.8% in the control group completely stopped smoking. The mean number of cigarettes consumed fell by six in the treatment group, compared with three in the control group. These differences were statistically significant. At eight months, the differences between the groups were maintained and altogether 10% in the treatment group, compared with 5% in the control group, had stopped smoking. Although these differences were statistically significant, one might have hoped for a greater effect judging from the result of our pilot study (Patel & Carruthers, 1977).

Plasma renin activity and aldosterone were analyzed in a sub-sample of 54 subjects. There were significantly greater reductions in both the parameters in the treatment group at eight weeks, but not at eight months ($P = <.05$). There was no correlation between the changes in blood pressure and changes in PRA in either group, but there were significant correlations between changes in aldosterone and changes in both systolic and diastolic blood pressure at eight weeks. Plasma renin activity is mediated through beta-adrenoreceptors, while it is possible that blood pressure was reduced by a central mechanism involving both alpha and beta adrenoreceptors as well as hormonal changes.

It was impractical to perform the study blind, when the subjects are expected to change their behaviors consciously. Therefore, attempts were made to standardize the measurements as closely as possible using a trained nurse to take blood pressure measurements with random-zero sphygmomanometer. This does not eliminate the bias completely, but it does reduce it. Smoking was assessed by questionnaire rather than confronting the subjects face to face in order to minimize the tendency to please the experimentor. The treatment group may still have exaggerated their reduction. Therefore, caution is in order.

It is unlikely that the blindly carried-out laboratory measurements of cholesterol, PRA and aldosterone were biased. This increases our confidence that observed blood pressure changes were neither the result of observer bias, nor merely short-term reactions at the time of measurement. However, reduction in biochemical measures were significant at eight

Table 3. Reductions in risk factors after
eight weeks and eight months

Reduction in risk factors	At Eight Weeks		At Eight Months	
	Biofeedback Group	Control Group	Biofeedback Group	Control Group
	Mean ± S.E.	Mean ± S.E.	Mean =+ S.E.	Mean =+ S.E.
Systolic B.P. (mmHg)				
Whole Group	*** 13.8 ± 1.34	4.0 ± 1.30	*** 15.3 ± 1.55	6.1 ± 1.56
Initial B.P. High ⩾ 140 mmHg	*** 18.0 ± 1.72	8.8 ± 1.42	** 20.2 ± 1.99	11.2 ± 1.70
Diastolic B.P. (mmHg)				
Whole Group	*** 7.2 ± 0.91	1.4 ± 0.81	*** 6.8 ± 0.09	0.63 ± 0.91
Initial B.P. High ⩾ 90 mmHg	*** 10.6 ± 1.37	3.8 ± 1.04	*** 11.5 ± 1.32	3.2 ± 1.32
Cigarette Smoking				
% reporting reduced smoking	*** 67.9 ± 5.19	39.1 ± 6.10	*** 67.5 ± 5.34	37.5 ± 6.47
Av. No. of cig/day less	** 5.8 ± 0.73	2.6 ± 0.75	* 4.8 ± 0.77	2.3 ± 0.74
Plasma Cholesterol (mmol/1)				
Whole Group	0.71 ± 0.12 *	0.53 ± 0.11	0.63 ± 0.09	0.57 ± 0.10
Initial choles. High ⩾ 6.3 mmol/1	0.90 ± 0.12	0.52 ± 0.12	0.77 ± 0.10	0.56 ± 0.11

Between group difference   Significance test (two tailed)
  *   .01 < P < .05
 **   .001 < P < .01
***   P < .001

weeks only. This suggests that mechanisms of acute blood pressure reduction might be different from those responsible for long-term reductions. It is a common experience that blood pressures of patients on long-term antihypertensive medications may not revert to their original levels for some weeks or months after the medications are stopped. This may be because of some adaptive changes which have been demonstrated in experimental studies (Folkow, Hallback, Lundgren, Silvertsson, & Weiss, 1973; Vaughan-Williams, Hassan, Florats, Sleight, & Jones, 1980). It is possible that a partial reversal of some of the factors responsible for maintaining and perpetuating hypertension and its persistence by limited practice of regular relaxation may be responsible for long-term maintenance of blood pressure reduction.

Conclusions

Although the hypothesis that psychosocial stress is a causative factor is not yet proven, there is enough evidence from the studies carried out so far that it is at least an important risk factor. Assuming that the greater reductions in blood pressure cigarette smoking and cholesterol achieved in this study would reduce the mortality from CHD, we can calculate the potential reduction in mortality using the multiple logistic function from the London Whitehall Study (Marmot et al., 1978). The figure at eight weeks is 21% reduction in the predicted risk of CHD death. At eight months, the figure is 18%.

This may not sound very impressive, but when one considers the fact that subjects in this study had elevations of risk factors too mild to warrant the hazards of pharmacological intervention and yet serious enough to increase their risk of dying from CHD, the results obtained may not be a mean achievement. It is possible that reduction in risk may occur through paths other than through the conventional risk factors, although it is not possible to estimate its magnitude from the present study. In fact, the results are quite encouraging when comparisons are made with other multiple risk factor intervention studies using conventional methods (Puska et al., 1979; Farquhar et al., 1977; Rose et al., 1980). The predicted reductions in mortality from CHD in these studies followed up to five-year periods have been estimated from 9 - 17.4%. Follow-up period in our study has been comparatively short. Therefore, we must remain cautious in making claims. However, there is no reason why relaxation therapy cannot be combined with conventional therapies so that the future intervention trials not only show large enough differences in morbidity and mortality to be detected by the available statistical means, but also make an impressive contribution in eradicating the great burden of mortality from CHD in our communities.

REFERENCES

Banquet, J. P. Spectral analysis of the EEG in meditation. Electroencephalography and Clinical Neurophysiology, 1973, 35, 143-151.

Breckenridge, A., & Dollery, C. T. Changing pattern of death
     in hypertension. British Heart Journal, 1969, 31, 387.
Committee of Principal Investigators. A co-operative trial in
     the primary prevention of ischemic heart disease. British
     Heart Journal, 1978, 40, 1069-1118.
Farqhar, J. W., MacCoby, N., & Wood, P. D. Community education
     for cardiovascular health. Lancet, I, 1192-1195.
Folkow, B., Hallback, M., Lundgren, Y., Silvertsson, R., &
     Weiss, L. Importance of adaptive changes in vascular
     design for establishment of primary hypertension. Studied
     in men and in spontaneously hypertensive rats.
     Circulation Research Supplement, 1973, Vol 1 to Vol.
     32-33, 2-16.
Friedman, M., Rosenman, R. H., & Carrol, V. Changes in the
     serum cholesterol and blood clotting time in men subjected
     to cyclic variation of occupational stress. Circulation,
     1958, 17, 852-861.
Friedman, M., & Rosenman, R. H. Type A Behavior and Your Heart.
     New York, Alfred A. Knopp, 1974.
Gellhorn, E. The influence of curare on hypothalamic
     excitability and the electroencephalogram.
     Electroencephalography and Clinical Neurophysiology,
     1957, 27, 697-703.
Gordon, T., Garcia-Palmier, M. R., Kagan, A., Kennel, W. B.,
     & Schiffman, J. Differences in coronary heart disease in
     Framingham, Honolulu, and Puerto Pico. Journal of Chronic
     Diseases, 1974, 27, 329-344.
Gutmann, H. C., & Benson, H. Interaction of environmental
     factors and systemic arterial blood pressure: A review.
     Medicine, 1971, 50, 543-553.
Hawthorne, V. M., Greaves, D. A., & Beevers, D. G. Blood
     pressure in a Scottish town. British Medical Journal,
     1974, 3, 600-603.
Henry, J. P., & Cassel, J. C. Psycho-social factors in
     essential hypertension. Recent epidemiological and animal
     experimental evidence. American Journal of Epidemiology,
     1969, 90, 171-200.
Hess, W. R. In J. R. Hughes (Ed.), Functional organization of
     diencephalon. Grune and Stratton, 1957.
Hinkle, L. E., & Wolf, S. A summary of experimental evidence
     relating life stress to diabetes mellitus. Journal of the
     Mount Sinai Hospital, 1952, 19, 537-570.
Hodes, R. Electroencephalographic synchronization resulting
     from reduced proprioceptive drive caused by neuromuscular
     blocking agents.
Hypertension Dection and Follow up Program Cooperative Group.
     Five year findings of the hypertension detection and

follow up program, reduction in mortality of persons with
high blood pressure, including mild hypertension. Journal
of the American Medical Association, 1979, 242, 2562-2571.

Intersociety Commission for Heart Disease. Primary prevention
of the atherosclerotic diseases. Circulation, 1970, 42
A55-A95.

Keys, A. Coronary heart disease in seven countries.
Supplement I to Circulation, 1970, 41 & 42.

Lind, A. R., Taylor, & S. H., Humphreys, P. W. The circulatory
effects of sustained voluntary muscle contraction.
Clinical Science, 1964, 27, 229-244.

McCubbin, J. W., Green, J. H., & Page, I. H. Baroreceptor
function in chronic renal hypertension.
Circulation Research, 1956, 4, 205-210.

Management Committee. The Australian therapeutic trial in mild
hypertension. Lancet, I:1261-1267.

Mann, G. V. Diet-heart: End of an era. New England Journal of
Medicine, 1977, 297, 644-650.

Marmot, M. G., Rose, G., Shipley, M., & Hamilton, P. J. S.
Employment grade and coronary heart disease in British
Civil Servants. Journal of Epidemiology and Community
Health, 1978, 32, 244-249.

Marmot, M., & Winkelstein, W. Epidemiologic observations on
intervention trials for prevention of coronary heart
disease. American Journal of Epidemiology, 1975, 101,
177-181.

Miall, W. E., & Chinn, S. Screening for hypertension: Some
epidemiological observation. British Medical Journal,
1974, 3, 595-600.

Magoun, H. W. The Waking Brain, Charles Thomas, Sprinfield,
Illinois, Second Edition, 1963.

Nixon, P. Human function curve with special reference to
cardiovascular disorders. Practitioner,
1976, 217, 765-770.

Patel, C. H. Yoga and biofeedback in the management of
hypertension. Lancet, II:1053-1055.

Patel, C. H. & North, W. R. S. Randomized controlled trial of
yoga and biofeedback in the management of hypertension.
Lancet, II, 93-95.

Patel, C. H. Twelve-month follow up of yoga and biofeedback in
the management of hypertension. Lancet, I, 62-65.

Patel, C. H. Reduction of serum cholesterol and blood pressure
in hypertensive patients by behavior modification.
Journal of the Royal College of General Practitioners,
1976, 26, 211-215.

Patel, C. H. Biofeedback-aided behavioral methods in the
management of hypertension. M.D. Thesis, University of
London, 1976.

Patel, C. H. Biofeedback-aided relaxation and meditation in
the management of hypertension. Biofeedback and

Self-Regulation, 1977, 2, 1-44.
Patel, C. H., & Carruthers, M. Coronary risk factor reduction
    through biofeedback-aided relaxation and meditation.
    Journal of the Royal College of General Practitioners,
    1977, 27, 401-405.
Patel, C., Marmot, M. M., Terry, D. J. Controlled trial of
    biofeedback-aided behavioral methods in reducing mild
    hypertension. British Medical Journal, in press.
Pickering, G. W. High Blood Pressure, Churchill Livingtone,
    London, 2nd Edition, 1968.
Puska, P., Tuomilehto, J., & Salenon, J. Changes in coronary
    risk factors during comprehensive five-year community
    program to control cardiovascular disease (North Karelia
    project). British Medical Journal, 1979, II, 1173-1178.
Reid, D., Hamilton, P. J. S., Keen, H., Brett, G. Z.,
    Jarrett, R. J., & Rose, G., Cardiorespiratory diseases and
    diabetes among middle-aged male civil servants, Lancet,
    1974, I, 469-473.
Rose, G., Heller, R. F., Pedee, H. T., & Christie, D. G. S.
    Heart disease prevention project: A randomized controlled
    trial in industry. British Medical Journal, 1980, 1,
    747-751.
Rose, G. Coronary heart disease: Check the "healthy" patient.
    Modern Medicine, 1976, 21, 6-11.
Shapiro, D., Tursky, B., & Schwartz, G. E. Control of blood
    pressure in man by operant conditioning.
    Circulation Research Supplement, 1970, 1.27-1.31.
Sleight, P. Baroreceptor function in hypertension. In
    Berglund, Lindgren & Soner, Pathophysiology and Management
    of Arterial Hypertension. Molndal, Sweden, 1975.
Thomas, C. B., & Murphy, E. A. Further studies on cholesterol
    levels in the John Hopkins medical students: The effect of
    stress at examination. Journal of Chronic Diseases,
    1958, 8, 661-
Vaughan-Williams, E. M., Hassan, M. O., Florats, J. S.,
    Sleight, P., & Jones, V. J. Adaptation of hypertensives
    to treatment with cardioselective and non-selective
    beta-blockers. Absence of correlation between bradycardia
    and blood pressure control and reduction in slope of QT/RR
    relation. British Heart Journal, 1980, 44, 437-487.
Veterans Administration Co-operative Study Group on
    Antihypertensive Agents. Effects of treatment on
    morbidity in hypertension. Results in patients with
    diastolic blood pressure averaging 115 through 129 mmHg.
    Journal of the American Medical Association, 1970, 213,
    1143-1152.
Wallace, R., & Benson, H. The physiology of meditation.
    Scientific American, 1972, 226, 84-90.
Wright, B. M., & Dore, C. F. A random-zero sphygmomanometer.
    Lancet, I, 337-338.

Self-Regulation

SECONDARY PREVENTION OF CORONARY HEART DISEASE

James A. Blumenthal and Robert Califf

Departments of Psychiatry and Medicine
Duke University Medical Center
Durham, North Carolina

Secondary prevention of coronary heart disease (CHD) is a term used to describe therapeutic efforts in patients with established disease. For patients who have sustained a myocardial infarction or who have developed clinical manifestations of coronary atherosclerosis, strategies for secondary prevention remain a subject of much controversy as well as burgeoning interest. Treatment of coronary disease has received so much attention because of its prevalence. Coronary disease affects 1½ million Americans annually and is the leading cause of death in the United States and many countries in Western Europe. Secondary prevention is controversial, in part, because the factors influencing the prognosis of patients with coronary disease are not fully understood, and the rationale for various interventions is not always evident.

In general, those risk factors that are predictive of initial coronary events are less important once the disease has developed. At Duke University Medical Center, for example, information has been gathered on over 2000 medically treated patients with coronary artery disease documented by coronary angiography. No significant relationship between such traditional risk factors as serum cholesterol, systolic or diastolic blood pressure, presence of diabetes or smoking status at the time of cardiac catheterization and subsequent mortality is evident. Similarly, the Coronary Drug Project Research Group (1980) reported that the traditional risk factors were only weakly related to subsequent mortality in patients after myocardial infarction.

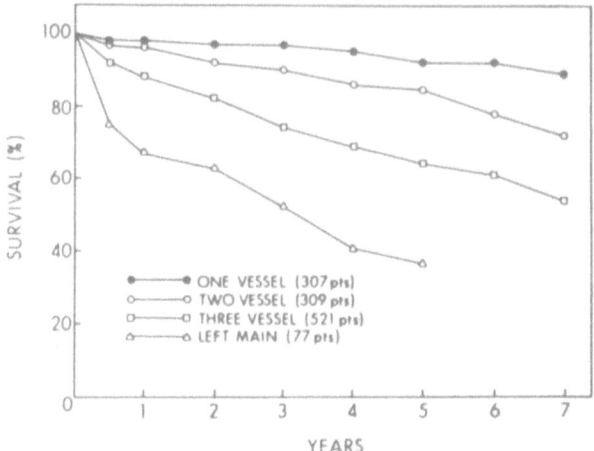

Figure 1.   Cumulative survival rates in patients with one-,
two-, three-vessel and left-main coronary
disease (from Harris et al., 1980a, p. 1264).

While traditional risk factors apparently have minimal
influence on future morbidity and mortality, the primary
determinants of survival appear to be based on the state of the
coronary vasculature, the degree of impairment of left
ventricular function, and the presence and type of clinical
symptoms (Burggraff & Parker, 1975). Figure 1 shows the
cumulative survival of patients with one-, two-, three-vessel
disease and left main coronary disease (Harris, Harrell, Lee,
Behar, & Rosati, 1980a). Left main disease and triple-vessel
disease have the worst prognosis.

Furthermore, the extent of coronary atherosclerosis, the
degree of left ventricular dysfunction, and the pattern of
symptoms of chest pain are additive in their effects on
mortality (Harris, Lee, Harrell, Behar & Rosati, 1980b). This
relationship can be seen in Figure 2.

Thus, the magnitude of anatomic involvement and the degree
of impairment of left ventricular function appear to be the most
important determinants of survival. Progressive chest pain is
an important short-term predictor of the likelihood of
subsequent ischemic events, possibly because the pain pattern is
a manifestation of a critical myocardial oxygen supply and
demand situation. The probability of surviving an ischemic

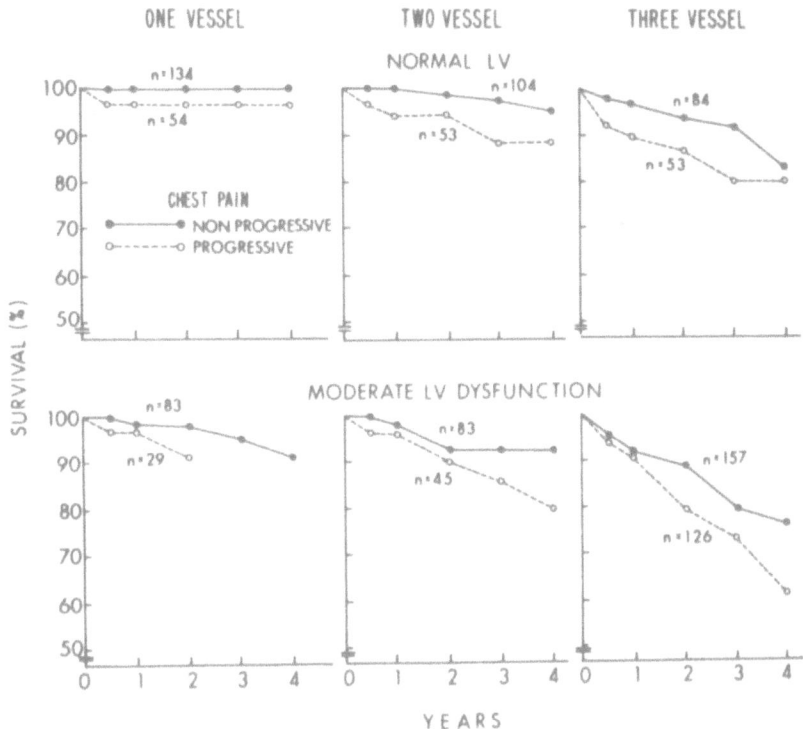

Figure 2.   Cumulative survival rates related to progressive
            and nonprogressive chest pain in patients with
            one-, two-, and three-vessel disease and normal
            and moderately impaired left ventricular (LV)
            function (from Harris et al., 1980a, p. 1266).

event is strongly influenced by the quality of left ventricular
function and the presence of stenosis of the left main coronary
artery. In the absence of left main coronary artery disease,
baseline ventricular function is the most important predictor of
survival over a short follow-up period (Harris et al., 1980b).
Over longer periods, however, the extent of anatomic involvement
and progression in the pattern of chest pain during follow-up
are likely to be the most important predictors of survival,
because these factors are strongly associated with the incidence

of non-fatal infarction. The results of studies of post-infarction patients have yielded similar results (Taylor, Humphries, Mellits, Pitt, Schulze, Griffith, & Achuff, 1980)

Traditionally, the effectiveness of secondary prevention has been measured by rates of cardiovascular morbidity, especially reinfarction, and mortality. The present review will consider three major intervention strategies: surgical, pharmacological and behavioral. The effectiveness and limitations of each treatment modality will be reviewed and directions for future research will be presented.

SURGICAL INTERVENTION

Coronary artery bypass grafting has become a popular form of treatment for coronary disease. Results of the major randomized and observational trials of coronary artery bypass grafting (CABG) were summarized at a recent National Institutes of Health Consensus Conference (Rahimtoola, 1981). For patients with significant stenosis of the left main coronary artery, survival is improved with surgical therapy (Takaro, Hultgren, Lipton, & Detre, 1976). Studies have also found increased survival in other subgroups of surgical patients depending upon considerations of anatomical involvement and left ventricular function. The European Cooperative Study, for example, reported that patients with stable angina, significant triple-vessel disease and normal left ventricular function have improved survival with surgery (European Coronary Surgery Group, 1979). In contrast, the National Unstable Angina Trial (1978) found no benefit in any subgroup in terms of survival of non-fatal myocardial infarction with surgery. Other clinical trials either have not adequately characterized the occurrence of non-fatal infarction or have performed so many statistical analyses that the results are difficult to interpret (Lee, McNeer, Starmer, Harris, & Rosati, 1980).

Numerous studies report improved symptoms and increased exercise tolerance in patients after CABG compared to patients without CABG (National Cooperative Study Group, 1979; Peduzzi & Hultgren, 1979). The one study that investigated the effect of CABG on asymptomatic patients after myocardial infarction found no difference in survival, although the sample size was too small to permit any definitive conclusions (Norris, Agnew, Brandt, Graham, Hill, Kerr, Lowe, Roche, Whitlock, & Barratt-Boyes, 1981).

While CABG has proven to be a useful procedure to ameliorate symptoms and reduce mortality in certain patients, available evidence indicates that coronary disease is usually progressive after surgery. Indeed, one recent study using life table techniques to estimate seven year post-CABG survival reported a 16% reinfarction rate, and a 57% rate of recurrent chest pain (Collins, Tisevova, Mudge, Cohn, & Koster, 1981). The recurrence of symptoms and cardiac events in such a large percentage of patients clearly indicates the need for concomitant treatment interventions when CABG is performed.

PHARMACOLOGICAL INTERVENTION

Beta-adrenergic blockade has been shown to improve the myocardial oxygen supply/demand ratio and may decrease infarct size during the coronary event. Numerous clinical trials with beta-blocking agents have been completed in post-infarction patients. Many of these trials have failed to demonstrate a benefit in terms of mortality or reinfarction (Reynolds & Whitlock, 1972; Baber, Evans, Howitt, Thomas, Wilson, Lewis, Daves, Handler, & Tuson, 1976; Wilcox, Roland, Banks, Hampton, & Mitchell, 1980). These negative trials have been criticized for failure to start therapy early enough, use of small doses, and short duration of therapy (Barber, Boyle, Chaturvedi, Singh, & Walsh, 1980). Several trials with Alprenolol have demonstrated a reduction in cardiovascular mortality, particularly in middle-aged patients (Anderson, Frederiksen, Jurgensen, Pederson, Bechsgaard, Hausen, Nielsen, Pedersen-Bjergaard, & Rasumussen, 1979). In the Multicentre International Study (1975) encouraging results were reported using Practolol. Unfortunately, due to adverse reactions, this drug is no longer in use for long-term therapy. More recently, long-term treatment with Timolol Maleate has been carefully studied (The Norwegian Multicenter Study Group, 1981). A sample of 1884 patients were randomized to drug or placebo with a follow-up period of 12 to 33 months. The study demonstrated a significant reduction in mortality and reinfarction when treatment with Timolol was started 7-28 days after the myocardial infarction. Additional reports from large cooperative studies using propranolol (β-Blocker Heart Attack Study Group, 1981) and metoprolol (Hjalmarson, Herlitz, Malek, Ryden, Vedin, Waldenstrom, Wedel, Elmfeldt, Holmberg, Nyberg, Swedberg, Waagstein, Waldenstrom, Wilhelmsen, & Wilhelmsen, 1981) have recently demonstrated a beneficial effect on survival in post-infarction patients.

Although the role of platelets in acute myocardial infarction and sudden death has not been clearly delineated, circumstantial evidence has accumulated indicating that platelet aggregation may be important. The rationale and results of clinical trials using antiplatelet agents have been extensively reviewed recently (Proceedings of the Workshop on Platelet-active Drugs, 1981). In seven large clinical trials using aspirin with or without persantine, no clear benefit was demonstrated with regard to subsequent mortality. Although differences may have been obscured by inadequate sample sizes in some studies, the maximum estimated benefit from any study was a 30% reduction in mortality. Arguments for further trials with larger sample sizes center around the fact that most studies may have used too large a dose of aspirin and the fact that the therapy has relatively few side effects (Genton, 1981). Another randomized trial using Sulfinpyrazone on 1558 patients after myocardial infarction found a significant reduction in total cardiac mortality and sudden death (The Anturane Reinfarction Trial Research Group, 1980). This study has been criticized by the Food and Drug Administration, however, and final results of a repeat analysis of the data are pending (Temple & Pledger, 1980).

The use of full anticoagulation therapy has been debated for years, especially in survivors of myocardial infarction. The preponderance of evidence is that any therapeutic benefit that occurs with anticoagulation is likely to be minimal in these patients (Soffer, 1976). However, recent evidence demonstrating the central role of thrombosis in acute myocardial infarction (Dewood, Spores, & Norski, 1980) has rekindled interest in anticoagulation to prevent myocardial infarction and in thrombolytic agents to treat myocardial infarction (Ganz, 1981).

Since the majority of cardiac deaths occur suddenly, it has been hypothesized that many of these deaths are due to primary "electrical" instability of the heart (Lown, Podrid, & Graboys, 1980). Support for this argument has come from a variety of sources, including the recent finding that the occurrence of ventricular arrhythmias is associated with increased risk of sudden death (Ruberman, Weinblatt, Goldberg, Frank, Chaudhary, & Shapiro, 1981). The obvious conclusion from the data is that antiarrhythmic therapy might prevent these cardiac deaths. In a review of seven trials of antiarrhythmic therapy prior to 1977, Bigger, Dresdale, Heissenhuttel, Weld, and Wit (1977) concluded that although several studies showed a trend towards increased survival with therapy, no definitive evidence is currently available. There has also been concern that the dosage levels of drugs used may not have been adequate. More recently, Myerberg and coworkers (1979) reported that therapeutic drug levels of antiarrhythmics were associated with enhanced survival of patients who had been resuscitated after an episode of ventricular fibrillation.

Although the important pharmacological result (i.e., prevention of death) of both sulfinpyrazone and beta-blockers may actually be an antiarrhythmic effect, no conclusive evidence has been produced that antiarrhythmics prevent cardiac death in the majority of coronary artery disease patients. Similarly, pharmacological intervention to reduce blood lipids has not been successful in reducing mortality or non-fatal infarction (Coronary Drug Project Research Group, 1979). Modification of lipid levels will continue to be an area of active study as new drugs are introduced with greater pharmacologic effects, particularly on high-density lipoproteins.

BEHAVIORAL INTERVENTION

Smoking. The benefits of nonsmoking have been established in numerous prospective and retrospective studies (U.S. Public Health Service, 1979). The benefits of smoking cessation have usually been inferred from the reduction in morbidity and mortality rates in ex-smokers relative to those who maintained their smoking habit (Doll & Hill, 1964; Kahn, 1966; Hammond & Garfinkel, 1969; Sparrow, Dawber, & Colson, 1978; Friedman, Petitti, Bawol, & Siegelaub, 1981).

Several studies have attempted to assess the benefits of stopping smoking in patients with established coronary disease. Sparrow and co-workers (1978) found an 18.8% mortality rate six years after a documented MI in a group of men who subsequently stopped smoking, compared to a 30.4% mortality rate in smokers who maintained their smoking habit or resumed smoking after their infarctions. This reduced risk of death among post-MI patients who stop smoking is consistent with data from other studies in the United States (Coronary Drug Project Research Group, 1979), Sweden (Wilhelmsen, Sanne, Elmfeldt, Grimby, Tibblin, & Wedel, 1975), and Ireland (Mulcahy, Hickey, Graham, & Macairt, 1977).

The literature on smoking cessation programs has received extensive review (Schwartz, 1969; Bernstein, 1969; Bradshaw, 1973; Evans, Henderson, Hill, & Rains, 1979; Lichtenstein & Danaher, 1978). In general, no one method or procedure appears more effective than any other, although certain aversive conditioning techniques, such as rapid smoking, may have potentially harmful side effects in patients with known CHD (Horan, Hackett, Nicholas, Linberg, Stone, & Lukaski, 1977; Hall, Sachs, & Hall, 1979). In general, successful temporary reduction or elimination of smoking behavior is relatively easy to achieve, while long-term abstinence is more difficult.

Available evidence suggests that about half of all patients who suffer an MI will stop smoking on their own or with minimal intervention, such as physician advice (Burt, Thornley, Illingworth, White, Shaw, & Turner, 1974; Weinblatt, Shapiro, & Frank, 1971; Werko, 1971; Russell, Wilson, Taylor, & Baker,

1979; Hay & Turbot, 1970; Croog & Richards, 1977). Data from
most studies, however, have relied on self-report or follow-up
questionnaire data which may be unreliable. The concurrent
measurement of such physiological variables as serum
thiocyanate, expired alveolar air carbon monoxide, and
carboxyhemoglobin levels may serve to corroborate self-reporting
of smoking behavior in future research. Instrumentation is
currently being developed which should add greater precision to
the evaluation of smoking cessation programs (Henningfield,
Stitzer & Griffiths, 1980).

In addition, there has been increased attention to
motivational aspects of smoking behavior. A number of models
have been proposed ranging from psychological (Tomkins, 1968;
Tamerin, 1972) to physiological (Eysenck, 1973; Schachter, 1978)
and situational (Best & Hakstian, 1978). No one model is
universally accepted. Future research should attempt to
evaluate individual differences in smoking motivation so as to
best tailor specific smoking intervention programs to the unique
needs of the patient. Moreover, the importance of environmental
factors which serve to elicit and reinforce smoking behavior
must be recognized. More careful attention to the factors
affecting maintenance of nonsmoking as well as to techniques to
help patients stop smoking will also be necessary.

Diet. Traditionally, patients with coronary disease are
encouraged to alter their usual diets. The rationale is largely
based on the association between cholesterol, diet, and
subsequent mortality in populations free of CHD such as those
studied in Framingham (Truett, Cornfield, & Kannel, 1967) and
Chicago (Shekelle, Shryock, Paul, Lepper, Stamler, Liv, &
Raynor, 1981). Reductions in salt intake, cholesterol, and
total caloric consumption are almost always suggested. However,
despite the correlational evidence that dietary intake of salt,
saturated fat, and cholesterol play an etiologic role in the
pathogenesis of coronary atherosclerosis and hypertension
(Epstein, 1965; Keys, 1966; Simborg, 1970), there is no
definitive evidence that modification of diet has any
significant impact upon morbidity or mortality of patients with
coronary disease. There is some suggestion that dietary
management may lead to regression of atheromatous lesions in
humans as well as animals (Wissler, 1980), although these data
are far too limited to draw any definitive conclusions.

Recent studies of patients with angina pectoris have shown
cholesterol to be unrelated to prognosis (Frank, 1968).
Similarly, the Coronary Drug Project (Canner & Halperin, 1978)
concluded that the impact of hyperlipidemia on mortality is much
less after a person has survived acute myocardial infarction

than before the clinical manifestation of the disease. While the relationship between diet and serum cholesterol is not clear, even if modification of diet resulted in substantial reductions in levels of serum cholesterol, large scale controlled studies would be necessary to document the effects of lowered blood lipids on subsequent mortality and morbidity (Coronary Drug Project Research Group, 1981). No such data are available at present.

Physical Exercise Training. There is currently a great deal of interest in the role of physical conditioning in the rehabilitation of the coronary patient. Numerous reviews are available to the interested reader (Haskell, 1974; Bruce, 1974; Amsterdam, Wilmore, & DeMaria, 1977; Mitchell, 1975; Froelicher, 1977; Scheur and Tipton, 1977; Greenberg, Arbeit, & Rubin, 1979).

The physiological effects of physical exercise training have been studied in both normal subjects and in patients with coronary artery disease (Clausen, 1976; Redwood, Rosing, & Epstein, 1972). In general, physical training refers to repetitive isotonic muscular exercise performed on a regular basis. Performance is often measured in terms of oxygen consumption (VO_2), and maximum oxygen consumption is considered a reproducible measure of work capacity and overall physical fitness.

Patients with coronary disease are often limited in their exercise capacity because their impaired coronary circulation cannot meet the increased myocardial oxygen demands during exercise. Compared to normal subjects, patients with coronary disease demonstrate a lower maximum cardiac output and maximum VO_2 (Clausen, 1976; Detry & Bruce, 1971; Bruce, Kusumi, Neiderberger, & Peterson, 1974; Rousseau, Brasseur, & Detry, 1973), a decreased heart rate response to exercise, and a decreased stroke volume at submaximum exercise workload (Clausen, 1976; Detry, 1971; Bruce et al., 1974; Rousseau et al., 1973; Clausen & Trap-Jensen, 1970). A subset of patients also develop myocardial ischemia before reaching their theoretical VO_2 maximum, and terminate exercise because of symptoms of pain. Therefore, one goal of physical training is to enable the individual to perform a given amount of work with less demand for the limited myocardial oxygen supply (Redwood et al., 1972; Detry et al., 1971; Sanne, 1973; Detry & Bruce, 1971; Clausen, Larsen, & Trap-Jensen, 1969). Exercise may also induce left ventricular dysfunction including decreased left ventricular ejection fraction and regional wall motion abnormalities (Upton, Rerych, Newman, Ports, & Jones, 1980; Rerych, Scholz, Newman, Sabiston, & James, 1978; Marshall, Berger, Costen, Freedman, 1977; Borer, Bachrach, Green Kent,

Stein, & Johnson, 1971). Such cardiac decompensation may result
in feelings of fatigue, weakness or dyspnea, which limit
exercise capacity

Studies of coronary patients have shown that increased
maximum oxygen consumption (Max. VO_2) can be achieved following
a program of regular exercise (Redwood et al., 1972). The
mechanism for this increase appears to be primarily the
improvement of the maximum arterial-venous oxygen extraction
($A-VO_2$), and not the enhancement of cardiac output (Rousseau et
al., 1973; Varnauskaw, Bergman, Houk, & Bjorntorp, 1966).
Hence, the increase in maximum work capacity is thought to be a
result of the increased efficiency of trained peripheral
muscles.   There is some evidence that exercise promotes
increased   coronary   collateral   vascularization   in   animals
(Redwood et al., 1972), although this finding has not been
reproduced in human studies (Kennedy, Spiekerman, Lindsay,
Mankin,   Frye,   &   McAllister,   1976;   Fergusun,   Petitclerc,
Choquette, Chaniotis, Gauthier, Hout, Allard, Jankowski, &
Campeau, 1974). Methodological problems make it difficult to
document collateralization in humans. Symptom-limited testing
has also demonstrated that maximum heart rates can be increased
in patients with angina (Redwood et al., 1972; Detry et al.,
1971). Submaximal workloads can also be increased. Decreased
resting heart rates and lowered heart rate and blood pressure
response   to   physical   activity   have   been   well   established
(Clausen & Trap-Jensen, 1970; Clausen et al., 1973) and
therefore the rate pressure product (i.e., myocardial oxygen
demand) is reduced.

In addition to these hemodynamic effects, other beneficial
effects include a reduction in risk factors such as weight,
serum triglycerides and cholesterol, free fatty acids, and
improved glucose tolerance (Clausen et al., 1969; Bjernulf,
Boberg, & Froberg, 1975). Several studies have demonstrated
that angina does not occur until higher double or triple
products are reached, thus suggesting increased myocardial
oxygen consumption (Redwood et al., 1972; Detry & Bruce, 1971).
However, this finding is not universally accepted (Sim & Neill,
1974). While even patients with poor left ventricular function
can achieve significant improvements in functional capacity
(Conn, Williams, & Wallace, in press), extensive training does
not appear to enhance left ventricular function (Conn et al., in
press; Cobb, Williams, McEwan, Jones, & Wallace, in press).

Despite the improved cardiovascular fitness and reduction
in risk factors as a result of exercise training, evidence for
prolongation of life or a reduction in reinfarctions in patients
with documented CHD remains equivocal. Several studies have

morbidity among patients who engage in regular exercise (Kentala, 1972; Wilhelmsen et al., 1975; Kellerman, 1973), although methodological problems and relatively small differences between exercise groups and controls have left the issue unsettled. For example, Figure 3 shows the results from a large scale randomized clinical trial (The National Exercise and Heart Disease Program, 1981). The cumulative three-year total mortality rate was 7.3% for the Control group and 4.6% for the Exercise group. The three-year rate for recurrent infarctions was 7.0% and 5.3% for Control and Exercise groups respectively. Kellerman (1973) studied patients who participated in brief (4-month) and extended (12-42 month) exercise programs. Mortality rates were lowest in the extended group, and there was no significant difference between the brief group and no-exercise controls. These findings suggest that benefits of exercise may be achieved only by continued compliance with the exercise regimen. For example, studies with high dropout rates showed no differences in morbidity or mortality between exercise groups and controls (Kentala, 1972; Wilhelmsen et al., 1975).

The problem of non-compliance in cardiac rehabilitation in itself represents a major obstacle to providing effective medical care in patients with coronary disease. Interestingly, those patients who comply with treatment may be at lower risk, even if the treatment is a placebo (The Coronary Drug Project Research Group, 1980). Compliance rates in most programs are disappointingly low, however. In one report of factors influencing prognosis following myocardial infarction (Kavenaugh Shephard, Chrisholm, Qureshi, & Kennedy, 1979), compliance was the most important single determinant of subsequent fatal and non-fatal reinfarctions. Moreover, sustained physical activity was associated with five-fold improvement in the risk ratio for cardiac events (Shephard, Corey, Kavanaugh, & 1981).

Efforts have been made to identify psychological and social characteristics of program dropouts. Oldridge (1979) found dropouts to have had significantly more recurrent infarctions, to have smoked cigarettes, to exhibit Type A behavior, to work in blue-collar occupations, and to be more sedentary than those who remained in the program. Other retrospective studies (Andrew & Parker, 1979; Bruce et al., 1976) have emphasized socio-economic factors such as interference with work or financial cost. In a recent study integrating psychological and physiological variables, Blumenthal and coworkers (1981) found that scores on the Ego strength and Social introversion scales of the MMPI, and left ventricular ejection fraction (LVEF) determined by radionuclide angiography, were highly significant predictors of compliance behavior among a sample of patients referred for cardiac rehabilitation at Duke University.

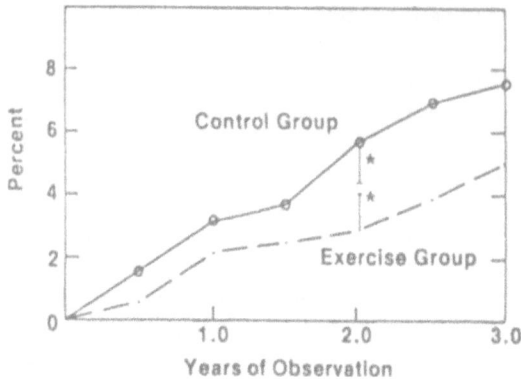

Figure 3.   Cumulative morality rates for control (N = 328)
            and exercise (N = 323) subjects in the National
            Exercise and Heart Disease Project (National
            Exercise and Heart Disease Project, 1981, p 41).

The kind and amount of exercise recommended for patients
with CHD is based upon a complete medical evaluation and is
always individualized to meet the unique needs of the patients.
In general, however, the basic principles of an exercise program
are similar in normal subjects and in patients with CHD.   The
typical prescription is for repetitive aerobic exercise of
larger muscles (walking, swimming, bicycling, etc.) for a
specific frequency (usually 3-5 days/week), duration (usually
20-45 minutes), and intensity (usually 70-85% maximum heart rate
achieved during symptom-limited treadmill testing).

The issue of exercise safety has received growing attention
and has been the subject of some controversy.   While some
studies emphasize the relative safety of medically supervised
physical exercise (Haskell, 1978; Kentala, 1972; Rechnitzer,
1979), others report medical complications and potential
frequency of fatal events (Hakkila, 1973; Mead, Pyfer, Trombold,
& Frederick, 1976).   The major objection to exercise is the
amount of sudden death that may occur with vigorous exercise
(Friedman et al., 1973).   It is generally believed that direct
medical supervision, (i.e., if medical personnel and emergency
defibrillation equipment are present and cardiopulmonary
resuscitation is performed in the event of ventricular
fibrillation (Haskell, 1978)) is adequate precaution for most
patients.   Guidelines for unsupervised exercise are currently
being developed (Williams et al., 1981).

Psychological.   While there has developed a rather
extensive literature on the psychological, social, and
behavioral aspects of coronary disease (Jenkins, 1971; 1976) the
number of intervention studies in patients with CHD is fairly

limited (Blanchard & Miller, 1977). In general, studies demonstrate that psychological treatment can facilitate early return to work (Thockcloth, Ho, & Wright, 1973), reduce emotional disturbance (Gruen, 1975; Adsett & Bruhn, 1968; Bilodeau & Hackett, 1971), and minimize post-infarction complications (Gruen, 1975; Rahe, O'Neil, Hagan and Arthur, 1975). In one of the few longitudinal studies, Rahe and coworkers (1979) compared the effects of brief psychotherapy with routine medical care in forty-four post-MI patients. Although some data suggested improved morbidity and mortality rates for the treatment group, the sample was too small to permit any general conclusions. There were also no differences between treatment and control groups with respect to a reduction in risk factors; weight loss, reduced cigarette consumption and lowered blood lipids occurred in only a minority of subjects. While some elements of Type A behavior, e.g. overwork and time urgency, were reduced in the treatment groups, no objective measure of the Type A behavior pattern was included in the study design.

Recently, efforts have been directed at modifying the Type A behavior pattern. Rosenman and Friedman (1977) describe their approach as involving group counseling, self-observation of Type A behaviors, the use of self-instruction and other cognitive techniques, relaxation training, recreational planning, and behavioral contracting. Initial anecdotal reports suggest a reduction in Type A behaviors, although no data regarding specific decrements in the Type A behavior pattern, or morbidity and mortality statistics are available.

Several studies of healthy adults have reported changes in Type A behavior as a result of such diverse approaches as psychodynamic and behavior therapy (Roskies, Spevack, Surkis, Cohen, & Gilman, 1978; Roskies, Kearney, Spevack, Surkis, Cohen, & Gilman, 1979), aerobic exercise (Blumenthal, Williams, Wallace, & Williams, 1980), behavior therapy and group support (Rahe et al., 1975; Rahe, Ward, & Hayes, 1979), and stress management training (Suinn, 1975). However, only a few studies have reported modification of Type A behavior in coronary patients. Suinn (1974) has reported outcome data using a combination of various behavioral techniques including relaxation training and stress management training. Ten post-MI patients were treated and compared with ten patients who received routine medical care. After only six sessions of treatment, the group receiving stress management training showed significantly greater reductions in serum cholesterol and triglycerides than the control group. No measures of Type A behavior were reported, however. Similar results have been subsequently reported (Suinn & Bloom, 1978; Suinn, Brock, &

Edie, 1975). Jenni and Wollersheim (1979) compared anxiety management training and cognitive therapy with a waiting list control group. Self-ratings of Type A behaviors were reduced for extreme Type A patients in the cognitive therapy group, although neither treatment affected blood pressure or cholesterol levels.

In one of few controlled clinical trials, Ibrahim and co-workers (1974) compared fifty-eight post-MI patients seen in weekly sessions of group therapy for one year, and sixty matched controls who were only periodically assessed. No statistically significant differences were shown in either physiological (e.g. blood pressure, serum lipids, glucose, uric acid, smoking behavior, weight) or psychological (e.g. mood, interpersonal orientation, autonomy, dominance) variables. The one-year survival was 10% higher among treatment patients than among the controls. However, this difference was also not statistically significant. The differences between survival rates were greatest among the severely ill patients (93% vs. 74%), with the advantage to the treatment group narrowly missing statistical significance.

In one of the more recent developments, cardiac arrhythmias, particularly those characterized by premature ventricular contractions (PVCs), have been shown to be reduced by employing biofeedback of heart rate responses (Engel & Bleecher, 1974). For example, Weiss & Engel (1971) were able to demonstrate a significant reduction in the presence of PVCs in four of eight patients treated by heart rate feedback. Other investigators have also reported successful outcomes using biofeedback for PVCs (Pickering & Gorham, 1975; Benson, 1975; Lown et al., 1976), sinus tachycardia (Blanchard & Abel, 1976; Scott, Blanchard, Edmunson, & Young, 1973), and cardiac neurosis (Wickramasekera, 1974).

SUMMARY

The efficacy of secondary prevention of coronary disease, i.e. the treatment of patients with established CHD, is traditionally evaluated by considering recurrence rates of non-fatal cardiac events, and by assessing subsequent mortality rates.

While large-scale clinical trials are necessary to evaluate the potential benefits of a given intervention, the methodological and practical problems of large-scale research should not be underestimated. Moreover, the existence of multiple studies must be considered when evaluating the literature. For example, at least 15 clinical trials with beta-blockers have been reported. The probability of at least one study finding a significant difference by chance alone is .50. When multiple subgroups within each study are investigated, the problem is compounded. For this reason, no clinical trial in this area should be accepted as definitive until replicated. Furthermore, the number of patients entered into a clinical trial is a vital determinant of the probability of success. With inadequate patient numbers, a true therapeutic benefit may not be detected (Type II error). A recent review indicated that this problem is substantial in the clinical literature (Freiman, Chalmers, Smith, & Kuebler, 1978).

Surgical intervention is an effective means of preventing death in selected patients with coronary disease. Although a number of approaches to pharmacologic therapy appear to be promising, no particular type of drug has been proved to be efficacious in the broad spectrum of coronary disease patients. Beta-adrenergic blocking agents appear to be beneficial in post-infarction patients. Additional clinical trials of anti-arrhythmics are needed.

In general, there is little conclusive evidence to support the efficacy of any behavioral intervention in the reduction of cardiac mortality. While this issue has been debated, there does not appear to be any definitive evidence that any psychological or behavioral intervention (i.e. nonmedical or surgical) can significantly improve survival in patients who have coronary disease. Most behavioral intervention efforts appear to be directed at reducing risk factors associated with the initial development of disease and increasing functional capacity. Modification of diet has traditionally been an essential component of most rehabilitation programs, including reduction in salt intake, cholesterol, and total caloric consumption. However, there is no conclusive evidence that modification of diet or loss of weight significantly affects the rate of death in patients with established coronary disease.

Smoking cessation also receives considerable emphasis in most cardiac rehabilitation programs. However, the majority of patients stop smoking on their own after sustaining a myocardial infarction. While stopping smoking has been shown to reduce morbidity-mortality rates, behavioral intervention efforts to eliminate smoking has not been systematically evaluated in cardiac rehabilitation settings.

The use of physical conditioning programs in patients with coronary artery disease has received the most attention and has the most encouraging results. While available evidence does not permit optimism about prolonging life, the data clearly document

improved functional capacity even in patients with severe
cardiac impairment.

Finally, the use of psychotherapy and behavior modification
has generally been shown to be effective in increasing
psychological well-being in post-MI patients. Some studies have
reported a reduction in risk factors and related cardiac
complications (e.g. arrhythmias), although there is no
conclusive evidence that any form of psychological or behavioral
intervention promotes longevity or reduces morbidity.

We conclude that the traditional concept of secondary
prevention should be expanded beyond the exclusive consideration
of reinfarction and mortality. Improving the quality of life,
including psychological well-being, and improving the ability to
perform various vocational and recreational activities should
also be recognized as important goals of cardiac rehabilitation.
A multifaceted program, including exercise, smoking cessation,
and stress management is the most prudent approach to secondary
prevention at this time. These behavioral measures should be
combined with appropriate surgical and pharmacological
interventions when indicated.

REFERENCES

Adsett, C. A., & Bruhn, J. G. Short term group psychotherapy for
    post-myocardial infarction patients and their wives.
    Canadian Medical Association Journal, 1968, 99, 577-584.

Amsterdam, E. A., Wilmore, J. H., & DeMaria A. N. Exercise in
    Cardiovascular Health and Disease. New York: Yorke Medical
    Books, 1977.

Anderson, M. P., Frederiksen, J., Jurgensen, H. J., Pederson,
    F., Bechsgaard, P., Hausen, D. A., Nielsen, B.,
    Pedersen-Bjergaard, O., & Rasumussen, S. L. Effects of
    Alprenolol on mortality among patients with definite or
    suspected acute myocardial infarction. Lancet, 1979, 2,
    865-868.

Andrew, G. M., & Parker, J. O. Factors related to dropout of
    post myocardial infarction patients from exercise programs.
    Medical Science Sports & Exercise, 1979, 11, 376-378.

Anturane Reinfarction Trial Research Group. Sulfinpyrazone in
    the prevention of sudden death after myocardial infarction.
    New England Journal of Medicine, 1980, 302, 250-256.

B-blocker Heart Attack Study Group, The B-blocker Heart Attack
    Trial. Journal of the American Medical Association, 1981,
    246, 2073-2074.

Baber, N. S., Evans, D. W., Howitt, G., Thomas, M., Wilson, C.,
    Lewis, J. A., Daves, P. M., Handler, K., & Tuson, R.
    Multicentre post-infarction trial of Propranolol in 49

hospitals in the United Kingdom, Italy and Yugoslavia. British Heart Journal, 1980, 44, 96-100.

Barber, J. M, Boyle, D. M. A., Chaturvedi, N. S., Singh, N., & Walsh, M. J. Practolol in acute myocardial infarction Acta Medical Scandinavica (Supplement), 1976, 587, 213-219.

Benson, H., Alexander, S., & Feldman, C. L. Decreased premature ventricular contractions through the use of the relaxation response in patients with stable ischemic heart disease. Lancet, 1975, 2, 380-382.

Bernstein, D. A. Modification of smoking behavior: An evaluative review. Psychological Bulletin, 1969, 71, 418-440.

Best, J. A., & Hakstian, A. R. A situation specific model for smoking behavior. Addictive Behavior, 1978, 3, 79-92.

Bigger, J. T., Dresdale, R. J., Heissenhuttel, R. H., Weld, F. M., & Wit, A. L. Ventricular arrhythmias in ischemic heart disease: Mechanism, prevalence, significance and management. Progress in Cardiovascular Disease, 1977, 19, 255-295.

Bilodeau, C. B., & Hackett, T. P. Issues raised in a group setting by patients recovering from myocardial infarction. American Journal of Psychology, 1971, 128, 73-78.

Bjernulf, A., Boberg, J., & Froberg, S. Physical training after myocardial infarction: Metabolic effects during short and prolonged exercise before and after physical training in male patients after myocardial infarction. Scandanavian Journal of Clinical Laboratory Investigations, 1975, 33, 173-185.

Blanchard, E. B., & Abel, G. G. An experimental study of the biofeedback treatment of a rape-induced psychophysiological cardiovascular disorder. Behavior Therapy, 1976, 7, 113-119.

Blanchard, E. B., & Miller, S. T. Psychological treatment of cardiovascular disease. Archives of General Psychiatry, 1977, 34, 1402-1413.

Blumenthal, J. A., Williams, R. S., Wallace, A. G., & Williams, R. B. Effects of exercise on the Type A (Coronary Prone) behavior pattern. Psychosomatic Medicine, 1980, 42, 289-296.

Blumenthal, J. A., Williams, R. S., Wallace, A. G., Williams, R. B, & Needles, T. Physiological and psychological variables predict adherence to prescribed exercise therapy in patients recovering from myocardial infarction. Unpublished manuscript, 1982.

Borer, J. S., Bacharach, S. L., Green, M. V., Kent, K. M. E., Stein, S. E., & Johnson, G. S. Real time radionuclide cineangiography in the man - invasive evaluation of global and regional left ventricular function at rest and during exercise in patients with coronary artery disease. New

England Journal of Medicine, 1971, 296, 839-844.

Bradshaw, P. W. The problem of cigarette smoking and its
    control. International Journal of Addiction, 1973, 8,
    353-371.

Bruce, E. H., Frederick, R., Bruce, R. A., & Fisher, C. D.
    Comparison of active participants and dropouts in CAPRI
    Cardiopulmonary Rehabilitation Programs. American Journal
    of Cardiology, 1976, 37, 53-60.

Bruce, R. A. The benefits of physical training for patients
    with coronary heart disease. In F. J. Gelfinger, R. V.
    Elbert, M. Finland, & A. S. Relman (Eds.), Controversy in
    Internal Medicine. Philadelphia: W.B. Saunders Company,
    1974.

Bruce, R. A, Kusumi, F., Neidergerger, M., & Peterson, J. L.
    Cardiovascular mechanisms of functional aerobic impairment
    in patients with coronary heart disease. Circulation, 1974,
    49, 696-702.

Burggraff, G. W., & Parker, J. O. Prognosis in coronary artery
    disease: Angiographic, hemodynamic, and clinical factors.
    Circulation, 1975, 51, 146-156.

Burt, A., Thornley, P., Illingworth, D., White, P., Shaw, T. R.
    D., & Turner, R. Stopping smoking after myocardial
    infarction. Lancet, 1974, 1, 304-306.

Canner, P. L., & Halperin, M. Implications of findings in the
    Coronary Drug Project for Secondary Prevention Trials of
    Lipid   Lowering   Drugs.   18th   Annual   Conference   on
    Cardiovascular Disease and Epidemiology, Orlando, Florida,
    1978.

Clausen, J. P. Circulatory adjustments to dynamic exercise and
    effects of physical training in normal subjects and in
    patients with coronary artery disease.   Progress in
    Cardiovascular Disease, 1976, 18, 459-495.

Clausen, J. P., Klausen, K., Rasmussen, B., & Trap-Jensen, J.
    Central and peripheral circulatory changes after training
    of the arms or legs. American Journal of Physiology, 1973,
    225, 675-682.

Clausen, J. P., Larsen, O. A., & Trap-Jensen, J. Physical
    training in the management of coronary artery disease.
    Circulation, 1969, 40, 143-154.

Clausen, J .P., & Trap-Jensen, J. Effects of training of the
    distribution of cardiac output in patients with coronary
    artery disease. Circulation, 1970, 42, 611-624.

Cobb, F. R., Williams, R. S., McEwan, P., Jones, R. H., &
    Wallace, A. G. Effects of exercise training on ventricular
    function in patients with recent myocardial infarction.
    Circulation, in press.

Collins, J. J., Tisevova, J., Mudge, G. H., Cohn, L. H., &
    Koster, J. K. Probability of survival, myocardial
    infarction and angina at seven years after coronary bypass

surgery. Circulation, 1981, 95, (Abstract).

Conn, E. H., Williams, R. S., & Wallace, A. G. Exercise
    responses before and after physical conditioning in
    patients with severely depressed left ventricular function.
    American Journal of Cardiology, in press.

Coronary Drug Project Research Group. Cigarette smoking as a
    risk factor in men with a prior history of myocardial
    infarction. Journal of Chronic Diseases, 1979, 32,
    415-425.

Coronary Drug Project Research Group. Influence of adherence to
    treatment and response of cholesterol on mortality in the
    Coronary Drug Project. New England Journal of Medicine,
    1980, 303, 1038-

Coronary Drug Project Research Group. Treatable risk factors -
    hypercholesterolemia, smoking, and hypertension after
    myocardial infarction. Primary Care, 1980, 7, 175-179.

Coronary Drug Project Research Group. Implications of findings
    in the Coronary Drug Project for secondary prevential
    trials in coronary heart disease. Circulation, 1981, 63,
    1342-1349.

Croog, S. H., & Richards, N. P. Health beliefs and smoking
    patterns in heart patients and their wives: a longitudinal
    study. American Journal of Public Health, 1977, 67,
    921-930.

Detry, J.-M., & Bruce, R. A. Effects of physical training on
    exertional S-T segment depression in coronary heart disease
    Circulation, 1971, 44, 390-396.

Detry, J.-M., Rousseau, M., Vandenbraucke, G., & Kusumi, F.
    Increased arteriovenous oxygen difference after physical
    training in coronary heart disease. Circulation, 1971, 44,
    109-118.

DeWood, M. A., Spores, J., Norski, R. Prevalence of total
    coronary occlusion during the early hours of transmural
    myocardial infarction. New England Journal of Medicine,
    1980, 303, 897-902.

Doll, R., & Hill, A. B. Mortality in relation to smoking: ten
    years' observations of British doctors. British Medical
    Journal, 1964, 1, 1460-1467.

Engel, B. T., & Bleecher, E. R. Application of operant
    conditioning techniques to the control of cardiac
    arrhythmias. In P. A. Obrist, A. H. Black, J. Brener, & L.
    V. DiCara (Eds.), Cardiovascular Physiology. Chicago:
    Aldine Publishing Company, 1974.

Epstein, F. The epidemiology of coronary heart disease: A
    review. Journal of Chronic Diseases, 1965, 18, 735-774.

European Coronary Surgery Group. Coronary artery surgery in
    stable angina pectoris: survival at two years. Lancet,
    1979, 889, 93.

Evans, R. I., Henderson, A. H., Hill, P. C., & Rains, B. E.

Current psychological, social, and educational programs in
control and prevention of smoking: A critical methodological review. Atherosclerosis Review, 1979, 6, 203-

Eysenck, H. J. Personality and the maintenance of the smoking
habit. In W. L. Dunn (Ed.), Smoking Behavior: Motives and
Incentives. Washington, D.C.: Winston & Sons, 1973.

Fergusun, R. J., Pititclerc, R., Chaquette, G., Chaniotis, L.,
Gauthier, P., Huot, R., Allard, C., Jankowski, L., &
Campeau, L. Effect of physical training on treadmill
exercise capacity, collateral circulation and progression
of coronary disease. American Journal of Cardiology,
1974, 34, 764-769.

Frank, C. W. The course of coronary heart disease: Factors
related to prognosis. Bulletin of the New York Academy of
Medicine, 1968, 44, 899-915.

Freiman, J. A., Chalmers, J. C., Smith, H., & Kuebler, R. R.
The importance of beta, the Type II error, in the design
and interpretation of the randomized control trial.
New England Journal of Medicine, 1978, 299, 690-694.

Friedman, M., Manwaring, J. H., Rosenman, R. H., Donlon, G.,
Ortega, P., & Grube, S. M. Instantaneous and sudden deaths:
clinical and pathological differentiation in coronary
artery disease. Journal of the American Medical
Association, 1973, 224, 1319-1328.

Friedman, G. D., Petitti, D. B., Bawol, R. D., & Siegelaub, A.
B. Mortality in cigarette smokers and quitters: Effect of
baseline differences. New England Journal of Medicine,
1981, 304, 1407-1410.

Froelicher, V. F. Does exercise conditioning delay progression
of myocardial ischemia in coronary atherosclerotic heart
disease. Cardiovascular Clinics, 1977, 8, 11-31.

Ganz, W. Intracoronary thrombosis in evolving myocardial
infarction (Editorial). Annals of Internal Medicine, 1981,
95, 500-502.

Genton, E. A perspective on platelet-suppressant drug treatment
in coronary artery and cerebrovascular disease.
Circulation, 1981, 62: (PartII) 111-120.

Gruen, W. Effects of brief psychotherapy during the
hospitalization period or the recovery process in heart
attacks. Journal of Consulting and Clinical
Psychology, 1975, 43, 223-232.

Greenberg, M. A., Arbeit, S., & Rubin, I. L. The role of
physical training in patients with coronary artery disease.
American Heart Journal, 1979, 97, 527-534.

Hakkila, J. Complications during physical rehabilitation of
coronary patients. Giornale Italiano di Cardiologia, 1973,
3, 362-

Hall, R. G., Sachs, D. P. L., & Hall, S. M. Medical risks and
therapeutic effectiveness of rapid smoking.

Behavior Therapy, 1979, 10, 249-259.

Hammond, E. C, & Garfinkel, L. Coronary heart disease, stroke and aortic aneurysm: factors in the etiology. Archives of Environmental Health, 1969, 19, 167-182.

Harris, P. J., Harrell, F. E., Lee, K. L., Behar, V. S., & Rosati, R. A. Survival in medically treated coronary artery disease. Circulation, 1980a, 60, 1259-1269.

Harris, P. J., Lee, K. L., Harrell, F. E., Behar, V. S., & Rosati, R.A. Outcome in medically treated coronary artery disease, Ischemic events: Non-fatal infarction and death. Circulation, 1980b, 62, 718-726.

Haskell, W. L. Physical activity after myocardial infarction, American Journal of Cardiology, 1974, 33, 776-

Haskell, W. L. Cardiovascular complications during exercise training of cardiac patients. Circulation, 1978, 57, 920-924.

Hay, D. R., & Turbot, S. Changes in smoking habits in men under 65 years after myocardial infarction and coronary insufficiency. British Heart Journal, 1970, 32, 738-740.

Henningfield, J. E., Stitzer, M. L., & Griffiths, . Expired air carbon monoxide accumulation and elimination as a function of number of cigarettes smoked. Addictive Behavior, 1980, 5, 265-272.

Horan, J. J., Hackett, G., Nicholas, W. C., Linberg, S. E, Stone, C. I., & Lukaski, H. C. Rapid smoking: a controversy note. Journal of Consulting and Clinical Psychology, 1977, 45, 341-343.

Hjalmarson, A., Herlitz, J., Malek, I., Ryden, L., Vedin, A., Waldenstrom, A., Wedel, H., Elmfeldt, D., Holmberg, S., Nyberg, G., Swedberg, K., Waagstein, F., Waldenstrom, Wilhelmsen, L., & Wilhelmsen, C. Effect on mortality of Metoprolol in acute myocardial infarction. Lancet, 1981, 2, 8251.

Ibrahim, M. A., Feldman, J. G., Sultz, H. A., Staiman, M. G., Young, L. J., & Dean, D. Management after myocardial infarction: a controlled trial of the effect of group psychotherapy. Psychiatry in Medicine, 1974, 5, 253-268.

Jenkins, C. D. Psychologic and social precursors in coronary disease. New England Journal of Medicne, 1971, 284, 244-307.

Jenkins, C. D. Recent evidence supporting psychological and social risk factors for coronary disease. New England Journal of Medicine, 294, 987-994; 1033-1038.

Jenni, M. A., & Wallersheim, J. P. Cognitive therapy, stress management training, and the Type A behavior pattern. Cognitive Therapy & Research, 3, 61-75.

Kahn, H. A. The Dorn study of smoking and mortality among U.S. veterans: Report on eight and one-half years of observation. In W. Haenszel (Ed.), Epidemiological

approaches to the study of cancer and other chronic
diseases. National Cancer Institute Monograph, Bethesda,
Maryland: Public Health Service, 1966.

Kavenaugh, T., Shephard, R. J., Chjrisholm, A. W., Qureshi, S.,
& Kennedy, J. Prognostic indexes for patients with
ischemic heart disease enrolled in an exercise-centered
rehabilitation program. American Journal of Cardiology,
1979, 44, 1230-1240.

Kellerman, J. J. Physical conditioning in patients after
myocardial infarction: results of a comparative study and
nine years followup. Schweizerische Medizinische
Wochenschrift, 1973, 103, 79-85.

Kennedy, C. V., Spiekerman, R. E., Lindsay, M. I., Mankin, H.
T., Frye, R. L. & McCallister, B. D. One year graduated
exercise program for patients with angina pectoris:
Evaluation by physiologic studies and coronary
arteriography. Mayo Clinic Proceeding, 1976, 51, 231-236.

Kentala, E. Physical fitness and feasibility of physical
rehabilitation after myocardial infarction in men of
working age. Annals of Clinical Research, 1972, 9,
(Supplement), 1-84.

Keys, A. The individual risk of coronary heart disease.
Annals of the New York Academy of Sciences, 1966,
134, 1046-1056.

Lee, K. L., McNeer, J. F., Starmer, C. F., Harris, P. J., &
Rosati, R. A. Clincal judgement and statistics: Lesions
from a simulated randomized trial in coronary artery
disease. Circulation, 1980, 61, 508-515.

Lichtenstein, E., & Danaher, B. G. Modification of smoking
behavior. A critical analysis of theory, research, and
practice. In M. Hersen, R. M., Eisler, & P. M. Miller
(Eds.), Progress in Behavior Modification, (Vol. 3), New
York: Academic Press, 1978.

Lown, B., Podrid, P. J., & Graboys, T. B. Sudden cardiac death
management of the patient at risk. Current Problems in
Cardiology, 1980, 4-

Lown, B., Temtl, J. B., Reich, P. Basis for recurring
ventricular fibrillation in the absence of coronary heart
disease and its management. New England Journal of
Medicine, 1976, 294, 623-629.

Marshall, R. C., Berger, H. J., Costen, J. C., Freedman, G. S.,
Wolberg, J., Cohen, L. S., Gottschalk, A., & Zaret, B. L.
Assessment of cardiac performance with quanitative
radionuclide angiocardiography: Sequential left
ventricular ejection fraction, normalized left
ventricular ejection rate, and regional wall motion.
Circulation, 1977, 56, 820-829.

Mead, W. F., Pyfer, H. R., Trombold, J. C., & Frederick, R. L.
Successful resuscitation of mean simultaneous cases of

cardiac arrest with a review of fifteen cases occurring during supervised exercise. Circulation, 1976, 53, 187-189.

Mitchell, J. H. Exercise training in the treatment of coronary heart disease. Advances in Internal Medicine, 1975, 20, 249-272.

Mitchell, J. R. A. Secondary prevention of myocardial infarction - the present state of the art. British Medical Journal, 1980, 280, 1128-1130.

Mulcahy, R., Hickey, N., Graham, I. M., & Macairt, J. Factors affecting the five-year survival rate of men following acute coronary heart disease. American Heart Journal, 1977, 93, 556-559.

Multicentre International Study. Improvement in prognosis of myocardial infarction by long-term beta-adrenoreceptor blockade using Practolol. British Medical Journal, 1975, 3, 735-740.

Myerberg, R. J., Conde, C., Sheps, D. S., Appel, R. A., Kiem, I., Sung, R. J., & Castellanos, A. Antiarrhythmic drug therapy in survivals of prehospital cardiac arrest: Comparison of effects of chronic ventricular arrhythmias and recurrent cardiac arrest. Circulation, 59, 855-863.

National Cooperative Study Group on Unstable Angina Pectoris. Unstable Angina Pectoris: In-hospital experience and initial follow-up results. American Journal of Cardiology, 1978, 42, 839-848.

National Cooperative Study Group on Unstable Angina Pectoris Medical or surgical therapy for angina pectoris. Cardiovascular Medicine, 1979, 1059.

National Exercise and Heart Disease Program. Effects of a prescribed supervised exercise program on mortality and cardiovascular morbidity in patients after a myocardial infarction. American Journal of Cardiology, 1981, 48, 39-46.

Norris, R. N., Agnew, T. M., Brandt, P. W., Graham, K. J., Hill, D. G., Kerr, A. R., Lowe, J. B., Roche, A. H. G., Whitlock, R. M. C., & Barratt-Boyes, B.G. Coronary surgery after recurrent myocardial infarction: Progress of a trial comparing surgical with non-surgical management for asymptomatic patients with advanced coronary disease, Circulation, 1981, 63, 785-792.

Norwegian Multicenter Study Group. Timolol-induced reduction in mortality and reinfarction in patients surviving acute myocardial infarction. New England Journal of Medicine, 304, 801-807.

Oldridge, N. B. Compliance of post-myocardial infarction patients to exercise program. Medical Science in Sports and Exercise, 1979, 11, 373-375.

Peduzzi, P., & Hultgren, H. N. Effects of medical versus

surgical treatment of symptoms in stable angina pectoris. The Veterans Administration Cooperative Study of surgery for coronary arterial occlusive disease. Circulation, 1979, 60, 888-900.

Pickering, T., & Gorham, G. Learned heart rate controlled by a patient with ventricular parasystolic rhythm. Lancet, 1975, 1, 252-253.

Proceedings of the Workshop on Platelet-Active Drugs in the Secondary Prevention of Cardiovascular Events. Circulation, 1981, 60, 1-135.

Rahe, R. H., O'Neil, T., Hagan, A., & Arthur, R. J. Brief group therapy following myocardial infarction: Eighteen month follow-up of a controlled trial. Psychiatry in Medicine, 1975, 6, 349-358.

Rahe, R. H., Ward, H. W., & Hayes, V. Brief group therapy in myocardial infarction rehabilitation three to four-year follow-up of a controlled trial. Psychosomatic Medicine, 1979, 41, 229-242.

Rahimtoola, S. H. A consensus on coronary bypass. New England Journal of Medicine, 1981, 94, 272-273.

Rechnitzer, P. A. The effects of training-reinfarction and death: An interim report. Medical Science in Sports and Exercise, 1979, 11, 382.

Redwood, D. R., Rosing, D. R., & Epstein, S.E. Circulatory and symptomatic effects of physical training in patients with coronary artery disease and angina pectoris. New England Journal of Medicine, 286, 959-965.

Rerych, S. K., Scholz, P. M., Newman, G. E., Sabiston, D. C., & James, R. N. Cardiac function at rest and during exercise in normals and in patients with coronary heart disease: Evaluation by radionuclide angiocardiography. Annals of Surgery, 1978, 187, 449-463.

Reynolds, J. L., & Whitlock, R. M. Effects of beta-adrenergic receptor blockers in myocardial infarction treated for one-year from onset. British Heart Journal, 34, 252-259.

Rosenman, R. H., & Friedman, M. Modifying Type A behavior pattern. Journal of Psychosomatic Research, 1977, 21, 323-331.

Roskies, E., Kearney, H., Spevack, M., Surkes, A., Cohen, C., & Gilman, S. Generalizability and durability of treatment effects in an intervention program for coronary-prone (Type A) managers. Journal of Behavioral Medicine, 1979, 2, 195-207.

Roskies, E., Spevack, M., Surkis, A., Cohen, C., & Gilman, S. Changing the coronary-prone (Type A) behavior pattern in a nonclinical population. Journal of Behavioral Medicine, 1978, 1, 201-216.

Rousseau, M. F., Brasseur, L. A., & Detry, J-M. R. Hemodynamic determinants of maximal oxygen intake in patients with healed myocardial infarction: Influence of physical training. Circulation, 1973, 48, 943-949.

Ruberman, W., Weinblatt, E., Goldberg, J. D., Frank, C. W., Chaudhary, B. S., & Shapiro, S.   Ventricular premature complexes and sudden death after myocardial infarction. Circulation, 1981, 64, 297-305.

Russell, M. A. H., Wilson, C., Taylor, C., & Baker, C. D. Effect of general practitioner's advice against smoking. British Medical Journal, 1979, 2, 231-235.

Sanne, H.   Exercise tolerance and physical training of non-selected patients after myocardial infarction.   Acta Medica Scandinavia, (Supplement), 1973, 1.

Schachter, S.   Pharmacological and psychological determinants of smoking. Annals of Internal Medicine, 1978, 88, 104-114.

Scheur, J., & Tipton, C. M. Cardiovascular adjustments to physical training.   Annual Review of Physiology, 1977, 39, 221-251.

Schwartz, J. L. A critical review and evaluation of smoking control methods. Public Health Reports, 1969, 84, 483-506.

Scott, R. W., Blanchard, E. B., Edmunson, E. D., & Young, L. D. A shaping procedure for heart-rate control in chronic tachycardia. Perceptual and Motor Skills, 1973, 37, 327-338.

Shekelle, R. B., Shryrock, A. M., Paul, O., Lepper, M., Stamler, J. Liv, S., & Raynor, W. J., 1981, Diet, serum cholesterol, and death from coronary heart disease. New England Journal of Medicine, 1981, 304, 65-70.

Shephard, R. J., Corey, P., Kavanaugh, T. Exercise compliance and the prevention of a recurrence of myocardial infarction. Medical Science in Sports and Exercise, 1981, 13, 1-5.

Sim, D. N., & Neill, W. W. Investigation of the physiologic basis for increased exercise threshold for angina pectoris after physical conditioning, Journal of Clinical Investigations, 1974, 54, 763-770.

Simborg, D. W. The status of risk factors and coronary heart disease. Journal of Chronic Diseases, 1970, 22, 515-552.

Soffer, A. Editorial comment. Archives of Internal Medicine, 1976, 136, 1229-1230.

Sparrow, D., Dawber, T. R., & Colson, T. The influence of cigarette smoking on prognosis after a first myocardial infarction. Journal of Chronic Diseases, 1978, 31, 425-432.

Suinn, R. M. Behavior therapy for cardiac patients. Behavior Therapy, 1974, 5, 569-571.

Suinn, R. M. The cardiac stress management program for Type A patients. Cardiovascular Rehabilitation, 1975, 5, 13-15.

Suinn, R. M., & Bloom, L. J. Anxiety management training for pattern A behavior. Journal of Behavioral Medicine, 1978, 1, 25-36.

Suinn, R. M., Brock, L., & Edie, C. A.   Behavior therapy for
    Type A patients. American Journal of Cardiology, 1975, 36,
    269.

Takaro, T., Hultgren, H. N., Lipton, M. J., & Detre, K. M.   The
    VA Cooperative randomized study of surgery for coronary
    arterial occlusive disease II.   Subgroup with significant
    left main lesions. Circulation, 1976, 54, 107-

Tamerin, J. S.   The psychodynamics of quitting smoking in a
    group. American Journal of Psychiatry, 1972, 129, 589-595.

Taylor, G. J., Humphries, J. O., Mellits, E. D., Pitt, B.,
    Schulze, R. A., Griffith, L. S. C., & Achuff, S. C.
    Predictors of clinical course, coronary anatomy and left
    ventricular function after recovery from acute
    myocardial infarction. Circulation, 1980, 62, 960-970.

Temple, R., & Pledger, G. W.   The FDA's critique of the anturane
    reinfarction trial. New England Journal of Medicine,
    1980, 303, 1488-1492.

Tomkins, S. S.   A modified model of smoking behavior.   In E.
    Borgatta & R. Evans (Eds.), Smoking, Health and Behavior.
    Chicago: Aldine, 1968.

Thockcloth, R. M., Ho, S. O., & Wright, W.   Is cardiac
    rehabilitation really necessary.   Medical Journal of
    Australia, 1973, 2, 669-674.

Truett, J., Cornfield, J., & Kannel, W.   A multivariate analysis
    of the risk of coronary disease.   Journal of Chronic
    Diseases, 1967, 20, 511-

U.S. Public Health Service.   Smoking and health: A report of the
    Surgeon General, DHEW, Publication #(PHS) 79-50066, 1979.

Upton, M. T., Rerych, S. K., Newman, G. E., Ports, S., Cobb, F.
    R., & Jones, R. H.   Detection of abnormalities in left
    ventricular function during exercise before angina and ST
    segment depression. Circulation, 1980, 62, 341-349.

Varnauskaw, E., Bergman, H., Houk, P., & Bjorntorp, P.
    Haemodynamic effects of physical training in coronary
    patients. Lancet, 1966, 2, 8-

Weinblatt, E., Shapiro, S., & Frank, C. W.   Changes in personal
    characteristics of men over five years following first
    diagnosis of coronary heart disease.   American Journal of
    Public Health, 1971, 61, 831-

Weiss, T., & Engel, B. T.   Operant conditioning ofheart rate in
    patients with premature ventricular contractions.
    Psychosomatic Medicine, 1971, 33, 301-331.

Werko, L.   Can we prevent heart disease.   Annals of Internal
    Medicine, 1971, 74, 278-288.

Wickramasekera, I.   Heart rate feedback and the management of
    cardiac neurosis. Journal of Abnormal Psychology, 1974,
    83, 578-580.

Wilcox, R. G., Roland, J. M., Banks, D. C., Hampton, J. R., &

Mitchell, J. R. A. Randomized trial comparing propranolol with atenolol in immediate treatment of suspected myocardial infarction. British Medical Journal, 1980, 280, 885-888.

Wilhelmsen, L., Sanne, H., Elmfeldt, D., Grimby, G., Tibblin, G., & Wedel, H. A controlled trial of physical training after myocardial infarction. Preventive Medicine, 1975, 4, 491-508.

Wilhelmsen, L., Vedin, J. A., Elmfeldt, D., Tibbin, G., & Wilhelmssen, L. Smoking and myocardial infarction. Lancet, 1975, 1, 415-420.

Williams, R. S., Miller, H., Koisch, F. P., Ribisl, P., & Graden, H. Guidelines for unsupervised exercise in patient with ischemic heart disease. Cardiac Rehabilitation, 1981, 1, 213-217.

Wissler, R. W. Nutrition, plasma lipids, and atherosclerosis. In R. M. Laver & R. B. Shekelle (Eds.), Childhood Prevention of Atherosclerosis and Hypertension. New York: Raven Press, 1980.

TYPE A INTERVENTION: FINDING THE DISEASE TO FIT THE CURES

Ethel Roskies

University of Montreal

Montreal, Canada

Less than a decade ago, Friedman and Rosenman (1974) first raised the possibility of modifying Type A behavior to reduce coronary risk. Since then, type A intervention has become one of the "hot" topics in the newly emerging specialty of behavioral medicine. The Society of Behavioral Medicine chose a symposia on type A treatment as the feature of its first annual meeting in 1979, while the same year the American Psychological Association, the American Psychosomatic Society, the American Public Health Association and the Association for the Advancement of Behavior Therapy all provided forums devoted to this issue. Reflecting this widespread interest, the editors of that year's Annual Review of Behavior Therapy describe the available treatment studies in detail and predict "greatly increased research activity in this important area of behavioral medicine in the next few years" (Franks & Wilson, 1979, p. 382).

Even more frenetic than the research activity has been the proliferation of clinical efforts directed towards this new health problem. Exact figures on clinical practice are difficult to obtain, but it is indicative that the new crop of popular stress management books has begun to include sections on Type A modification (Girdano & Everly, 1979; Goldberg, 1978). The media have also become attracted to the topic, with a recent feature article in The New York Times and an interview with Dr. Friedman on a national television show. And should a therapist feel inadequate in beginning Type A treatment, there are a multitude of workshops ready to show him how. Even a questionnaire designed simply to diagnose Type A is now being promoted by the Psychological Corporation in full-page advertisements as "an effective way to outsmart heart disease".

71

Table 1

Summary of Type A Intervention Studies,

Techniques and Objectives

Authors and date	Treatment techniques	Treatment objectives	Treatment format
Suinn, 1975	Combination of Anxiety Management Training and Visuomotor Behavioral Rehearsal	Stress management	Group 5 sessions
Rosenman and Friedman, 1977	Psychoanalytically oriented psychotherapy	General change in Type A pattern: philoso-phical and behavioral	Group 18 mos
Roskies et al., 1978	Behavior Therapy (Rel-axation); Brief Psycho-therapy	For B.T.: to learn to respond to perception of loss of control with rel-axation, rather than frantic activity. For psychotherapy: insight	Group 14 sessions
Suinn & Bloom, 1978	Anxiety Management Training	Stress management	Group 6 sessions
Jenni & Wollersheim, 1979	Stress management Cognitive therapy (rational-emotive)	Reducing stress associa-ted with type A pattern	Group 6 sessions
Rahe, Ward & Hayes, 1979	Brief group therapy	General cardiac reha-bilitation, altering coronary-prone behavior	Group 6 sessions
Roskies et al., 1979	Same as Roskies et al., 1978		
Blumenthal et al., 1980	Adult fitness program	Reduction of CHD risk	Group 30 sessions
Friedman, 1980	Medical information + modification of type A cognitions, behaviors and environmental stimuli	Modify value system, diminish time urgency, reduce free-floating hostility	Group 5 yers. program weekly mos 1 + 2 biweekly mos 4 - 6 montly mos 7 - 60
Roskies, in press	Cognitive behavior therapy (relaxation, R.E.T., communication skills, problem-solving, stress inoculation)	Modification of frequency, intensity and duration of auto-nomic and endocrine arousal	Group 13 sessions

Unfortunately, this enthusiasm for embarking on Type A treatment has been accompanied by considerable confusion about what, or even who, we are trying to treat. The handful of studies published to date contain a bewilderingly large

Table 2

Summary of Type A Intervention Studies

Sample Characteristics

Authors and date	Sample size	Measurement of Type A	Type A status	Clinical status	Age	Sex	Occupation	Other Characteristics
Suinn, 1975	10	?	All As	Immediate post MI	?	?	?	
Rosenman and Friedman, 1977	12	S.I.	All As	All CHD patients	?	M	?	
Roskies et al., 1978	27	S.I.	All A_1	No overt CHD	39-59 $x = 47.6$	M	Professionals and Managers	Non-smokers
Suinn and Bloom, 1978	14	JAS	All As	No overt CHD	24-55 $x = 38.0$	12 M 2 F	Professionals and Managers	
Jenni and Wollersheim, 1979	42	S.I.	All As	7M S post MI Others?	29-58 $x = 42.5$	27 M 15 F	16 M Managers Others?	Non-smokers
Rahe, Ward and Hayes 1979	54	None	?	All post first MI	$x = 52.5$	M	Military Officers	
Roskies et al., 1979	31	S.I.	All A_1	No overt CHD	39-59 $x = 48.1$	M	Professionals	
Blumenthal, 1980	46	JAS	21 A 25 B	No overt CHD	25-61 $x = 42.6$	20 M 26 F	?	
Friedman, 1980	1,035	S.I.	98% Type A	All 1-6 MI	$x = 53.0$	946 M 89 F	Varied. 23% supervisory or managerial	Non-smokers
Roskies, in press	66	S.I.	47% A_1 40% A_2	No overt CHD	$x = 41.3$	M	Middle managers	

S.I.   Standardized Interview
JAS   Jenkins Activity Survey
?   Not specified

assortment of treatment approaches -- all the way from psychoanalytically oriented psychotherapy to physical fitness training -- but beyond vague references to "improved stress management" or "reduced coronary risk", there is no description of specifically what problems these remedies are expected to alleviate, nor how (see Table 1). The individuals subjected to these diverse treatments also do not form a distinct clinical group, varying in age, sex, occupation, experience of heart disease, and even in whether and how they were identified as As (see Table 2). Most confusing of all, the universal claim of therapeutic success is based on such different and contradictory results as lowered, raised and unchanged serum cholesterol levels, lowered and unchanged anxiety scores, lowered and unchanged scores on the Jenkins Activity Survey and so on (Table 3). Based on the studies reported to date, one can only conclude that Type A treatment consists of a variety of cures for a disease that remains to be defined!

There are a number of reasons that help to explain the tremendous interest in Type A intervention, as well as the failure to date to pay much attention to exactly what we are trying to treat. Type A treatment is attractive to behavioral scientists working in the health field because this pattern provides an unusually clear demonstration of the etiological importance of behavior, successfully and independently predicting the future emergence of a major somatic disease. Moreover, the pattern is also an excellent example of the Type of health problem for which traditional pharmaceutical and surgical remedies have little relevance. For behavioral scientists eager to prove to their medical colleagues that psychological treatments can be of practical value in the prevention and treatment of disease, Type A constitutes the ideal "disease". Given these circumstances, one can easily understand the pressure to quickly produce a suitable treatment.

The therapeutic target toward which these enthusiastic treatment efforts are being directed, however, is much more complex and ambiguous than the more traditional lifestyle health problems, such as smoking, alcoholism, or lack of exercise. It would be naive to claim that it is easy to modify any deeply engrained habit pattern, but at least we have a fairly clear idea of the behavioral changes required for the smokers, the drinkers and the sedentary to improve their health status. For the Type A individual, in contrast, we do not know what exactly makes him or her at higher risk for heart disease and, consequently, what must be changed if this risk is to be reduced.

The problem is not lack of data concerning what distinguishes a Type A individual from a Type B one. On the contrary, a host of recent studies has detailed the behavioral and physiological characteristics differentiating the two types: Type A persons as a group speak louder, faster and more explosively than their Type B counterparts (Schucker & Jacobs, 1977); they show greater cardiovascular and biochemical reactivity to certain types of challenge (Dembroski, MacDougall and Shields, 1977; Dembroski, MacDougall, Shields, Pettito and Jushene, 1978; Dembroski, MacDougall, Herd and Shields, 1979; Friedman, Byers, Diamant and Rosenman, 1975; Glass, Krakoff,

TABLE 3

Evaluation of Treatment Effects

Authors and Date	Comparison Group	Outcome Measures	Treatment Results
Suinn, 1975	Pre vs Post T vs C	Cholesterol Triglycerides	T greater reduction in choles-terol and triglycerides than C
Rosenman and Friedman, 1977	Pre vs Post	Clinical impressions	?
Roskies et al., 1978	Pre vs Post BT vs Psycho	Cholesterol, Trigly-cerides, BP, STAI-S STAI-T, Satisfaction, Goldberg, Time Pressure	Cholesterol, BP, Satisfaction, Goldberg, Time pressure significantly reduced in both groups. No difference between groups
Suinn & Bloom, 1978	Pre vs Post T vs C	Cholesterol, Triglyce-rides, BP, STAI-S, STAI-T, JAS	T significant change in one com-ponent of JAS (Hard-Driving), STAI-S and STAI-T
Jenni & Wollersheim,	Pre vs Post SM vs CT vs C	Bortner, Cholesterol, STAI-S, STAI-T	Increase in cholesterol for SM group. Decrease in STAI-S for two treatment groups
Rahe, Ward & Hayes, 1979	Pre vs Post vs Follow-up T vs C	Recurrent cardiac complications, Return to work, Coronary risk factors, Coronary prone behavior	T greater reduction in coronary morbidity and mortality, higher return to work, decrease in over-work, and time urgency
Roskies et al., 1979	Pre vs Post vs Follow-up BT vs Psycho	Cholesterol, Triglyce-rides, BP, STAI-S, STAI-T, Satisfaction, Goldberg, Time pressure	Reduction in cholesterol, BP, Satisfaction, Goldberg, Time pressure. Better maintenance in behavior therapy groups
Blumenthal et al., 1980	Pre vs Post As vs Bs	JAS, BP, Serum lipids, Body weight, Plasmino-gen activator release, Treadmill performance	All Ss decrease in BP, body weight, increase in treadmill performance and plasminogen activator release. Type As decrease in JAS. Type A females and Type B males increase in HDL
Friedman, 1980	Pre vs Post T vs C	Coronary mortality	Study in progress
Roskies, in press	Pre vs Post T vs C	JAS, Satisfaction, Goldberg, Time pressure, Anger, BP and catecholamine reactivity	T greater increase in satisfac-tion, decrease in Goldberg

T = Treatment group
C = Controls
BP = Blood pressure
STAI = Spielberger State-Trait Anxiety Inventory
Goldberg = General Health Questionnaire (self-report of psycho-physiological symptoms)
Bortner = Self rating of type A behavior

Contrada, Hilton, Kehoe, Mannucci, Collins, Snow, Elting, 1980); they report less contentment on their jobs, but also feel more able to make a change should they desire it (Howard, Cunningham

and Rechnitzer, 1977); on a treadmill task, they work closer to
the limits of their endurance, but, even as they do so, they are
more likely to suppress feelings of fatigue (Carver, Coleman and
Glass, 1976); they prefer to wait with others prior to working
on a stressful task, but while actually doing the task they
prefer to work alone (Dembroski & MacDougall, 1978). What this
catalogue of presumed Type A characteristics does not do,
unfortunately, is discriminate between those attributes which
are simply chance findings in a specific sample, those which are
generalizable to most Type A people but are irrelevant to
increased risk of CHD, and those which are the true pathogenic
ones (Roskies, 1980).

The lack of specificity in the Type A syndrome might be
easier to tolerate if it were confined to a small proportion of
the population, all with a high risk for heart disease. Under
these circumstances, we could at least be confident of our
choice of who to treat. However, recent samples of symptom-free
individuals in North America show a preponderance of As with
prevalence rates ranging from 50-76% (Chesney, Black, Chadwick
and Rosenman, in press; Howard, Cunningham and Rechnitzer, 1976;
MacDougall, Dembroski and Musante, 1979). Obviously, only a
small proportion of these Type As are likely to develop heart
disease. Thus, the hapless therapist who embarks on
indiscriminate treatment of Type As will not only find him or
herself in the ridiculous position of seeking to treat most of
the population of North America, but he or she will also have to
bear in mind that most of this treatment is irrelevant to
coronary risk. Given the fact that many characteristics of the
Type A pattern have considerable personal and social value,
indiscriminate therapeutic intervention may even be harmful (cf.
Roskies, in press a + b).

It is possible, of course, to continue using Type A
treatment as an all-purpose nostrum, non-specific both as to
treatment objectives and target populations. We must then
accept that it is unlikely to be considered as anything more
than another passing treatment fad. If, on the other hand, the
aim is to attain scientific and clinical credibility, then we
must begin the task of achieving greater treatment specificity.
Difficult as it may be, we must narrow down who and what we are
seeking to treat. This paper seeks to make a contribution to
this process by sketching some of the treatment alternatives.

Type As at High Risk for Coronary Heart Disease

One obvious approach to avoiding unnecessary treatment is
to select for therapy only those Type A individuals known to be
at high risk either because of a history of heart disease, or
because they manifest other risk factors in conjunction with the

Type A pattern. Type A individuals who have already suffered a heart attack have convincingly demonstrated their vulnerability to heart disease. Moreover, Type A continues to operate as a risk factor for them, since they have twice as many subsequent incidents as do Type B coronary survivors (Jenkins, Zyzanski and Rosenman, 1976). In Type A individuals who do not yet manifest overt heart disease, the simultaneous presence of multiple traditional risk factors (age, sex, family history, smoking, diabetes, hypertension, cholesterol) makes the individual in question a likely candidate for a future coronary event, and hence suitable for treatment.

This selection of individuals on the basis of their CHD risk certainly justifies the decision to intervene, but, unfortunately, does not resolve the more fundamental issue of precisely what to intervene upon. Given a Type A who has suffered a heart attack, does one seek to teach him or her to speak less loudly and explosively, to become more aware of fatigue, to use fewer self references in speech, or to control his or her blood pressure when placed in a competitive situation? The choice of therapeutic objectives is even more complicated when other risk factors are present, since it may be unethical to leave smoking and hypertension untouched, for instance, but confusing to engage in multiple risk factor intervention, if the aim is to evaluate the efficacy of modifying Type A.

The strongest advocate of restricting Type A intervention to individuals who have already suffered a heart attack is Friedman, who is currently directing a large-scale intervention project with this population. Based on preliminary reports (Friedman, 1979, 1980), it would appear that this program uses a broad spectrum treatment approach, seeking to modify the multiple behavioral manifestations of the Type A pattern (e.g. time pressure, hostility), the situations that provoke them, and the values that make these persons susceptible to intense competition. Moreover, while the program does not accept current smokers, it includes some intervention with other lifestyle risk factors. Should this project succeed in its aim of demonstrating reduced coronary mortality in the treated group, this would provide strong evidence of the value of treating Type As. It may be more difficult to distinguish whether the benefits observed resulted from generally improved health habits, or from modification of specific Type A characteristics, and if so, which ones.

Type As in High Risk Situations

Studies   demonstrating   behavioral   and   physiological differences between As and Bs typically show that these

differences are not manifest in the resting condition, but only
become apparent when both groups are exposed to appropriate
situations of threat and challenge. As Rosenman and Friedman
have repeatedly emphasized, Type A is not simply an innate
personality trait, but instead reflects a response, by a
predisposed individual, to certain types of environmental
pressure (Friedman & Rosenman, 1974; Rosenman, 1974, 1977).
Following this line of reasoning, Type As who live or work in
pressured situations can be considered suitable targets for
intervention because of the frequency with which these
situations are likely to call forth pathogenic A
characteristics. And since the environment has been indicted as
the agent provocateur, there is a clearcut remedy of simply
removing the individual from the harmful environment.

There are two main problems with this approach, one
conceptual and the other practical. On the conceptual level, we
do not know exactly what it is that makes an environment
challening for the Type A. For instance, a recent study
(Frankenhaeuser, Lundberg & Forsman, 1980) found that depriving
a Type A of work could produce physiological arousal remarkably
similar to that observed during intense effort. Outside the
laboratory, the data on the interaction between Type A
predisposition and environmental provocation are even more
confusing. There are correlational studies showing Type As to
be more prevalent at higher occupational levels (Caplan, Cobbs,
French, Harrison and Pinneau, 1975; Mettlin, 1976; Zyzanski,
1977; Shekelle, Schoenberger and Stamler, 1976), but these do
not distinguish whether it is the job level that produces the
behavior, or the Type A who seeks out that job level.

More relevant to the understanding of the effects of
pressured work situations is the intriguing finding of Friedman,
Rosenman and Carrol (1958) of increased cholesterol levels in
accountants at tax time. Even here, however, cholesterol was
more strongly related to events recorded in the individual's
personal diary than to the objective tax deadline. To further
complicate matters, in a recent study of air traffic
controllers, Rose, Jenkins and Hurst (1978) found that
individuals with higher Type A scores generally had higher rates
of illness than Type Bs in the same jobs, but, paradoxically,
were less likely to suffer from the illness most characteristic
of this occupation -- hypertension. Caplan and his colleagues
(1975), on the other hand, found no relationship between Type A,
job stress, or the physiological correlates of stress.

Until we have much firmer data showing which environments
are safe for Type As, it would be unethical to suggest drastic
environmental change as a means of reducing their coronary risk.
The ethics of this treatment approach need not concern us

unduly, however, because few Type As would expose themselves to
it. Individuals who hold demanding jobs, for instance, or who
are advancing rapidly in their careers, will be reluctant to
abandon the psychological and material rewards currently
provided by their occupations for the sake of health benefits
far in the future. Much more likely, they will decide that the
cure is worse than the disease: Better to envisage the prospect
of a shortened life span than to endure a long life devoid of
the very considerations that give it meaning.

Type As with Coronary-Prone Characteristics

A third approach to narrowing down therapy objectives is to
select for treatment Type As considered to be at high coronary
risk because they manifest a more "virulent" form of the
pattern. The task here is to isolate those specific components
within the global pattern that produce heart disease.
Individuals manifesting these pathogenic components could then
be selected for treatment, with the intervention focused
directly on removing or diminishing the harmful elements.

An early attempt to isolate specific components, by factor
analysis of the Jenkins Activity Survey, proved disappointing
since the subscales so derived, in contrast to the global score,
were not predictive of heart disease (Zyzanski & Jenkins, 1970).
More recently, Matthews, Glass, Rosenman and Borner (1977)
selected a subsample from the Western Collaborative Group Study
and conducted an item analysis of their interview records to
isolate the predictors of later heart disease. Among the five
factors derived only two significantly distinguished cases from
controls. These were competitive drive (manifested by explosive
voice modulation, vigorous answers and potential for hostility)
and impatience (manifested by irritation at waiting in lines).

Before accepting these findings as definitive, it would be
desirable to cross-validate them on a new sample. Should they
prove robust, it would then be possible to select individuals
who manifested high competitive drive and impatience and focus
therapy on modifying these specific behaviors. Ideally,
evaluation of the efficacy of treatment would be done by waiting
until sufficient new cases of heart disease had been generated
to permit statistical analysis of differences between treated
and control groups. A less expensive and time-consuming
alternative, at least for preliminary studies, is to use the
interview itself as a process measure. If individuals following
treatment showed a decrease in these predictor characteristics
greater than that observed in a control group, then one would
have a basis for embarking on a large-scale prospective study.

In contrast to this empirical approach, another way of reducing the myriad manifestations of Type A into specific therapeutic goals is by model-building. An explanatory model of the factors giving rise to the pattern and the mechanisms linking behavior to the disease end point can help us focus treatment on an underlying psychological need, for instance, or a common physiological pathway. The problem here is to isolate process factors that are known to be related both to Type A and to heart disease. For instance, Glass' hypothesis, that Type As behave the way they do because they have a strong need to control their environment, might permit us to select for treatment Type As who have an unusually strong need for control and to seek to modify this need during treatment.

Even more interesting is the current exploration of the possibility that Type As' greater vulnerability to heart disease might be attributable, at least in part, to their hyperreactive sympathetic adrenomedullary systems (SAM). It has consistently been found that As, as a group, show significantly greater autonomic and endocrine reactivity to a wide variety of stressors than do Bs. There is also a growing body of evidence implicating repeated sympathetic arousal, with its associated endocrine activity, in the development of arteriosclerosis and, eventually, heart disease (Davis, 1974; Gilmore, 1974; Krantz, Glass, Schaeffer and Davia, in press; Obrist, 1976; Williams, 1975). Presumably, therefore, any reduction in the frequency, intensity and duration of arousal manifested by a Type A could be taken as a sign both of decreased Type Aness and diminished coronary risk.

What makes this option particularly exciting in the search for specificity is that not all Type As demonstrate the same degree of physiological reactivity (Dembroski et al., 1978; Roskies, in press b). Moreover, Dembroski and his colleagues found that individuals who received higher clinical ratings of hostility and competitiveness were those most likely to show a strong cardiovascular response, even in a low challenge situation. It will be recalled that it was these same factors of competitiveness and hostility that were found to be the best predictors of eventual heart disease in the Western Collaborative Group Study interviews (Matthews et al., 1977). One can therefore speculate that Type As who manifest greater cardiovascular reactivity, for whatever reason, are the ones most likely to be coronary prone, and would achieve the greatest benefit -- in terms of the coronary risk associated with the presence of pattern A -- from treatment designed to reduce this reactivity.

It is this approach that my colleagues and I have adopted in our attempt to define precise treatment goals and develop meaningful evaluation measures (Brochocka, 1981; Roskies, Spevack, Surkis, Cohen & Gilman, 1978; Roskies, Kearney, Spevack, Surkis, Cohen & Gilman, 1979; Roskies, 1979; Roskies, Seraganian % Oseasohn, 1981; Roskies, in press a + b; Roskies & Avard, in press). The effort to operationalize "reactivity" and to develop a methodology for repeatedly testing a host of autonomic and endocrine variables has not been an easy one, and we have not yet succeeded in demonstrating significant differences in reactivity attributable to treatment. It may be that we shall have to eventually abandon this approach either because we will have to admit failure in obtaining good measures of reactivity, or because no treatment that we can devise can significantly modify these physiological characteristics, or even because research yet to come points to a better way of understanding the relationship between Type A and heart disease. For the time being, however, this approach presents the clearest understanding we have of how to select Type As for treatment and what to treat them for. And unless we are willing to take the risk of being proven clearly wrong, we also lose the chance of being at least partially right.

ACKNOWLEDGMENTS

This work has been partially supported by grants from the Conseil de la Recherche en Sante du Quebec, and the Department of Health and Welfare, Ottawa.

REFERENCES

Blumenthal, J. A., Williams, R. S., Williams, R. B. & Wallace, A. G. Effects of exercise on the Type A (coronary-prone) behavior pattern. Psychosomatic Medicine, 1980, 42, 289-296.

Brochocka, J. Evaluation d'un traitement chex les sujets de Type A dans le milieu de travail. Memoire de maitrise en psychologie, Universite de Montreal, 1981.

Caplan, R. D., Dobbs, S., French, J. R. P., Harrison, R. V. & Pinneau, S. R. Job demands and worker health, U.S. Department of Health, Education and Welfare, Publication No. (NIOSH) 75, 1975.

Carver, C. S., Coleman, A. E & Glass, D. C. The coronary-prone behavior pattern and suppression of fatigue on a treadmill test. Journal of Personality and Social Psychology, 1976, 33, 460-466.

Chesney, M. A., Black, G. W., Chadwick, J. H. & Rosenman, R. H. Psychological correlates of the coronary-prone behavior pattern. Journal of Behavioral Medicine, in press.

Davis, R. Stress and hemostatic mechanisms. In R. S. Eliot (eds.) Stress and the heart, New York: Futura, 1974.

Dembroski, T. & MacDougall, J. Stress effects on affiliation preferences among subjects possessing the Type A coronary-prone behavior pattern, Journal of Personality and Social Psychology, 1978, 36, 23-33.

Dembroski, T. M., MacDougall, J. M. & Shields, J. L. Physiologic reactions to social challenge in persons evidencing the Type A coronary-prone behavior pattern. Journal of Human Stress, 1977, 3, 2-10.

Dembroski, T. M., MacDougall, J. M., Shields, J. L., Pettito, J. & Lushene, R. Components of the Type A coronary-prone behavior pattern and cardiovascular responses to psychomotor performance challenge. Journal of Behavioral Medicine, 1978, 1, 159-176.

Dembroski, T. M., MacDougall, J. M., Herd, J. A. & Shields, J. L. Effect of level of challenge on pressor and heart rate responses in Type A and B subjects. Journal of Applied Social Psychology, 1979, 9, 209-228.

Frankenhaeuser, M., Lundberg, U. & Forsman, L. Note on arousing Type A persons by depriving them of work. Journal of Psychosomatic Research, 1980, 24, 45-47.

Franks, C. M. & Wilson, G. T. (eds.) Annual review of behavior therapy: Theory and practice. New York: Brunner/Mazel, 1979.

Friedman, M. The modification of Type A behavior in post-infarction patients. American Heart Journal, 1979, 97, 551-560.

Friedman, M. Progress report on the Recurrent Coronary Prevention Project, 1980.

Friedman, M. & Rosenman, R. H. Type A behavior and your heart. Greenwich, Conn.: Fawcett, 1974.

Friedman, M., Byers, S. O., Diamant J. & Rosenman, R. H. Plasma catecholamine response of coronary-prone subjects (Type A) to a specific challenge. Metabolism, 1975, 4, 205-210.

Friedman, M., Rosenman, R. H. & Carroll, V. Changes in serum cholesterol and blood clotting time in men subjected to cyclic variation of occurpational stress. Circulation, 1958, 17, 852-861.

Gilmore, J. P. Physiology of stress. In R. S. Eliot (ed.) Stress and the heart. New York: Futura, 1974.

Girdano, D. & Everly, G. Controlling stress and tension: A holistic approach. Englewood Cliffs, N.J.: Prentice Hall, 1979.

Glass, D. C. Behavior patterns, stress and coronary disease. Hillsdale, N.J.: Erlbaum, 1977.

Glass, D. C., Krakoff, L. R., Contrada, R., Hilton, W. F., Kehoe, K., Mannucci, E. G., Collins, C., Snow, B. & Elting, E. Effect of harassment and competition upon vardiovascul and catecholamine responses in Type A and Type B individuals. Psychophysiology, 1980, 17, 453-463.

Goldberg, P. Executive health. New York: McGraw-Hill, 1978.

Howard, J. H., Cunningham, D. A. & Rechnitzer, P. A. Health

patterns associated with Type A behavior: A managerial
population. Journal of Human Stress, 1976, 2, 24-33.

Howard, J. H., Cunningham, D. A. & Rechnitzer, P. A. Work
patterns associated with Type A behavior: A managerial
population. Human Relations, 1977, 30, 825-836.

Jenni, M. A. & Wollersheim, J. P. Cognitive therapy, stress
management training and the Type A behavior pattern.
Cognitive Therapy and Research, 1979, 3, 61-75.

Jenkins, C. D., Zyzanski, S. J. & Rosenman, R. H. Risk of new
myocardial infarction in middle-aged men with manifest
coronary heart disease. Circulation, 1976, 53, 342-347.

Krantz, D., Glass, D. C., Schaeffer, M. A. & Davia, J.E.
Behavior patterns and coronary disease: A criticial
evaluation. In J. T. Cacioppo and R.E. Petty (eds.) Focus
on cardiovascular psychophysiology. New York: Guilford
Press, in press.

MacDougall, J. M., Dembroski, T. M. & Musante, L. The
structured interview and questionnaire methods of assessing
coronary-prone behavior in male and female college
students. Journal of Behavioral Medicine, 1979, 2, 71-83.

Matthews, K. A., Glass, D. C., Rosenman, R. H. & Bortner, R. W.
Competitive drive, pattern A, and coronary heart disease.
A further analysis of some data from the Western
Collaborative Group Study. Journal of Chronic Disease,
1977, 30, 489-498.

Mettlin, C. Occupational careers and the prevention of
coronary-prone behavior. Social Science and Medicine,
1976, 10, 367-373.

Obrist, P. A. The cardiovascular – behavior interaction – as it
appears today. Psychophysiology, 1976, 13, 95-107.

Rahe, R. H., Ward, H. W. & Hayes, V. Brief group psychotherapy
in myocardial infarction rehabilitaton: Three-to-four year
follow-up of a controlled trial. Psychosomatic Medicine,
1979, 41, 229-241.

Rose, R. M., Jenkins, C. D. & Hurst, M. W. Air traffic
controller healthchange study: A prospective study of
physical, psychological and work-related changes.
Published by the authors, 1978.

Rosenman, R. H. The role of behavior patterns and neurogenic
factors in the pathogenesis of coronary heart disease. In
R. S. Eliot (ed.), Stress and the heart. New York:
Futura, 1974.

Rosenman, R. H. History and definition of the Type A
coronary-prone behavior pattern. IN T. Dembroski (ed.)
Proceedings of the forum on coronary-prone behavior.
Washington, D.C., Dept. of Health, Education and Welfare
Publication, NO. (NIH) 78-1451, 1977,

Rosenman, R. H. & Friedman, M. Modifying Type A behavior
pattern. Journal of Psychosomatic Research, 1977, 21,
323-333.

Roskies, E. Evaluating improvement in the coronary-prone (Type A) behavior pattern. In D. J. Osborne, M. M. Gruneberg and J.R. Eiser (eds.), Research in psychology and medicine, Vol. 1. New York: Academic Press, 1979.

Roskies, E. Considerations in developing a treatment program for the coronary-prone (Type A) behavior pattern. In P. Davidson and S.M. Davidson (eds.) Behavioral medicine: Changing health lifestyles. New York: Brunner/Mazel, 1980.

Roskies, E. Stress management for Type A individuals. In D. Meichenbaum and M. Jaremki (eds.) Stress prevention and management: A cognitive-behavioral approach. New York: Plenum, in press.

Roskies, E. Modification of coronary-risk behavior. In D. Krantzl, A. Baum, and J. E. Singer (eds.) Handbook of Psychology and Health. Hillsdale, N.J. Lawrence Erlbaum, in press.

Roskies, E. & Avard, J. Teaching healthy managers to control their coronary-prone (Type A) behavior. In K. Blankstein & J. Polivy (eds.), Self-control and self-modification of emotional behaviors. New York: Plenum, in press.

Roskies, E., Spevack, M., Surkis, A., Cohen, C. & Gilman, S. Changing the coronary-prone (Type A) behavior pattern in a non-clinical population. Journal of Behavioral Medicine, 1978, 1, 201-215.

Roskies, E., Kearney, H., Spevack, M., Surkis, A., Cohen, C. & Gilman, S. Generalizability and durability of treatment effects in Journal of Behavioral Medicine, 1979, 2, 195-207.

Roskies, E., Seraganian, P. & Oseasohn, R. Changing Type A in a non-clinical population: Project III. Research proposal submitted to Health and Welfare, July, 1981.

Schucker, B. & Jacobs, D.R. Assessment of behavioral risk for coronary disease by voice characteristics. Psychosomatic Medicine, 1977, 39, 219-228.

Shekelle, R. B., Schoenberger, J. A. & Stamler, J. Correlates of the JAS Type A behavior pattern score. Journal of Chronic Diseases, 1976, 29, 381-394.

Suinn, R.M. The cardiac stress management program for Type A patients. Cardiac Rehabilitation, 1975, 5, 13-15.

Suinn, R.M. & Bloom, L.J. Anxiety management training for Pattern A behavior. Journal of Behavioral Medicine, 1978, 1, 25-37.

Williams, R.B. Physiological mechanisms underlying the association between psychsocial factors and coronary disease. In W.D. Gentry & R.B. Williams (eds.) Psychological aspects of myocardial infarction and coronary care. St. Louis: Mosby, 1975.

Zyzanski, S.J. Associations of the coronary-prone behavior pattern. In T. D. Dembroski (ed.) Proceedings of the forum on coronary-prone behavior. Washington, D.C. Department

of Health, Education and Welfare Publication No. (NIH)
78-1451, 1977.

Zyzanski, S.J. & Jenkins, C.D. Basic dimensions within the
coronary-prone behavior pattern.   Journal of Chronic
Diseases, 1970, 22, 781-795.

BEHAVIORAL TREATMENT OF RAYNAUD'S DISEASE

Richard S. Surwit

Department of Psychiatry
Duke University Medical Center
Durham, North Carolina

In 1862 Maurice Raynaud published his now famous thesis in which he first described the syndrome today known as Raynaud's disease. Three interrelated phenomena were considered part of his new diagnostic category: local syncope or sudden blanching and numbness of the digits; cyanosis, during which time the pallor previously observed evolves into a blue color characteristic of deoxygenated tissue; and reactive hyperemia, characterized by the spread of red oxygenated blood through the upper layers of the epidermis. This last phase is often accompanied by burning and tingling and lasts until the skin returns to its normal pink color. In severe cases, patients experience chronic vasoconstriction or such frequent episodes of cyanosis that gangrene or small nutritive lesions and ulcerations often appear at the distal end of the digits. While vasospasms are usually confined to the digits of the hands and feet, they can occasionally appear on parts of the face as well. Cold stimulation is the most reliable eliciting stimulus, although emotional stress has also been reported to produce these attacks (Spittell, 1972). This is so because the mechanism of peripheral vasoconstriction is mediated by the sympathetic nervous system. Severe manifestations of this condition are not common, but Lewis (1949) has estimated that it affects approximately 20% of most young people in its mildest forms. Clinical Raynaud's disease is found to occur five times more often in women than in men, the time of onset occurring in the first and second decades of life. When this syndrome results from an identifiable pathological process, it is known as Raynaud's phenomenon.

The pathophysiology of "idiopathic" Raynaud's disease is not completely understood. While Raynaud himself attributed the malady to sympathetic overreactivity, Lewis (1949) maintained that the problem resulted from a local fault in the peripheral digital vessels. He collected evidence showing that changes in environmental temperature could have specific effects on the part of the digits stimulated by cold. Lewis did not believe that the patients he examined who were suffering from Raynaud's disease were abnormally "nervous." This led him to downplay the contribution of emotional and central nervous system activity upon the manifestations of this disorder.

Mittelmann and Wolff (1939) demonstrated that emotional stress could reduce the digital blood flow as measured by skin temperature in both normals and Raynaud's patients. In Raynaud's patients, however, these changes in temperature were accompanied by the blanching-cyanotic-edemic color change and pain typical of the vasospastic disorder. They reported that temperature changes were not in and of themselves sufficient cause for the vasospasm. Rather, the attacks seemed to occur most reliably when emotional stress and low environmental temperature interacted. In addition, they failed to find emotional stimuli effective in producing this reaction after sympathectomy. Graham (1955) was also able to demonstrate the vasoconstrictive effects of disturbing interviews on the skin temperature of both patients with Raynaud's disease as well as normal subjects. In addition, he was able to isolate hostility and anxiety as the emotion most often responsible for this reaction. In a subsequent study, Graham, Stern and Winokur (1958) demonstrated that by suggesting these emotions to normal subjects under hypnosis, vasoconstriction in the digits could be produced.

Although this evidence strongly implies that emotional stimuli are at least a contributing factor in the elicitation of Raynaud's disease, the local fault hypothesis of Lewis cannot be immediately ruled out. Mendlowitz and Naftchi (1959) have suggested that Raynaud's disease might be dichotomized into two separate disorders, one in which the vasculature is normal and vasomotor tone is heightened by sympathetic overreactivity and another in which normal vasomotor tone produces an overreaction in pathological local vasculature. Thus, much of the dispute as to the etiology of Raynaud's disease is probably attributable to diagnostic confusion. Allen and Brown (1932) produced one of the first attempts to improve on the nosological categories established by Maurice Raynaud. They suggested the following criteria for the diagnosis of true idiopathic Raynaud's disease: 1] intermittent attacks of discoloration of the extremities, 2] absence of evidence of organic arterial occlusion, 3] symmetric or bilateral distribution, 4] trophic changes when present, limited to skin and never consisting of gross gangrene, 5] the disease must have been present for at least two years, and 6] there should be no evidence of any other disease that could produce the symptoms secondarily. These criteria were recently revised by Spittell (1972) to include cold and emotion as the stimuli for vasospastic attacks.

Case Studies. Shapiro and Schwartz (1972) conducted the first study in which patients suffering from Raynaud's disease were trained to increase peripheral blood flow. In this report, two patients were provided with biofeedback of blood volume changes as recorded by a photo-plethysmograph. Feedback of blood volume changes occurring in the finger was provided to one patient, while the other patient was provided with feedback of blood volume changes recorded from the toes. The treatment was moderately successful for one patient who reported a reduction in the severity of Raynaud's symptoms.

Surwit (1973) investigated the use of skin temperature biofeedback in the treatment of a case of Raynaud's disease. The patient in this study was a 21-year-old female who reported vasospasms occurring in both hands and feet. The patient previously had bilateral cervical and lumbar sympathectomies. The former surgical procedure was unsuccessful and she underwent a considerable number of vasospasms in her fingers each year. She initially trained in autogenic and progressive relaxation techniques and was then provided with a series of 52 laboratory feedback sessions spaced over a nine-month period. During training, feedback was provided by a computer CRT which displayed a cumulative record of skin temperature while an audible bell underscored each $.1^{\circ}C$ increase in temperature. Over four months of training the patient's basal hand temperature rose from $23.3^{\circ}C$ to $26.6^{\circ}C$ and a concomitant decrease in the frequency of Raynaud's attacks was reported. Recently the patient was contacted for a nine-year follow-up. She indicated that she continues to practice voluntary control and claims a continued maintenance of therapeutic effect.

Jacobson, Hackett, Surman and Silverberg (1973) explored the utility of hypnosis and temperature biofeedback. In this case, the patient showed very little improvement in the hypnosis portion of training. However, when the patient was provided with temperature biofeedback designed to teach him to increase finger temperature relative to forehead temperature, a marked reduction in the frequency of Raynaud's attacks occurred. These gains were maintained at eight-month follow-up. As contrasted with the earlier work of Surwit (1973), less emphasis was placed on laboratory training and much more emphasis on the importance of the patient's practice with self-control techniques at home. No follow-up was reported.

Sundermann and Delk (1978) reported the case of a 40-year-old patient with a 15-year history of Raynaud's disease. The patient was provided with temperature biofeedback to teach her to increase hand temperature. During biofeedback sessions, the patient was encouraged to use whatever technique would provide temperature increases. She arrived at a strategy of subvocally quoting biblical scriptures, and by the end of training, actually brought the Bible to the sessions. The patient had a long course of treatment - three sessions a week for 13 months. In addition to the biofeedback and subvocal biblical quoting, the patient was also on medication. Although the authors reported that this patient showed gradual temperature changes over the course of treatment, the unorthodox combination of treatment procedures makes it very difficult to determine what elements are responsible for change.

Blanchard and Haynes (1975) have conducted the most systematic and controlled single case study published to date. In this study, changes in skin temperature and the frequency of Raynaud's attacks were evaluated under three conditions: 1) no treatment baseline, 2) self-control technique in which the patient was asked to try to increase her hand temperature any way that she could, and 3) biofeedback training to increase hand relative to forehead temperature. The results provided strong support for the utility of temperature biofeedback. During the biofeedback sessions, the patient showed an ability to increase her hand temperature an average of 3.4°F, whereas no consistent changes in temperature were noted under any of the other conditions. The authors also reported a gradual increase in basal finger temperature from 79 to 91.1°F. Reductions in the frequency of vasospasms were achieved concurrent with temperature biofeedback training. Follow-up evaluations at two and four months revealed maintenance of treatment gains. Although control over skin temperature had deteriorated by seven months post-treatment, acquisition of learned control was reinstated after five additional training sessions. No long-term follow-up was available.

The results of these case studies seem to suggest that temperature biofeedback training may be helpful in the treatment of Raynaud's disease. However, the primary source of outcome data in all these studies is self-report. Self-report is a notoriously inaccurate and biased method of evaluating treatment (Keefe, Kopel & Gordon, 1978). Self-report can be influenced strongly by the demand characteristics of the situation. In an effort to provide a more stringent measure of learned control of peripheral vasodilatation, a number of researchers have suggested that one examine the ability of the patients to maintain hand temperature under cold ambient stress. Taub (1977) described using such a procedure with Raynaud's patients.

In this study, patients who had been trained in temperature self-control were fitted with a "cold suit" and instructed to attempt to maintain hand temperature. This suit was designed in such a way that cold water of specified temperatures could be rapidly circulated so that the patient's whole body could be stressed. Taub has reported on the results achieved with only one patient so far. This patient was able to actually increase hand temperature from 88 to 89.5°F, while the temperature of the cold suit was decreased from 80 to 60°F. Unfortunately, no data are provided on the clinical damage observed in this patient.

In each of the other case studies presented, the number of subjects was small and no statistical treatment of the data was presented. Because only cases in which such treatment techniques are successful are published, the number of failures goes unreported. Without a no-treatment or other appropriate control group, the therapeutic gains reported cannot be honestly attributed to treatment. Finally, in each study multiple treatment techniques were used. When treatment effects are analyzed over such a long time period, the possibility of carryover effects from one treatment to another is strong. Although patients may show change during temperature biofeedback sessions, these changes may be mediated by a cognitive strategy previously taught to the patient, for example, self-hypnosis or autogenic training. It seems fair to conclude, however, that these studies suggest that behavioral techniques such as biofeedback may have an important role to play in the management of patients with Raynaud's symptoms. Recently, controlled group outcome studies have attempted to more systematically address the potential contribution particular behavioral techniques may provide.

Controlled Group Outcome Studies. In the past two years a series of controlled studies investigating the behavioral treatment of Raynaud's disease have been conducted. These studies have attempted to identify the relative contribution of particular behavioral techniques in facilitating self-control of skin temperature and reducing Raynaud's attacks. In the first study of this series (Surwit, Pilon & Fenton, 1978), three major questions were addressed. Can Raynaud's disease be treated behaviorally; does biofeedback have any advantage over simplier relaxation procedures; and does training need to be performed under laboratory conditions in order for patients to benefit?

Thirty female patients diagnosed as suffering from idiopathic Raynaud's disease were trained to control their digital skin temperature using either autogenic training or a combination of autogenic training and skin temperature feedback. Training was conducted either in a laboratory or in three group sessions supplemented by extensive home practice. All subjects

were exposed to an initial cold stress procedure in which they
were seated in an experimental chamber while the ambient
temperature was slowly dropped from 26 to 17°C over 72 minutes.
Skin temperature was monitored during the procedure. This
procedure was given to half the subjects immediately before and
immediately following a four-week training sequence. The
remaining half of the sample were exposed to an additional cold
stress challenge prior to treatment as a control for possible
habituation effects. The results of this study are illustrated
in Figures 1 and 2. All subjects, regardless of which condition
they were trained in, showed a significant improvement in their
ability to maintain digital skin temperature both relative to
their initial cold stress and relative to the second cold stress
given to the half of the sample not immediately treated.
Patients who served as a no-treatment control not only failed to
show improvement during the second test, but actually
deteriorated in performance. In addition to this objective
finding, all treated patients reported significant reductions in
the frequency of vasospastic attacks over the four-week
treatment period. No additional benefits could be observed for
those subjects receiving skin temperature biofeedback or for
those subjects whose training was conducted in the laboratory.

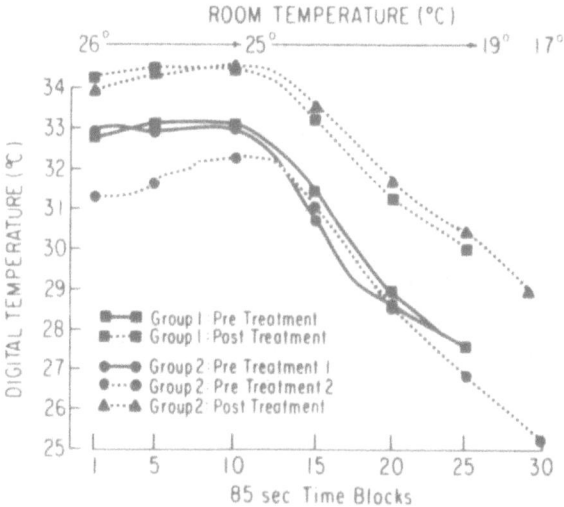

Figure 1.   Mean digital temperature during pre- and post-
             treatment stress tests. Recording of skin
             temperature begun after a 10-minute stabiliza-
             tion period (From R. S. Surwit, R. N. Pilon,
             & C. H. Fenton. Behavioral treatment of
             Raynaud's disease. Journal of Behavioral
             Medicine, 1978, 1, 329).

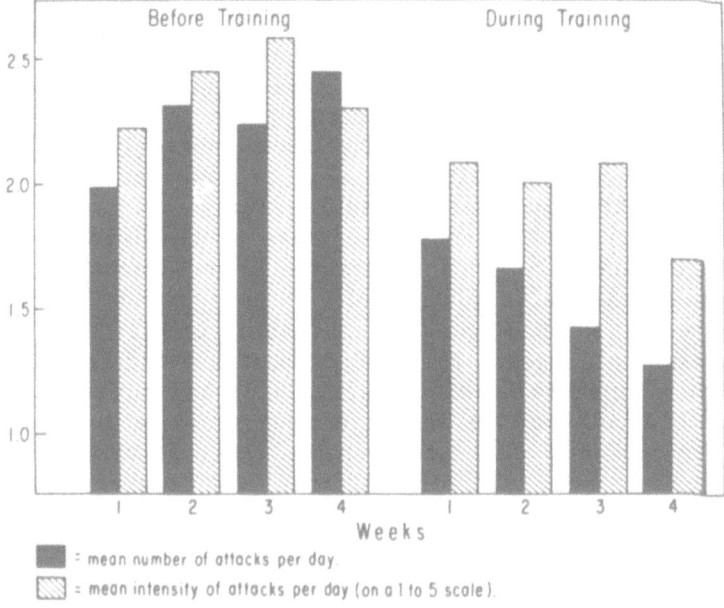

Figure 2. Mean number and intensity of attacks per day reported by all subjects during the four weeks immediately preceding training and the four weeks of training. The reduction in the number of attacks across weeks of treatment was significant. (From R. S. Surwit, R. N. Pilon, & C. H. Fenton. Behavioral treatment of Raynaud's disease. Journal of Behavioral Medicine, 1978, 1, 331. Solid bars indicate mean number of attacks per day. Ruled bars indicate mean intensity of attacks per day (on a 1-5 scale).

The second study of this series (Keefe, Surwit, & Pilon, 1980) served as a partial replication as well as an extension of the study just described. This study attempted to provide a more rigorous test of home biofeedback training by having patients on home practice regimens use more sophisticated and sensitive temperature feedback equipment than had been used in the prior study. This study also compared the efficacy of autogenic training (which focuses specifically upon sensations of warmth and heaviness in the hands) to general relaxation training (which focuses upon reducing muscular tension generally throughout the body). In addition, this study used four laboratory cold stress challenges like those described in the previous study given at week 1 of a four-week baseline and during weeks 1, 3, and 5 of training. Twenty-one patients were randomly assigned to one of three treatment conditions. The first group received progressive muscle relaxation and home practice instructions, the second group received autogenic training and home practice instructions, while the third group received autogenic training and skin temperature feedback with autogenic instructions and portable skin temperature feedback equipment. The results confirmed those of the initial study in that all patients improved regardless of treatment (see Figure 3). Data gathered from the cold stress procedures indicated that subjects improved gradually and significantly over the four cold stress challenges. This improvement was not felt to be due to habituation since Surwit et al. (1978) had demonstrated that patients did not improve in performance without training. The gradual improvement of the response to the cold stress procedure suggests that some learning process was taking place. As previously reported, all treated patients also experienced an approximate 40% reduction in the frequency of vasospastic attacks. This reduction in symptoms was obtained at the same time as a significant drop in outdoor temperature was occurring.

Another group of investigators have obtained similar results. Jacobson, Manschreck and Silverberg (1979) gave 12 patients suffering from idiopathic Raynaud's disease 12 sessions of progressive muscle relaxation over a six-week period. Half of the patients were also given auditory and visual skin temperature feedback during the training sessions. Skin temperature during training as well as patients' self-reports of improvement were collected. Both groups showed significant increases in skin temperature during training with larger skin temperature increases shown by the group not receiving feedback. All subjects rated themselves moderately to markedly improved at one month with seven subjects continuing to report improvement at two years. No data was collected on objective hand temperature changes at follow-up.

In the final study (Keefe, Surwit & Pilon, 1979) the maintenance of treatment gains was evaluated. Nineteen patients who had undergone behavioral training in the initial study (Surwit et al., 1978) were asked to keep a daily log of frequency and severity of vasospastic attacks and to fill in a follow-up questionnaire dealing with their satisfaction with various elements in the treatment regimen. One year after initial treatment, these patients were given an additional cold stress challenge. Thus, as before, both objective and subjective indices of symptom improvement were obtained. The results of the study are fascinating in that they appear on the surface to be contradictory. One year post-treatment, patients reported an average of 1.2 vasospasms per day compared to 1.3

attacks per day immediately following treatment one year
earlier. However, the ability of patients to maintain digital
temperature in the face of cold stress had significantly
deteriorated and was virtually identical to their initial
treatment performance (see Figures 4 and 5). The contradiction
implied by these two sets of data can be explained by examining
data from the follow-up questionnaires administered to all
patients. Most patients had stopped practicing the behavioral
techniques they were taught during the spring months following
their initial training. These patients had not returned to
levels of practice comparable to those they were engaged in
during the initial treatment. Thus, while response to performance during cold stress was seen as related to practice, the
patient's subjective reports of improvement seem to be under the
control of other variables.

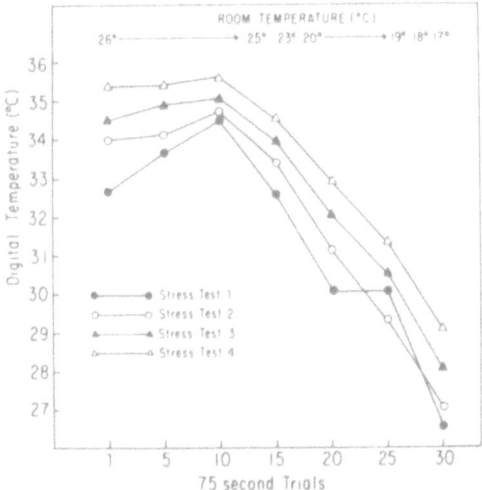

Figure 3.   Mean digital temperature during pre-treatment
            stress tests (Stress Test 1) and post-treatment
            stress tests (Stress Test 2-4). Recording of
            skin temperature began after a 10-minute
            stabilization period (From F. J. Keefe, R. S.
            Surwit, & R. N. Pilon. Biofeedback, autogenic
            training and progressive relaxation in the
            treatment of Raynaud's disease. Journal of
            Applied Behavior Analysis, 1980, 13, 7).

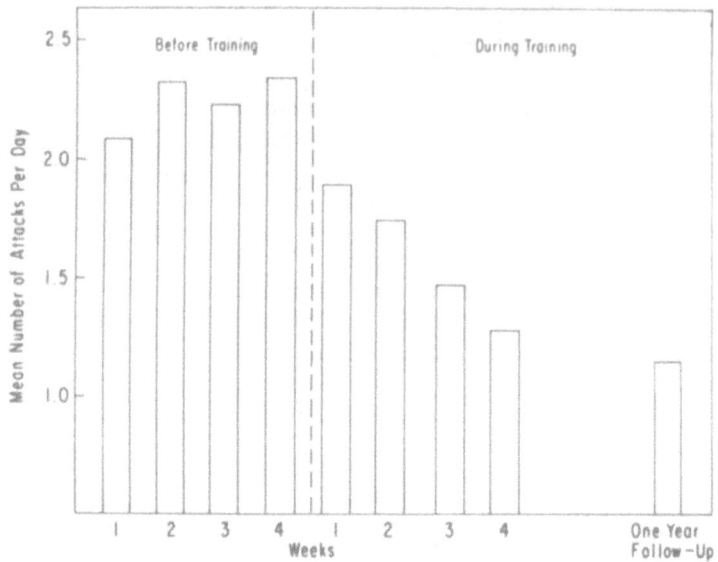

Figure 4.   Mean frequency of vasospastic attacks recorded
            during the four weeks before training, the four
            weeks post-training, and a 1-week 1-year follow-
            up.   (From F. J. Keefe, R. S. Surwit, & R. N.
            Pilon.   A one year follow-up of Raynaud's
            patients treated with behavioral therapy
            techniques.   Journal of Behavioral Medicine,
            1979, 2, 389).

Two other explanations can be offered to reconcile the discrepancy between subjects continuing to report fewer vasospastic attacks despite a deterioration of their ability to tolerate the cold stress challenge. First, it is possible that subjects were simply trying to please the investigators in reporting fewer vasospastic attacks. It is well known that different variables control verbal and nonverbal behavior (Keefe, Kopel & Gordon, 1979). However, it is also possible that subjects did retain some control of their ability to voluntarily vasodilate -- enough to prevent vasospasms but not enough to meet the challenge of the cold stress test. Mittelmann and Wolff (1939) demonstrated that a drop in digital temperature alone is not sufficient to bring on a vasospasm. Low digital temperature is a precursor to vasospasm, they noted, only if the subject is autonomically aroused.

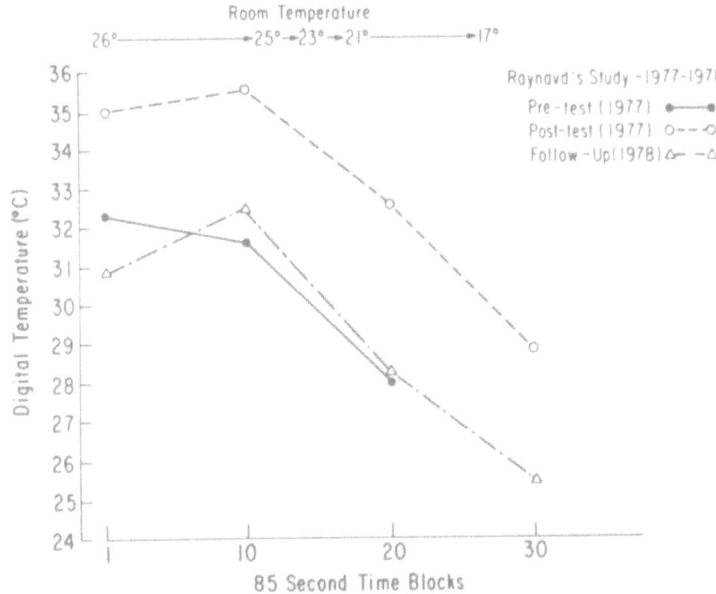

Figure 5. Mean digital temperature (°C) during cold stress
stress tests conducted pre-treatment, post-
treatment, and at a 1-year followup. (From
F. J. Keefe, R. S. Surwit, & R. N. Pilon.
A one year follow-up of Raynaud's patients
treated with behavioral therapy techniques.
Journal of Behavioral Medicine, 1979, 2, 388).

A recent study by Surwit, Bradner, Fenton and Pilon (1979)
sheds some light on which behavioral variables may be important
in predicting the response of patients to a behavioral program
designed to treat Raynaud's disease. These investigators found
that subjects' improvement, as measured by increasing skin
temperature during the cold stress test, could be predicted by
subjects' responses to a simple paper-and-pencil test. Thirty
subjects, trained in voluntary vasomotor with skin temperature
feedback and autogenic training were given the Psychological
Screening Inventory (Lanyon, 1973). Those subjects scoring high
on the Alienation scale of this inventory were found to show no
improvement in skin temperature in response to cold stress,
while those subjects with low scores showed a net increase of
4.5°C during the test after training. According to Lanyon
(1973) high scores on the Alienation scale are associated with
high scores on those MMPI scales related to serious psychopathology (Schizophrenia, Infrequency, Paranoia, Hypomania).
People with high scores on this scale perceive themselves as not
responsible for or in control of their own lives. Thus, this
scale appears sensitive to a feeling of self-control which seems

important for success in treatments based on exercise of self-control.

CONCLUSIONS

The above review allows us to draw the following conclusions about the utility of behavioral procedures in the treatment of Raynaud's disease. Investigators have described both subjective reductions of symptomatology and objective evidence of increased blood flow under conditions of cold stress subsequent to training. Typically, patients report up to 50% reduction in symptom frequency following training, with the increase of resting digital temperature approximately 3-4°C. These results are impressive and parallel the best clinical effects of many medical as well as surgical interventions. However, there does not appear to be any specific advantage of one behavioral technique over another. In the three controlled group outcome studies just reviewed, there is typically no difference in efficacy between different relaxation techniques or relaxation techniques supplemented with biofeedback.

REFERENCES

Allen, E. V., & Brown, G. E. Raynaud's disease: A critical review of the minimal requisite for diagnosis. American Journ of Medical Sciences, 1932, 183, 187-200.

Blanchard, E. B., & Haynes, M. R. Biofeedback treatment of a case of Raynaud's disease. Journal of Behavior Therapy and Experimental Psychiatry, 1975, 6, 230-234.

Graham, D. T. Cutaneous vascular reactions in Raynaud's disease and in states of hostility, anxiety, and depression. Psychosomatic Medicine, 1955, 17, 200-207.

Graham, D. T., Stern, J. A., & Winokur, C. Experimental investigation of the specificity of attitude hypothesis in psychosomatic disease. Psychosomatic Medicine, 1958, 20, 446-457.

Jacobson, A. M., Hackett, T. P., Surman, O. S., & Silverberg, E. L. Raynaud's phenomenon: Treatment with hypnotic and operant technique. Journal of the American Medical Association, 1973, 225, 739-740.

Jacobson, A. M., Manschreck, T. C., & Silverberg, E. L. Behavioral treatment for Raynaud's disease: A comparative study with long-term follow-up. American Journal of Psychiatry, 1979, 136, 844-846.

Keefe, F. J., Kopel, S., & Gordon, S. A Practical Guide to Behavioral Assessment. New York: Springer, 1978.

Keefe, F. J., Surwit, R. S., & Pilon, R. N. A one year follow-up of Raynaud's patients treated with behavioral therapy techniques. Journal of Behavioral Medicine, 1979, 2, 385-391.

Keefe, F. J., Surwit, R. S., & Pilon, R. N. Biofeedback, autogenic training and progressive relaxation in the treatment of Raynaud's disease. Journal of Applied Behavior Analysis, 1980, 13, 3-11.

Lanyon, R. I. Psychological Screening Inventory: Manual. Goshen, N.Y.: Research Psychologists' Press, 1973.

Lewis, T. Vascular Disorders of the Limbs: Described for Practitioners and Students. London: MacMillan, 1949.

Mendlowitz, M., & Naftchi, N. The digital circulation in Raynaud's disease. American Journal of Cardiology, 1959, 4, 580-584.

Mittelmann, B., & Wolff, H. G. Affective states and skin temperature: Experimental study of subjects with "cold hands" and Raynaud's syndrome. Psychosomatic Medicine, 1939, 1, 271-292.

Shapiro, D., & Schwartz, G. E. Biofeedback and visceral learning: Clinical applications. Seminars in Psychiatry, 1972, 4, 171-184.

Spittell, J. A., Jr. Raynaud's phenomenon and allied vasospastic conditions. In J. F. Fairbairn, J. C. Juergens, & A Spittell (Eds.), Allen-Barber-Hines Peripheral Vascular Diseases. (4th ed.). Philadelphia: Saunders, 1972.

Sundermann, R. H., & Delk, J. L. Treatment of Raynaud's disease with temperature biofeedback. Southern Medicine Journal, 1978, 71, 340-342.

Surwit, R. S. Raynaud's disease. In L. Birk (Ed.), Biofeedback: Behavioral medicine New York

Surwit, R. S., Pilon, R. N., & Fenton, C. H. Behavioral treatment of Raynaud's disease. Journal of Behavioral Medicine, 1978, 1, 323-335.

Surwit, R. S., Bradner, M. B., Fenton, C. H., & Pilon, R. N. Individual differences in response to the behavioral treatment of Raynaud's disease. Journal of Consulting and Clinical Psychology, 1979, 47, 363-367.

Surwit, R. S., & Fenton, C. H. Feedback and instructions in the control of digital skin temperature. Psychophysiology, 1980, 17, 129-132.

Taub, E. Self regulation of human tissue temperature. In G. E. Schwartz & J. Beatty (Eds.), Biofeedback: Theory and research, New York: Academic Press, 1977.

PATHOPHYSIOLOGY AND BEHAVIORAL TREATMENT OF SCLERODERMA

Robert Freedman and Paul Wenig

Lafayette Clinic and Wayne State University

Detroit, Michigan

Pathophysiology

Scleroderma (progressive systemic sclerosis or PSS) is a connective tissue disorder distinguished by symmetric thickening of the skin, by abnormalities of the microvasculature and large blood vessels, and by degenerative fibrotic changes in the viscera, muscles and joints (Rodnan, 1979a). Although the etiology of the disease is unknown, mechanisms involving the vascular, immune, and connective tissue systems have been proposed (LeRoy, 1981). The vasospastic attacks of Raynaud's phenomenon, consisting of pallor, cyanosis, and rubor in the fingers and sometimes the toes, occur in over 95% of patients with progressive systemic sclerosis (Rodnan, 1979a).

The term scleroderma refers specifically to a hardening or thickening of the skin. Since this change can occur in a variety of conditions, criteria for the classification of PSS have been established (American Rheumatism Association, 1981). In a multicenter study of 797 patients, proximal scleroderma, i.e. "bilateral and symmetrical sclerodermatous changes in any area proximal to the metacarpophalangeal or metatarsophalangeal joints" was found in 91% of the cases diagnosed as having definite systemic sclerosis but in only .2% of the comparison cases (systemic lupus erythematosus, polymyositis/dermatomyositis, Raynaud's phenomenon in the absence of these disorders). Proximal scleroderma has therefore been designated as the major criterion for systemic sclerosis. In the absence of this symptom, patients having two or more minor criteria (sclerodactyly, digital pitting scars of the fingertips or loss of distal finger pad substance, or bilateral basilar pulmonary

101

fibrosis on chest roentgenogram) are also classified as having PSS. Additional criteria are needed to differentiate systemic sclerosis from other so-called "overlap" disorders.

Estimates of the annual incidence of PSS range from 2.7 (Medsger & Masi, 1971) to 12 (Kurland, Hauser, Ferguson & Holley, 1969) new cases per million population, increasing with age. It is rare in children and more common in women than men, particularly women of childbearing age (Medsger & Masi, 1979). Survival with PSS is clearly related to the type and severity of visceral involvement, the overall 5-year survival rate being 50%. In a study of 358 hospitalized male veterans (Medsger & Masi, 1973), all 17 with renal involvement were dead within 10 months. Twenty-five percent of patients with cardiac involvement lived for 5 years and those with pulmonary involvement in the absence of cardiac involvement had a 45% 5-year survival rate. This rate was more favorable for women (60%) than for men (40%) and worse for blacks than for whites (Medsger et al., 1971).

Visceral Changes

It has been suggested that the pathogenesis of the visceral lesions in PSS is related to decreased nutritional blood flow caused by widespread degeneration of the intima of small arteries (Campbell & LeRoy, 1975). Abnormalities of the pulmonary vascular tree in PSS (Guttaduaria, Ellman & Kaplan, 1979) most often include arteriolar intimal proliferation, medial hypertrophy, and myxomatous changes in the vessel walls. Pulmonary hypertension is a frequent result of these vascular changes and may progress rapidly within a short period of time. The pulmonary vessels may have increased reactivity to cold, suggested by the finding that scleroderma patients have lower diffusion capacities in cold compared to warm weather (Emmanuel, Saroja, Gopinathan & Kaplan, 1976). In the kidney, blood flow decreases when peripheral vasospasms are provoked by local cooling (Cannon, Hassar, Case, Casarella, Sommers & LeRoy, 1974) and the incidence of renal involvement with scleroderma increases during cold weather (Gladman, Gordon, Urowitz & Levy, 1976). Vasomotor instability may also exist in the heart in scleroderma. In one study 45% of scleroderma patients with unobstructed coronary arteries had contraction band necrosis, a form of ischemic damage which develops when myocardium is reperfused after a transient interruption in blood flow (Bulkley, Ridolfi, Salyer & Hutchins, 1976). Esophageal involvement occurs in up to 80% of patients with PSS, consisting of reduced or absent peristalsis in the distal portion and diminished pressure in the lower sphincter (Turner, Lipshutz, Miller, Rittenberg, Schumacher & Cohen, 1973). It may occur in the absence of obvious cutaneous or other visceral involvement, but is closely related to disease of the stomach and large and small intestines, where similar pathological changes are found, i.e. atrophy and fibrosis of the muscularis (Myers, 1979).

Cutaneous Changes

Three stages have been described in the development of scleroderma: edematous, indurative, and atrophic (Rodnan, Lipinski, and Luksick, 1979). The initial phase consists of painless, pitting edema in the hands and fingers which may also involve all extremities as well as the face. In patients developing diffuse scleroderma (PSS) the swelling is gradually replaced by thickening and tightening of the skin, progressing proximally and becoming increasingly taut, shiny and hardened (Rodnan, 1979b). Contraction of the tissues leads to a loss of wrinkles and skin folds and the lips become thin and tightly pursed. Telangiectases, consisting of dilated capillary loops and venules, appear on the fingers, hands, and face. Progressive loss of the bone and soft tissue of the fingertips may result in the dissolution of the terminal phalanges. Subcutaneous calcinosis is particularly common in women (Rodnan, 1979a). Skin biopsies taken from indurated forearm skin of PSS patients show an increase of compact collagen fibers in the reticular dermis, along with hyalinization and obliteration of small blood vessels and loss of normal dermal appendages (Rodnan et. al.,1979). The thickness and weight of these biopsies are significantly greater in PSS patients than in normal controls and are proportional to the increase in dermal collagen content.

Immunological Abnormalities

The occurrence of immune abnormalities in PSS is indicated by the presence of antinuclear antibodies in 40% to 90% of the patients (Rodnan, 1979a). The most common patterns of nuclear fluorescence are those of fine or large speckles or of thread with occasional diffuse or homogeneous staining. Usually the titres are relatively low, particularly compared to those found in SLE (Rodnan, 1979a). There seems to be little correlation between the presence or titre of antinuclear antibodies and the duration or clinical severity of PSS (Rodnan, 1979a). Given the lack of specificity with which antinuclear antibodies occur in systemic sclerosis, work is currently underway to find marker antibodies which will specifically differentiate it from other diseases.

Microvascular Abnormalities

The pattern of cutaneous microvessels is markedly altered in patients with scleroderma. Using widefield microscopy to observe the nailfold and skin of the hand, Maricq and LeRoy

(1973) have found enlarged and distorted capillary loops and a reduction in their number in patients with PSS. Recently, this pattern was found in 82% of patients with PSS, and in 54% of those with mixed connective tissue disease (MCTD), but rarely in patients with systemic lupus erythematous or Raynaud's disease (Maricq, LeRoy, D'Angelo, Medsger, Rodnan, Sharp & Wolfe, 1980). The rarity of these abnormalities in the latter two disorders despite the presence of Raynaud's phenomenon suggests that they are not an expression of the Raynaud's phenomenon frequently seen in scleroderma and MCTD. It is hoped that this technique will be of prognostic value in specifying those patients who will eventually develop scleroderma but has not yet shown diagnostic skin changes. Capillary microscopy observations of PSS patients have also revealed frequent capillary hemorrhages and shown that the characteristic telangiectases found in the hand are clusters of dilated capillaries (Maricq & LeRoy, 1979).

Histological examination of digital arteries of scleroderma patients has shown proliferation of new connective tissue in the intima, subendothelial fibrin deposits, thickening of the basement membrane, reduplication and fraying of the internal elastic lamina and fibrous thickening of the adventitia (Rodnan, 1979a). These changes eventually lead to narrowing or obstruction of the lumen. Similarly, angiographic studies have revealed injury to the digital arteries with narrowing, obstruction, and absence of collateral circulation (Rosch, Porter, and Gralino, 1977).

Abnormalities of Peripheral Blood Flow

It has been previously noted that Raynaud's phenomenon occurs in the vast majority of patients with scleroderma. Symptoms typically begin with bilateral pallor and/or cyanosis of the fingers, which feel cold and numb. Upon rewarming, reactive hyperemia often occurs with concomitant rubor and tingling, burning, or painful sensations. Attacks are precipitated by various combinations of local cold, environmental cold, and emotional stress and have an average duration of 15 minutes (Freedman, Lynn, and Ianni, in press).

Many studies have shown that the patients with Raynaud's phenomenon have subnormal levels of peripheral blood flow, although the distinction between primary and secondary forms of the disorder has sometimes been obscured. Patients with Raynaud's phenomenon and scleroderma have lower basal digital temperatures than normal controls (Fries, 1969), reduced finger capillary blood flow in both warm and cool environments (Coffman & Cohen, 1971), a delay in skin rewarming after cooling (LeRoy, Downey, and Cannon, 1971), and increased blood viscosity at low shear rates (McGrath, Peek, and Penny, 1977). Plethysmographic

studies have shown abnormally reduced pulse volume during
maximum arterial dilation (Dabich, Bookstein, Sweifler &
Zarafonetis, 1972) and during cooling (Wouda, 1977). Complete
standstill of capillary blood flow during cooling was found in
scleroderma patients, but not controls (Maricq, Downey, and
LeRoy, 1976). Using the Xenon diffusion technique to measure
cutaneous blood flow in scleroderma patients, Nielsen (1978)
found an increased reduction in blood flow due to cooling,
although Kristensen (1980) found decreased reactivity to
vascular occlusion and to changes in transmural pressure.

Etiology

It is not known if the vasospasms of Raynaud's phenomenon
are the cause or effect of the structural vascular changes found
in scleroderma. Lewis (1929) believed that local structural
defects rendered the digital arteries susceptible to vasospasm;
subsequent work suggested that this might be manifest by
increased tissue concentrations of vasoconstrictor agents, at
least in patients with Raynaud's disease (Peacock, 1959). These
findings, however, were not replicated; Sapira, Rodnan, Scheib,
Klaniecki & Rizk (1972) and Surwit, Allen, Kuhn, Gilgor, Duvic,
Schanberg & Williams (1981) failed to find increased levels of
epinephrine or norepinephrine in venous blood of patients with
Raynaud's disease or scleroderma.

Theories of the etiology of systemic sclerosis have
attempted to integrate the vascular, immune, and connective
tissue abnormalities of this disease. Recent emphasis on the
vascular lesions of PSS has led to the hypothesis that sclerosis
is secondary to circulatory changes. LeRoy (1981) has proposed
that an unknown cytotoxic stimulus causes repeated disruption of
endothelial cells, leading to vascular permeability, plasma
leakage, fibrin deposition, collagen synthesis, and fibrosis.
In a similar theory, Fries (1979) suggests that vasoregulatory
failure causes widespread microvascular hypertension leading to
vascular leakage and subsequent fibrotic changes. These
hypotheses are consistent with evidence of abnormal blood flow
in the periphery and viscera and with the positive correlation
between the severity of digital capillary abnormalities and
extent of visceral involvement in PSS (Maricq, Spencer-Green,
and LeRoy, 1976). However, they cannot explain the fact that
many other diseases impair circulation to the point of gangrene
without causing sclerosis (Maricq & LeRoy, 1979), and that blood
flow is not impaired in sclerodermatous skin and subcutaneous
tissue of the forearm (Coffman, 1970).

An immune mechanism has been suspected as the trigger of
the vascular alterations of PSS. Significant numbers of
antinuclear and other antibodies are present in the sera of

scleroderma patients and those with recognized autoimmune
disorders such as systemic lupus erythematosus (Tan, Rodnan,
Garcia, Morin, Fritzler & Peebles, 1980). In addition, many
patients with graft-versus host disease due to bone marrow
transplantation develop a scleroderma-like skin disease (LeRoy,
1981). Cells from these organs synthesize increased amounts of
collagen compared to cell strains from control tissue (Rodnan,
1979a). It is likely, however, that extracutaneous factors
influence cutaneous involvement in scleroderma since autografts
of clinically normal abdominal skin transferred to the
PSS-involved forearm have become progressively sclerodermatous
while forearm skin transplanted to the abdominal site remained
thickened (Fries, Hoopes, and Shulman, 1971).

Treatment

     Traditional medical treatments for scleroderma have
generally proven unsatisfactory. Thoracic sympathectomy has
been used in the treatment of Raynaud's phenomenon in
scleroderma but is no longer advocated due to the transitory
nature of its effects (Rodnan, 1979a). Similar results have
been found with vasodilating drugs such as Reserpine.
Corticosteroids may loosen and thin swollen collagen bundles but
these changes are also temporary, and adverse side effects, such
as renal lesions, have been reported. A few patients with PSS
renal crisis have responded favorably to aggressive
antihypertensive treatment and have also shown remission of
sclerodermatous skin changes (Fries, 1979). Recommended medical
treatments have been mainly palliative, with patients instructed
to keep warm, eat balanced meals, reduce external stress and
avoid injury to the skin (LeRoy, 1981).

     Recent research has shown that normal persons can achieve
self-control of digital blood flow using skin temperature
biofeedback and other behavioral procedures, such as autogenic
training (e.g. Freedman & Ianni, 1981; Keefe, 1978). Raynaud's
disease patients treated with these techniques typically report
vasospastic symptom improvement and show evidence of increased
peripheral blood flow (see Chapter 6). More recently behavioral
methods have been employed to treat Raynaud's phenomenon in
patients with scleroderma. May and Weber (1976) treated 4
patients having Raynaud's phenomenon and scleroderma or systemic
lupus erythematosus with 16 sessions of combined skin
temperature biofeedback, autogenic training, and progressive
relaxation. These patients reported fewer vasospastic attacks
over the 8-week training period and showed increased finger
temperature during biofeedback. Unfortunately, these findings
were not statistically tested nor were they examined over
seasonal changes in weather. The combination of the 3

treatments makes it impossible to determine the effects of any one. Lastly, the specific number of patients having scleroderma was not given nor were the criteria by which they were classified. Adair and Theobald (1978) treated one severe case of scleroderma and Raynaud's phenomenon with finger temperature feedback and imagery techniques. The patient's baseline and maximum finger temperatures increased over the 10 training sessions and these gains were partially maintained at 6-month follow-up. The patient also showed reduced digital ulceration and required less pain medication. Vasospastic attack data were not reported. Like the previous study, this one suffers from the confounding of treatments and lack of statistical analysis.

Work completed thus far in our laboratory has focused on the use of temperature biofeedback without other concomitant treatments. In one study (Freedman, Lynn, Ianni, & Hale, 1981), 6 Raynaud's disease and 4 Raynaud's phenomenon patients (2 scleroderma, 1 vibration disease, 1 carpal tunnel syndrome), each received 12 laboratory sessions of finger temperature biofeedback. The frequency of vasospastic attacks was reduced to 7.5% of that reported during the pre-treatment baseline, and maintained for a 1-year follow-up period. Control of digital temperature was demonstrated during training, with Raynaud's phenomenon patients showing significantly greater temperature increases than those with Raynaud's disease. Correlations between finger temperature and heart rate, respiration rate, skin conductance level, and frontalis EMG were either nonsignificant or in the direction of increased physiological activity, suggesting that the results were not due to general physical relaxation. In another study we attempted to increase the likelihood that the effects of temperature biofeedback would generalize beyond the laboratory setting by introducing a cold stressor during biofeedback training (Freedman, Ianni, Lynn, & Hale, 1979). Six Raynaud's disease and 2 scleroderma patients received 12 sessions of finger temperature feedback; during the last 6 sessions the finger monitored for feedback rested on a thermoelectric stimulus whose temperature decreased from 30°C to 20°C. Patients again reported significant declines in vasospastic attack frequency and maintained these improvements at 1-year follow-up. Patients were tested for their ability to control finger temperature without feedback prior to and after training. Temperatures were significantly ($F = 17.78$, $p < .0001$) warmer in the post-treatment session ($X = 27.2°C$) than the pre-treatment session ($X = 24.5°C$), but rarely increased above baseline values. Temperature differences between the two sessions were significantly related to attack frequencies reported at the 1-year follow-up ($r = -.69$, $p < .05$) as were absolute temperatures during the final session ($r = -.88$, $p < .01$).

Results for patients meeting the classification criteria for PSS (American Rheumatism Association, 1981) and treated in our laboratory with temperature biofeedback are summarized in Table 1. Temperature changes reflect the average within-session differences between 16-minute baseline periods and 24-minute feedback periods. Attack frequencies were obtained from symptom report cards completed by the patients for one month prior to treatment and one year following treatment. The post-treatment and 1-year follow-up periods each occurred during February, the coldest month of the year in Detroit. It can be seen that all patients reported substantial decreases in vasospastic attack frequency; follow-up data are not yet available for patients 4 and 5.

Surwit (personal communication) has recently completed a study comparing the effects of prazosin HCL (Minipress) and autogenic training on Raynaud's phenomenon in PSS. Twenty scleroderma patients first received a cold stress test and were then given Minipress or placebo for one month. The cold stress test was then repeated; no changes were found. All subjects then received autogenic training for one month. During subsequent administration of the cold stress test, subjects who had received autogenic training and Minipress had significantly higher finger temperatures than those who had received training plus placebo. Thus, autogenic training increased the efficacy of the drug.

Comment

Although surgical and pharmacological treatments have been generally ineffective for Raynaud's disease and phenomenon, behavioral methods appear promising. Improvements in reported symptomatology and evidence of peripheral blood flow control have been found using biofeedback, autogenic training, progressive relaxation, or combinations of these procedures. To date, however, no behavioral treatment has proven superior to any other. It is not known if the changes in skin temperature and in reported symptoms are due to direct self-regulation of digital blood flow, to general relaxation, or to both. Results from studies employing biofeedback alone are most parsimoniously explained by the first hypothesis. The plausibility of this explanation is enhanced by the recent finding of an active beta-adrenergic vasodilator mechanism in the human finger (Cohen & Coffman, 1980). However, increased anxiety has been reported prior to vasospastic attacks (Freedman, Lynn, and Ianni, in press) and anecdotal evidence from our laboratory[1] suggests that stress may play a role in scleroderma. If stress is a factor in the provocation of vasospastic symptoms, the decreased physiological activity produced by all of the behavioral

Table 1

Biofeedback treatment of Raynaud's phenomenon in scleroderma

Subject Age, Sex	Duration (Years)	Viscera Involved	Temp. Change	Number of Attacks per Week		
				Pre	Post	1 Yr. F.U.
1. 19 F	7	E,H,L,K	+.43°C	28	0	0
2. 35 F	3	E,H,L	-.18°C	21	7	7
3. 19 M	5	E	+.72°C	30	7	14
4. 28 F	2	E	+1.43°C	20	7	-
5. 65 M	5	E	+.46°C	25	0	-
Means	4.4		+.57°C	24.8	4.2	3.0

Viscera: E=Esophagus, H=Heart, L=Lung, K=Kidney

procedures including thermal biofeedback (Freedman & Ianni, 1981) might well mitigate against it.

It is not known if similar pathophysiological mechanisms underlie the vasospasms of Raynaud's disease and Raynaud's phenomenon in PSS. Some evidence exists for SNS hypoactivity in scleroderma (Fries, 1969) and neuroendocrine differences have been found between this disorder and Raynaud's disease (Surwit et al., 1981). In our laboratory scleroderma patients produced average finger temperature increases of .57°C (Table 1) during biofeedback, similar to the responses of normal subjects but not to those of patients with Raynaud's disease. The latter group rarely raises digital temperature above baseline values but appears to inhibit larger temperature decreases than would have occurred without training (Freedman et al., 1981; Surwit & Fenton, 1980). This may be sufficient to prevent the closure of digital arteries and cessation of blood flow occurring during vasospastic attacks. It is important to determine if behavioral treatments lead to general changes in tonic levels of peripheral blood flow or if they provide patients with methods of inhibiting symptoms at discrete points in time. We are currently examining this issue using 24-hour ambulatory recordings of finger and ambient temperature in Raynaud's disease and scleroderma patients (Freedman et al., in press).

Research on the behavioral treatment of Raynaud's
phenomenon in scleroderma is clearly in an early stage of
development. Controlled studies must first be performed to
determine if behavioral treatments are superior to plausible
placebo conditions. If behavioral treatments prove superior,
additional studies should be undertaken to assess their relative
efficacy. The determination of outcome in this research may be
problematic since optimal procedures for the assessment of
Raynaud's phenomenon do not yet exist (Freedman et al., in
press). For example, the relationship between patient reports
of vasospastic symptoms and laboratory measures of peripheral
blood flow is not clear. Lastly, the role of Raynaud's
phenomenon in the symptom constellation of scleroderma is
currently unknown. If the sclerotic changes in PSS are
secondary to Raynaud's phenomenon, it is possible that
vasospastic symptom improvement, achieved through behavioral or
other methods, could lead to cutaneous improvement. If the
reverse is true, one would expect these treatment gains to be
limited to improvements in vasospastic symptoms alone. These
questions can only be answered through continued research on the
etiology and treatment of Raynaud's phenomenon and scleroderma.

REFERENCES

Adair, J. & Theobald, D. Raynaud's phenomenon: treatment of a
    severe case with biofeedback. Journal of the Indiana State
    Medical Association, 1978, 71, 990-993.
American Rheumatism Association. Preliminary criteria for the
    classification of systemic sclerosis (scleroderma).
    Bulletin on the Rheumatic Diseases, 1981, 31(1), 1-6.
Bulkley, B., Ridolfi, R., Salyer, W., & Hutchins, G. Myocardial
    lesions of progressive systemic sclerosis, a cause of
    cardiac dysfunction. Circulation, 1976, 53, 483-490.
Campbell, P, & LeRoy, E. Pathogenesis of systemic sclerosis: a
    vascular hypothesis. Seminars in Arthritis and Rheumatism,
    1975, 4, 351-368.
Cannon, P., Hassar, M., Case, D., Casarella, W., Sommers, S., &
    LeRoy, E. The relationship of hypertension and renal
    failure in scleroderma to structural and functional
    abnormalities of the renal cortical circulation. Medicine,
    1974, 53, 1-46.
Coffman, J. Skin blood flow in scleroderma. Journal of
    Laboratory and Clinical Medicine, 1970, 76, 480-484.
Coffman, J., & Cohen, A. Total and capillary fingertip blood
    flow in Raynaud's phenomenon. New England Journal of
    Medicine, 1971, 285, 259-263.
Cohen, R., & Coffman, J. Beta-adrenergic vasodilator mechanism
    in the finger. Clinical Research, 1980, 28, 161A.
Dabich, L., Bookstein, J., Sweifler, A., & Zarafonetis, J.
    Digital arteries in patients with scleroderma. Archives of

Internal Medicine, 1972, 130, 708-714.

Emmannel, G., Saroja, D., Gopinathan, K., & Kaplan, D.
   Environmental factors and the diffusing capacity of the
   lungs in progressive systemic sclerosis. Chest, 1976
   (suppl.) 2, 304-307.

Freedman, R., & Ianni, P. Voluntary control of skin
   temperature. Psychophysiology, 1981, 18, 197.

Freedman, R., Ianni, P., Hale, P., & Lynn, S. Treatment of
   Raynaud's phenomenon with biofeedback and cold
   sensitization. Psychophysiology, 1979, 16, 182.

Freedman, R., Lynn, S., & Ianni, P. Raynaud's disease. In F.
   Keefe & J. Blumenthal (eds.) Assessment   Strategies   in
   Behavioral Medicine. New York: Grune and Stratton, in
   press.

Freedman, R., Lynn, S., Ianni, P., & Hale, P. Biofeedback
   treatment of Raynaud's disease and phenomenon. Biofeedback
   and Self-Regulation, 1981, 6, 355-365.

Fries, J. Physiologic studies in systemic sclerosis
   (scleroderma). Archives of Internal Medicine, 1969, 123,
   22-25.

Fries, J. The microvascular pathogenesis of scleroderma: an
   hypothesis. Annals of Internal Medicine, 1979, 9, 788-789.

Fries, J., Hoopes, J., & Shulman, L. Reciprocal skin grafts in
   systemic   sclerosis   (scleroderma).   Arthritis   and
   Rheumatology, 1971 14, 571-578.

Gladman, D., Gordon, D., Urowitz, M., & Levy, H. Pericardial
   fluid   analysis   in   scleroderma. American   Journal   of
   Medicine, 1976, 60, 1064-1068.

Guttadauria, M., Ellman, H., & Kaplan, D. Progressive systemic
   sclerosis: pulmonary   involvement. Clinics   in   Rhematic
   Diseases, 1979, 5(1), 151-166.

Keefe, F.J. Biofeedback vs. instructional control in skin
   temperature. Journal of Behavior Medicine, 1978, 1,
   383-390.

Kristensen, J. Local regulation of blood flow in cutaneous and
   subcutaneous tissue in patients with generalized
   scleroderma. Acta Dermato-Venerologica, 1980 (suppl.) 90,
   1-39.

Kurland, L., Hauser, W., Ferguson, R., & Holley, K.
   Epidemiologic features of connective tissue diseases in
   Rochester, Minnesota. Mayo Clinic Proceedings, 1969, 44,
   649-663.

LeRoy, E. Scleroderma (systemic sclerosis). In W. Kelley, E.
   Harris, S. Ruddy, & C. Sledge,(eds.) Textbook of
   Rheumatology. Philadelphia: Saunders, 1981.

LeRoy, E., Downey, J., & Cannon, P. Skin capillary blood flow
   in scleroderma. Journal of Clinical Investigation, 1971,
   50, 930-939.

Lewis, T. Experiments relating to the peripheral mechanism
   involved in spasmodic arrest of the circulation in the

fingers, a variety of Raynaud's disease. Heart, 1929, 15,
    7-101.
McGrath, M., Peek, R., & Penny, R. Blood hyperviscosity with
    reduced skin blood flow in scleroderma, Annals of the
    Rheumatic Diseases, 1977, 36, 569-574.
Maricq, H., Downey, J., & LeRoy, E. Standstill of nailfold
    capillary blood flow during cooling in scleroderma and
    Raynaud's syndrome. Blood Vessels, 1976, 13, 338-349.
Maricq, H., & LeRoy, E. Patterns of finger capillary
    abnormalities in connective tissue disease by "widefield"
    microscopy. Arthritis and Rheumatism, 1973, 16, 619-628.
Maricq, H., and LeRoy, E. Progressive systemic sclerosis:
    disorders of the microcirculation. Clinics in Rheumatic
    Diseases, 1979, 5(1), 81-102.
Maricq, H., LeRoy, E., D'Angelo, W., Medsger, T., Rodnan, G.,
    Sharp, G., & Wolfe, J. Diagnostic potential of in vivo
    capillary microscopy in scleroderma and related disorders.
    Arthritis and Rheumatism, 1980, 61, 862-870.
May, D., & Weber, C. Temperature feedback training for symptom
    reduction in primary and secondary Raynaud's disease.
    Biofeedback and Self-Regulation, 1976, 1, 317.
Medsger, T., & Masi, A. Epidemiology of systemic sclerosis
    (scleroderma). Annals of Internal Medicine, 1971, 74,
    714-721.
Medsger, T., & Masi, A. Epidemilogy of progressive systemic
    sclerosis. Clinics in Rheumatic Diseases, 1979, 5(1):
    15-25.
Medsger, T., Masi, A., Rodnan, G., Benedek, T., & Robinson, H.
    Survival with systemic sclerosis (scleroderma) Annals of
    Internal Medicine, 1971, 75, 369-376.
Myers, A. Progressive systemic sclerosis: gastrointestinal
    involvement. Clinics in Rheumatic Diseases, 1979, 5(1),
    115-129.
Nilsen, K. Assessment of cold sensitivity in Raynaud's
    phenomenon associated with scleroderma. Microvascular
    Research, 1978, 15, 251-256.
Peacock, J.H. Peripheral venous blood concentrations of
    epinephrine and norepinephrine in primary Raynaud's
    disease. Circulation Research, 1959, 1, 821-827.
Rodnan, G. Progressive systemic sclerosis (scleroderma). In D.
    McCarty (ed.) Arthritis and Allied Conditions.
    Philadelphia: Lea and Febiger, 1979a.
Rodnan, G. Progressive systemic sclerosis: clinical feature and
    pathogenesis of cutaneous involvement (scleroderma).
    Clinics in Rheumatic Diseases, 1979b, 5(1), 49-79.
Rodnan, G., Lipinski, E., & Luksick, J. Skin thickness and
    collagen content in progressive systemic scleorsis and
    localized scleroderma. Arthritis and Rheumatism, 1979, 22,
    130-140.
Rosch, J., Porter, J., & Gralino, B. Cyrodynamic hand

angiography in the diagnosis and management of Raynaud's syndrome. Circulation, 1977, 55, 807-814.

Sapira, J., Rodnan, G., Scheib, E., Klaniecki, T., & Rizk, M. Studies of endogeneous catecholamines secondary to progressive systemic sclerosis. American Journal of Medicine, 1972, 52, 330-337.

Surwit, R., Allen, L., Kuhn, C., Gilgor, R., Duvic, M., Schanberg, J., & Williams, R. Neuroendocrine correlates of Raynaud's disease and Raynaud's phenomenon, Psychophysiology, 1981, 18, 204.

Surwit, R., & Fenton, C. Feedback and instructions in the control of digital skin temperature. Psychophysiology, 1980, 17, 129-132.

Tan, E., Rodnan, G., Garcia, I., Morin, Y., Fritzler, M., and Peebles, C. , Diversity of antinuclear antibodies in progressive systemic sclerosis, Arthritis & Rheumatism, 1980, 23, 617-625.

Turner, R., Lipshutz, W., Miller, W., Rittenberg, G., Schumacher, H., & Cohen, S. Esophageal dysfunction in collagen disease. American Journal of Medical Sciences, 1973, 265, 191-199.

Wouda, A. Raynaud's phenomenon. Acta Medica Scandanavica, 1977, 201, 519-523.

THE BEHAVIORAL TREATMENT OF THE SYMPTOMS OF ISCHEMIC HEART DISEASE

Derek W. Johnston

Department of Psychiatry
University of Oxford
Oxford, England

In this chapter I shall consider the behavioral treatment
of two symptoms or correlates of ischemic heart disease, cardiac
arrhythmias and angina pectoris. These symptoms can be disturbing, disabling or even life threatening, and effective
behavioral treatments could aid, supplement, or in some
instances replace existing medical and surgical treatments, and
be expected to lead to significant improvements in the quality
of life and life expectancy of many of the patients who suffer
from ischemic heart disease. So far, attempts to treat these
symptoms behaviorally have relied either on biofeedback or some
form of relaxation training. While the rationale for the use of
these techniques is obvious and appealing, they do not represent
the full strength of the behavioral approach. Therefore, after
reviewing the evidence for the efficacy of these techniques, I
shall outline the developments that I consider to be most
helpful in promoting a more comprehensive and imaginative
behavioral approach to the treatment of the symptoms of ischemic
heart disease.

Cardiac Arrhythmias

For a variety of reasons, including ischemic heart disease,
the regular rate and rhythm of the heart can be persistently
disrupted. In people with otherwise healthy hearts many arrhythmias are benign but in others they can indicate the presence
of serious heart disease, reduce the efficient functioning of
the heart, and cause discomfort and apprehension. In addition,
arrhythmias that are not fatal in themselves can be associated
with increased mortality, perhaps because they can lead to
catastrophic arrhythmias such as ventricular fibrillation.

Although many arrhythmias can occur in the diseased heart, it is convenient to concentrate on premature ventricular contractions (PVCs), since they are among the most common and are thought by some to have ominous implications in patients with evidence of ischemic heart disease. They therefore provide a clinically meaningful model for the study of arrhythmias in general.

There is considerable evidence that sympathetic stimulation can increase PVC frequency, and such increased sympathetic activity has been implicated in the effects of stress on PVCs; for example, Corbalan, Verrier and Lown (1974) studying dogs that had undergone experimental coronary occlusion, demonstrated that in their home cages such dogs showed no ventricular arrhythmias, but when placed in a stressful environment a variety of arrhythmias were observed. They have also shown that beta blockers can reduce or eliminate some manifestations of stress-induced ventricular instability (Matta, Lawler & Lown, 1976). Taggart, Gibbons and Somerville (1969), using ambulatory monitoring techniques, documented a dramatic increase in ventricular arrhythmias in patients with coronary heart disease exposed to the (presumed) stress of driving in London traffic. This association between psychological and environmental stressors and PVCs would appear to provide a promising basis for the behavioral treatment of such arrhythmias, but this has not yet been reflected in the therapies that have been examined. Instead, most investigators have used biofeedback as a direct symptomatic treatment that bypasses the need for a more complete analysis of the problem. It has been voluminously documented that healthy volunteers can control their heart rate, and that feedback can aid the process (Williamson & Blanchard, 1979). However, studies of rhythm, rather than rate, are less common. Lang (Hnatiow & Lang, 1965; Lang, Stroufe & Hastings, 1967; Sroufe, 1969) demonstrated in an early series of studies that subjects could reduce heart rate variability with analogue heart rate feedback. More recently, Johnston and Lethem (1981) have examined the role of feedback in enabling subjects to increase their heart rate slightly and hold it near a specific target. This involves control of both rate and rhythm. Volunteers quickly learned to do this and feedback proved to be essential for an adequate performance. These studies therefore suggest that the control of some aspects of rhythm is possible.

The major published study of the use of biofeedback to control PVCs is the well-known study by Weiss and Engel (1971). They studied the control of PVCs in eight patients, most of whom had ischemic heart disease. In addition, Engel and Bleeker (1974) report on one further patient. All patients were admitted to hospital and given many hours of heart rate feedback of the same general form. After a brief pre-treatment baseline, usually only one session, patients received training in the following fixed order -- heart rate increase, heart rate decrease, alternations of increase and decrease training, and finally training in the control of heart rhythm. While the patient response to training was very variable, the data suggest that at least three patients obtained some degree of control of PVC frequency. Because of the very brief baseline data obtained in the laboratory and the scanty extra-laboratory information, it is not possible to determine if patients actually reduced PVCs below their pre-treatment level.

The only other published reports of the use of biofeedback are case reports from Pickering and his colleagues. Pickering and Gorham (1975) studied an otherwise healthy woman with a ventricular parasystolic rhythm. The patient's PVCs were strongly tied to her absolute heart rate and increased in frequency as rate increased. She learned to increase her heart rate by approximately 25 beats per minute and at the same time suppress the PVCs that had previously occurred at these higher heart rate levels, so that in the course of the study her PVC threshold rose from 79 to 94 beats per minute. This transferred from the biofeedback situation to exercise induced tachycardia. Pickering and Miller (1977) described two very convincing individual case studies. The first was a 14-year-old boy with persistent bigeminy, i.e. each normal beat was followed by a PVC. He was given oscilloscope feedback of his ECG and learned to increase the percentage of normal (sinus) rhythm markedly, from 3.2% during baseline periods to 27.4% when feedback was provided. The subject could neither control nor detect his arrhythmia in the absence of feedback. The second patient had an infrequent arrhythmia that could be reliably produced by the Valsalva maneuver, i.e. the attempt to expel breath against a closed glottis. With 32 hours of training he learned to suppress this induced arrhythmia with high reliability, i.e. he was successful in 10 out of 10 trials compared to 0 out of 10 during baseline periods. He was able to suppress the arrhythmia without feedback, and while he did not have accurate perception of the return of sinus rhythm he could always detect the onset of bigeminal rhythm. Pickering made no attempt to assess the clinical significance of these results.

In addition to these published studies, Cheatle and Weiss (in press) describe a number of unpublished studies of fifteen patients with PVCs. In contrast to the published studies all these reports are negative. The most important study was an attempt by Weiss and colleagues to train twelve patients using a similar, but abbreviated, protocol to that described by Weiss and Engel (1971). No patients learned to reduce PVC frequency in 24 16-minute sessions of heart rate feedback. It is impossible to evaluate this unpublished study but the lack of success is disquieting. Cheatle and Weiss consider that it may reflect the very limited time that was spent in training compared to the earlier studies. However, it is not clear that the patients who gained control in the Weiss and Engel studies did so as a result of extensive training, and even if that were so it need not be assumed that extensive training is always required, particularly as Weiss and Engel have placed heavy reliance on the use of binary feedback of heart rate. This may not be the most efficient way of learning to control heart rate (see Williamson & Blanchard, 1979), and seems particularly inappropriate in the control of arrhythmias, since it would make it difficult for the subject to detect the occurrence of the arrhythmia. The use of analogue feedback which provides information on both rate and rhythm may be better. An additional point of technique that does not seem to have been adequately exploited is methods of developing the patient's awareness of the occurrence of a PVC. It seems very likely that for a long-term clinical change the patient should become aware of his arrhythmia in the absence of feedback. Neither Weiss and Engel nor Pickering specifically attempted to train their subjects in such awareness and, in fact, make only passing reference to the extent that subjects were aware of the occurrence of PVCs.

In addition to these studies of heart rate feedback, Benson, Alexander and Feldman (1975) have provided an uncontrolled study of a form of relaxation training in eleven patients with stable PVCs. All patients received a very brief (5 minutes) relaxation training procedure, described by Benson, Beary and Carol (1974), which they then practiced for 10 to 20 minutes twice daily for four weeks. PVC frequency was assessed during two 24-hour ambulatory monitoring assessments, before and after treatment, and was shown to have diminished both over the complete 24-hour period and more especially during sleep, when PVC frequency decreased from 125.5 per hour to 87.9 in the group as a whole. There were wide individual variations in the effectiveness of treatment but Benson et al. (1975) considered that eight out of eleven patients showed a reduction in PVC frequency.

While acknowledging the doubts that must be raised by the unpublished studies, the available data on the behavioral control of arrhythmias are sufficiently promising to suggest that further study of behavioral methods would be justified.

Angina Pectoris

Angina pectoris is the precordial pain that results when myocardial oxygen consumption (MVO_2) exceeds the available oxygen supply. This is usually the result of a reduction in coronary arterial flow caused by arteriosclerosis or coronary artery spasm. The pain is considered particularly unpleasant and disturbing and usually leads to the patient ceasing his ongoing activity until it remits. Angina pectoris is a particularly appropriate symptom for behavioral management since sufferers are only too aware of the presence of angina and the pain and fear associated with it might be expected to motivate patients to comply with treatment. It is also very clearly related to environmental events. The link with exercise is extensively documented and, indeed, forms the basis of the clinical assessment of angina (Goldstein & Epstein, 1972), and while the link with emotion is largely anecdotal it is almost universally agreed (Goldstein & Epstein, 1972; Sleight, 1977). Ishikawa, Tawara, Ohtsuku, Takeyama and Kobayashi (1971) have outlined a survey of the precipitants of angina in 100 patients. They report that 45 had angina induced by both exercise and emotion. They also claim that angina could be experimentally induced in some patients by the emotion engendered by discussion of past traumatic incidents in the patient's life. Other laboratory and ambulatory data supporting the association between emotion and angina have been provided by Robinson (1967), Taggart et al. (1969) and Little, Honour, Sleight and Stott (1973). In addition to this limited direct evidence of the effects of psychological stress on angina, it is surely the inevitable consequence of the well- known effects of emotion and environmental stressors on heart rate and blood pressure (and hence MVO_2) in both man and animals (Steptoe, 1981). It has also been suggested that emotional stress can induce coronary artery spasm (Schiffer, Hartley, Schulman & Abelman, 1980) as indexed by ECG changes, but studies measuring coronary flow directly cast some doubt on this interpretation of these changes (Taggart, Carruthers, Joseph, Kelly, Marcomichelakis, Noble, O'Neill & Somerville, 1979).

As was the case in the treatment of arrhythmias, the published reports of behavioral methods in the treatment of angina rely largely on the power of biofeedback and relaxation to reduce heart rate and blood pressure and hence MVO_2. Since the critical values of MVO_2 associated with ischemia are seldom reached when the patient is in a resting, relaxed state, the evidence on the control of heart rate and blood pressure at rest is of little significance in the treatment of angina. However, there is now a growing body of evidence that self- control of cardiovascular responses to physical and psychological stressors is possible and that biofeedback may be particularly effective in aiding such control. As the literature on the self- control of cardiovascular responses during psychological stress is reviewed by Steptoe (Chapter 12), this review is confined to the control of heart rate and blood pressure during exercise, the most common precipitant of angina pectoris.

Goldstein, Ross and Brady (1977) examined the effects of heart rate feedback on healthy volunteers exercising on a treadmill. Subjects receiving feedback produced smaller elevations in heart rate and systolic blood pressure than subjects who were simply exercising. Perski and Engel (1980) obtained somewhat similar results by providing heart rate feedback to subjects exercising on a bicycle ergometer, although they found the effect to be limited to heart rate and did not generalize to systolic blood pressure. Robinson (1967) has shown that MVO_2 is conveniently indexed by the product of heart rate and systolic blood pressure, the rate pressure product (RPP), and we have, therefore, attempted to provide subjects with feedback that relates to both these parameters of cardiovascular functioning by providing feedback on the time interval between heart beats, the interbeat interval (IBI), and pulse transit time (TT), a close correlate of systolic blood pressure (Marie, Lo, Van Jones & Johnston, Note 1; Steptoe, Smulyan & Gribben, 1976). In the first of two experiments we compared instructions to lower heart rate and blood pressure with IBI feedback and product (IBI X TT) feedback in healthy volunteers exercising on a bicycle ergometer (Lo & Johnston, Note 2). Feedback was more effective than instructions in increasing IBI and TT (i.e. lowering heart rate and systolic blood pressure). There was also a tendency for product feedback to be more effective than IBI feedback alone. In the second experiment (Lo & Johnston, Note 3), product feedback was compared with relaxatin training based on Benson's analysis of the relaxation response (Benson et al., 1974) and simply exercising with no attempt to control cardiovascular responses. Product feedback was very clearly better than either of the other two procedures in controlling IBI and TT. Rather surprisingly, relaxation training was slightly worse than exercise alone in the control of IBI and while it showed some initial advantage over exercise in the control of TT this was transitory. Therefore, four studies carried out in three different laboratories and using different forms of feedback and different forms of dynamic exercise, have all shown that feedback is effective in lowering heart rate and possibly systolic blood pressure. In addition, there is suggestive evidence that product feedback is better than heart rate feedback alone.

Clements and Shattock (1979) have shown that subjects given heart rate feedback can attenuate the heart rate response to isometric exercise. Marie and Johnston (Note 4) explored the effect of TT feedback on IBI and TT during a 50% isometric contraction produced using a hand dynometer. In two extensive and rather complex experiments no difference was found between TT feedback and simple instructions to lower blood pressure, neither during the isometric exercise nor afterwards in the recovery period. While the issue is far from closed, it is therefore currently unsafe to assume that cardiovascular feedback is as beneficial during all types of exercise as it is during dynamic exercise.

The use of feedback in patients with angina pectoris is still in a very undeveloped stage, but some preliminary studies have been reported at conferences or in the form of brief abstracts, (Underhill, Wills & Mansfield, 1977; Herring & Richlin, 1977; McCruskery, Engel, Gottlieb & Lacatta, 1978). The abstracts available on the first two studies are too brief to evaluate but it is worth noting that both claim positive outcomes from a variety of biofeedback procedures. Perski, Engel and McCruskery (in press) have reported an extension of their earlier paper with the addition of two more patients, making six in all. These patients were trained to slow their heart rate (at rest) and then tested on a standardized exercise test. Four of the six patients showed some increase in exercise tolerance. The group as a whole showed an increase in exercise duration and work load and this was associated with an increase in the rate pressure product and presumably MVO_2, indicating that the increase in exercise tolerance was not entirely due to a reduction in MVO_2. Since it is unlikely that coronary blood flow had increased, it is probable that the patient's pain or fatigue tolerance had changed. Johnston and Lo (Note 5; see also Johnston, 1981) carried out a pilot investigation of seven patients, each of whom received cardiovascular feedback (usually the product of TT and IBI) at rest and while exercising on a bicycle ergometer. They also received a variety of control conditions, including relaxation training and frontalis EMG feedback. Patients completed detailed diaries of anginal attacks and their use of glycerine trinitrate (GTN) for three weeks before and after treatment and carried out standardized exercise tests. Anginal frequency was reduced in six of the seven patients, GTN intake was reduced in all patients and exercise tolerance increased in six out of seven without an increase in RPP, i.e. MVO_2 was presumably diminished after treatment.

In addition to these uncontrolled studies of biofeedback, there is evidence that relaxation procedures can also be effective in patients with angina. Zamarra, Besseghini and Wittenberg (1977) compared ten patients who received Transcendental Meditation (TM) with six untreated patients. Patients appeared to have received an extensive medical work-up prior to the study and underwent standardized exercise testing on a number of occasions before the treatment and six to eight months after TM. In the TM group, exercise duration increased by 14.5%, work load by 11% and an ECG indicator of myocardial ischemia (ST depression) was delayed by 16%. The RPP was constant at the end of the test and significantly reduced in the early stages, suggesting that treatment had reduced MVO_2. There was no change in the exercise tolerance of the control subjects. This study is noteworthy for its use of a control group (albeit a minimal one) and the apparent persistence of the effects of TM over six to eight months. It is somewhat weakened by the fact that patients were not allocated to treatment randomly; the first twelve patients to enter the study were allocated to TM.

All the available studies of behavioral methods of treatment of angina have produced positive results: exercise tolerance has been increased, usually by a reduction in MVO_2 for a given work load, as has pain frequency and GTN use. However, the studies have only involved a handful of patients and have been methodologically crude, and can serve as little more than a stimulus for further action.

And Now?

The research on the treatment of arrhythmias and angina described above has been motivated primarily by the availability of a technique, usually biofeedback, plus an assumption that psychological factors influence the occurrence or severity of these cardiovascular symptoms. These psychological factors have not, however, greatly influenced the treatments, the various authors relying on their favored treatment without regard to the etiology of the condition. This may not be unreasonable in the case of a treatment such as biofeedback, which offers the promise of being a direct symptomatic treatment that can be applied irrespective of the condition's etiology and does not even carry with it the assumption that the etiology should have a significant psychological component. Indeed, the considerable success of biofeedback in the treatment of fecal incontinence is testimony to the power of biofeedback when used in this way. (See Chapter 17.) Nevertheless, in the treatment of cardiovascular symptoms this technique-driven approach to therapy has been much less successful and has encouraged the development of treatments of very limited power, and a more comprehensive approach to therapy is therefore called for.

Practicing clinicians using whatever treatment method they favor, physical or psychological, do not apply treatment blindly to patients simply on the basis of diagnosis. Good clinicians supplement diagnosis with an analysis of the patient's symptoms and problems, and treat those aspects of the complex picture that emerges which they consider significant and amenable to treatment. Such an approach should certainly be basic to behavioral medicine, which is, or should be, based on the tradition of behavior modification, with its strong emphasis on the explicit analysis of the patient's problems and an explicit statement of the proposed solutions and targets for treatment.

There are three main stages in behavior modification: [1] the analysis of the problem and the specification of the targets for treatment, [2] the development and implementation of a treatment plan, and [3] the measurement of the effectiveness of that implementation. If this method were applied to a patient with chest pain typical of angina, the analysis might proceed as follows. Firstly, the exact circumstances surrounding the occurrence of anginal pain would be determined. This would include detailed descriptions of the precipitants of pain, e.g. if exercise was a precipitant the therapist would seek to determine what types of exercise, and under what conditions, and if anxiety was a precipitant then anxiety over what, in whose presence, etc. The analysis would also consider the consequences of pain: these could include medication, rest, attention from an over-protective wife, the opportunity to avoid some unpleasant activity, etc. Behaviors reinforced by the lack of angina, i.e. maintained on an avoidance schedule such as reduced exercise or abuse of medication, would also be examined. This analysis would be carried out by detailed questioning of the patient (and possibly a relative) and also by the careful completion of diaries by the patient, specifying accurately his activities and psychological state at the time of interest. This analysis would generate a series of hypotheses linking the patient's anginal pain to features of the patient's external and internal environment. These hypotheses would then be tested by training the patient to implement the appropriate changes in his life. These may be very simple, e.g. walking more slowly after a heavy meal, or using medication more appropriately, or much more complex, perhaps using self-talk to interrupt the vicious cycle of angina leading to anxiety, leading to a worsening of angina, or instructions to the patient's spouse as to how to deal with the patient's angina in a way that does not reinforce it. They could also, of course, involve treatments that would be specific to angina, e.g. biofeedback training in the lowering of heart rate and blood pressure. Finally, the impact of these interventions would be assessed, i.e. the hypotheses tested by measuring if, in fact, the treatments have been successfully implemented, does the patient walk more slowly, feel less anxious and also, and more critically, is there a reduction in anginal frequency, severity and duration? The strategy I have proposed will be familiar to all behavioral clinicians. However, if the reader is not familiar with this approach, then I would suggest he or she refer to one of the standard texts on the practice of behavior modification such as Gambrill, 1977, in which such techniques have been exhaustively described. I would like to simply point out that the adoption of this approach increases the armamentarium of the clinician enormously, and makes the treatment of the symptoms of ischemic heart disease the legitimate concern of a much wider range of behavioral clinicians than simply those cognizant of biofeedback techniques.

In advocating this approach to the treatment of physical illness in general, it must be acknowledged that there are important differences between such symptoms and behavioral problems such as phobias or tics. It must always be remembered that in the vast majority of cases these symptoms do have an organic basis, and a behavioral analysis of the circumstances associated with their occurrence can only supplement a proper medical assessment and diagnosis of the underlying condition. In addition, there are special problems in applying these methods to cardiac symptoms. Behavioral methods. are usually applied to problems which are either public, such as tics, smoking or disruptive classroom behavior, or, if private, are obvious to the patient, such as anxiety or depression, and have some observable public correlates that help to validate the patient's self-report, such as trembling, tears or even a sad expression. This is much less likely to be the case in the treatment of disturbances of the cardiovascular system, and the patient's subjective report may have to be supplemented with ambulatory physiological measurement.

CONCLUSIONS

The evidence from clinical and laboratory studies of the symptoms of ischemic heart disease encourage the belief that behavioral methods of treatment could be used with advantage in these conditions. While only research and experience will determine if this is so, I consider that the likelihood of rapid progress towards significant clinical goals is greatest if a comprehensive behavioral approach is adopted and the lessons learned in the successful application of behavior modification to behavioral problems is applied to this new area.

REFERENCE NOTES

1. Marie, G., Lo, C.R., Van Jones, J. & Johnston, D.W. The relationship between transit time and arterial blood pressure during dynamic and isometric exercise. Manuscript in preparation.
2. Lo, C.R., & Johnston, D.W. Cardiovascular feedback during dynamic exercise. Manuscript in preparation.
3. Lo, C.R., & Johnston, D.W. A comparison of the feedback of the product of interbeat interval and pulse transit time with relaxation training at rest and during dynamic exercise. Manuscript in preparation.
4. Marie, G., & Johnston, D.W. Pulse transit time during isometric exercise. Manuscript in preparation.
5. Johnston, D.W., & Lo, C.R. The effects of heart rate and

blood pressure feedback on angina pectoris. Unpublished
manuscript, 1979.

REFERENCES

Benson, H., Alexander, S., & Feldman, C.L. Decreased premature
     ventricular contractions through the use of the relaxation
     response in patients with ischemic heart disease. The
     Lancet, 1975, 2, 380-382.
Benson, H., Beary, J.F., & Carol, M.P. The relaxation response.
     Psychiatry, 1974, 37, 37-46.
Cheatle, M.D., & Weiss, T., Biofeedback in heart rate control
     and the treatment of cardiac arrhythmias. In Clinical
     Biofeedback: Efficacy and Mechanisms. L. White   & B.
     Tursky, (eds.), Guilford Press, New York, (in press).
Clements, W.J., & Shattock, R.J. Voluntary heart rate control
     during static muscular effort. Psychophysiology, 1979, 16,
     327-332.
Corbalan, R., Verrier, R. and Lown, B. Psychologic stress and
     ventricular arrhythmias during myocardial infarction in
     conscious dogs. American Journal of Cardiology, 1974, 39,
     692-696.
Engel, B.T., & Bleecker, E.R. Application of operant
     conditioning techniques to the control of cardiac
     arrhythmias. In Cardiovascular psychophysiology. Current
     issues in response mechanisms, biofeedback and methodology.
     Chicago: Aldine, 1974.
Gambrill, E.D. Behavior Modification. Handbook of Assessment,
     Intervention and Evaluation. San Francisco: Josey-Bass
     Publishers, 1977.
Goldstein, D.S., Ross, R.S., & Brady, J.V. Biofeedback heart
     rate   training   during   exercise.   Biofeedback   and
     Self-Regulation, 1977, 2, 107-125.
Goldstein, R.E., & Epstein, S.E. Medical management of patients
     with angina pectoris. Progress in Cardiovascular Diseases,
     1972, 14, 360-397.
Herring, M., & Richlin, M. Effectiveness of self-relaxation
     training in reducing the severity of recurrent angina
     pectoris. Psychophysiology, 1977, 14, 101.
Hnatiow, M., & Lang, P.J. Learned stabilization of cardiac
     rate. Psychophysiology, 1965, 1, 330-336.
Ishikawa, H., Tawara, I., Ohtsuka, R.W., Takeyama, M., &
     Kobayashi, T. Psychosomatic study of angina pectoris.
     Psychosomatics, 1971, 12, 390-397.
Johnston, D.W. Exploiting the uniqueness of biofeedback. In J.
     Tiller & P. Martin (eds.), Behavioral Medicine. Sydney:
     Ciba-Geigy (in press).
Johnston, D.W., & Lethem, J. The production of specific
     decrease in interbeat interval and the motor skill analogy.
     Psychophysiology, 1981, 18, 288-300.

Lang, P.J., Sroufe, L.A., & Hastings, J.E. Effects of feedback
    and instructional set on the control of cardiac rate
    variability. Journal of Experimental Psychology, 1967,75,
    425-431.
Littler, W.A., Honour, A.J., Sleight, P., & Scott, F.D. Direct
    arterial pressure and the electrocardiogram in unrestricted
    patients with angina pectoris. Circulation, 1973, 148,
    125-134.
McCruskery, J.H., Engel, B.T., Gotlieb, S.M., & Lakatta, E.G.
    Operant conditioning of heart rate in patients with angina
    pectoris. Psychosomatic Medicine, 1978, 40, 89-90.
Matta, R.J.,Lawler, J.E., & Lown, B. Ventricular electrical
    instability in the conscious dog: effects of psychologic
    stress and beta adrenergic blockage. American Journal of
    Cardiology, 1976, 38, 594-598.
Perski, A., & Engel, B.T. The role of behavioral condi-tioning
    in the cardiovascular adjustment to exercise. Biofeedback
    and Self-Regulation, 1980, 5, 91-104.
Perski, A., Engel, B.T., & McCruskery, J.H. The modification of
    elicited cardiovascular responses by operant conditioning
    of heart rate. In J.T. Cacioppo & R.E. Petty (eds.). Focus
    on cardiovascular psychophysiology, (in press).
Pickering, T., & Gorham, G. Learned heart rate control by a
    patient with ventricular parasystolic rhythm. The Lancet,
    1975, 1, 252-253.
Pickering, T.D., & Miller, N.E. Learned voluntary control of
    heart rate and rhythm in two subjects with premature
    ventricular contractions. British Heart Journal, 1977, 39,
    152-159.
Robinson, B.F. Relation of heart rate and systolic blood
    pressure to the onset of pain in angina pectoris.
    Circulation, 1967, 35, 1073-1083.
Schiffer, F., Harley, L.H., Schulman, C.L., & Abelman, W.H.
    Evidence for emotionally induced coronary artery spasm in
    patients with angina pectoris. British Heart Journal,
    1980, 44, 62-66.
Sleight, P. Stress factors in angina pectoris. Scottish
    Medical Journal, 1977, 22, 34-37.
Sroufe, L.A. Learned stabilization of cardiac rate with
    respiration experimentally controlled. Journal of
    Experimental Psychology, 1969, 81, 391-393.
Steptoe, A. Psychological Factors in Cardiovascular Disease.
    London: Academic Press, 1981.
Steptoe, A., Smulyan, H., & Gribben, B. Pulse wave velocity and
    blood pressure: calibration and application.
    Psychophysiology, 1976, 13, 488-493.
Taggart, P., Gibbons, D., & Somerville, W. Some effects of
    motor car driving on the normal and abnormal heart.
    British Medical Journal, 1969, 4, 130-134.
Taggart,   P.,   Carrthers,   M.,   Joseph,   S.,   Kelly,   H.B.,

Marcomichelakis, J., Noble, D., O'Neill, & Somerville, W.
Electrocardiographic changes resembling myocardial ischemia
in asymptomatic men with abnormal coronary arterio-grams.
British Heart Journal, 1979, 41, 214-225.

Underhill, S.L., Wills, R.E., & Mansfield, L.W. Biofeedback
control to decrease heart rate for relief of angina
pectoris. Circulation, 1977, 55 & 56, Supplement 111,102.

Weiss, T., & Engel, B.T. Operant conditioning of heart rate in
patients with premature ventricular contractions.
Psychosomatic Medicine, 1971, 33, 301-321.

Williamson, D.A., & Blanchard, E.B. Heart rate and blood
pressure biofeedback: I. A review of the recent
experimental literature. Biofeedback and Self-Regulation,
1979, 4, 1-34.

Zammara, J.W., Besseghini, I., & Wittenberg, S. The effects of
the transcendental meditation program on the exercise
performance of patients with angina pectoris. In D.W.
Orme-Johnston & J.F. Farrow, (eds.) Scientific Research on
the Transcendental Meditation Program. Maharishi European
Research Press, 1977.

PATHOPHYSIOLOGY OF HYPERTENSION: CNS AND BEHAVIORAL COMPONENTS

Redford B. Williams, Jr.

Duke University Medical Center
Department of Psychiatry
Durham, North Carolina

Hypertension, or high blood pressure, is the most prevalent form of cardiovascular disease. Directly, its consequences include cerebral hemorrhage and congestive heart failure; while, indirectly, it serves as a major risk factor for atherosclerosis in both the cerebral and coronary vascular beds. Thus, the costs attributable to hypertension in terms of medical care and lost productivity are enormous. Large-scale prospective studies have now shown that these costs can be reduced. Early treatment of hypertension at all levels of severity has been shown to result in significant reduction of both mortality and morbidity associated with this disorder. A major problem in achieving this reduction, however, has been the low rates of patient compliance with antihypertensive pharmacologic regimens. Since the symptoms and discomfort associated with high blood pressure are minimal, particularly in the initial phase, and since the side effects of the medications prescribed to treat hypertension are often major (e.g. memory loss, depression, fatigue, impotence, etc.), it is not hard to understand why many patients choose not to take their medications. Now that the Hypertension Detection and Follow-up Program Cooperative Group (1979) has shown that adequate reduction of blood pressure is effective in reducing mortality even among patients with only mild hypertension, the problem of compliance will in all likelihood assume even greater prominence as we try to reduce the personal and societal costs of hypertension.

The topic of this Section is behavioral approaches to the treatment of hypertension. Behavioral approaches to the treatment of hypertension offer much promise in dealing with the problem of compliance with pharmacologic regimen which is

described above. To the extent that behavioral approaches can
achieve as little as a 5 mm Hg reduction in blood pressure -- a
goal which, as we shall learn in reading the papers in this
Section, appears within reach -- there is every reason to
believe that reductions in mortality and morbidity comparable to
those achieved with pharmacologic treatment are a realistic
goal. Since the side effects of behavioral treatments are
essentially non-existent when compared to pharmacologic
treatment, there is also every reason to believe that behavioral
approaches may offer at least a partial solution to the problem
of compliance with treatment regimen.

Before moving on to the chapters of this Section, each
dealing with some aspect of the behavioral approach to treatment
of hypertension, it will be the goal of this introductory
chapter to make the scientific case for such treatment
approaches. To the extent that environmental stresses and
challenges and psychological characteristics of the individual
can be shown to play a role in the etiology and pathogenesis of
hypertension, the rationale for using behavioral approaches to
treat this disorder is strengthened and specific targets for
behavioral interventions are identified. Since the brain is the
locus of the transduction of both environmental events and
psychological states into pathophysiology, it will be important
also to consider the evidence documenting the role of brain
mechanisms in the pathogenesis of hypertension.

Environmental Factors in Hypertension

Both animal and human studies have suggested that
environmental stresses and challenges can result in both acute
and sustained increases in blood pressure, as well as the
pathological vascular changes commonly associated with
hypertension.

Early studies (Rothlin, Cerletti & Ammenegger, 1955) showed
that rats exposed to loud noise developed significant blood
pressure elevations over the course of 1-9 months. However, no
pathological consequences of hypertension were noted in these
studies. Among rats genetically susceptible to developing
hypertension when fed a high sodium diet, Friedman and Iwai
(1979) were able to show that hypertension developed when these
animals were exposed to a conflict situation wherein they had to
experience a painful shock in order to obtain food. Thus, not
only environmental stresses, but also genetic characteristics
may both need to be present in order for hypertension to
develop. As we shall see later in this chapter, this
interaction of genetic and environmental influences may also be
important in the development of hypertension in humans.

Another form of environmental stress-induced hypertension in animals is that of chronic immobilization, involving the daily subjection of rats to two-hour immobilization periods over a 2-4 week span (Lamprecht, Williams & Kopin, 1973). Rats subjected to this paradigm gradually show a 50 mm Hg blood pressure increase over the 4-week span. Increased sympathetic nervous system activity appears involved in the initial phase of this form of hypertension, as indexed by increased levels of catecholamine forming enzymes in the adrenal medulla (Kvetnansky, Weise & Kopin, 1970), enhanced excretion of epinephrine and norepinephrine (Kvetnansky & Mikulaj, 1970) and increased serum levels of dopamine-beta-hydroxylase (Lamprecht et al., 1973).

The most comprehensively studied animal model of stress-induced hypertension, and that which appears to most closely approximate what might be occurring in humans, is the "mouse city" paradigm of Henry, Stephens and Santisteban (1975). The special population cages used in this model subject male mice to repeated confrontations as they compete for access to food and water, as well as females. As with the interaction of genetic and environmental influences noted above, in this model the early experience of the animals interacts with the environmental stress of the population cages in that hypertension is more likely to develop among mice raised in isolation prior to placing them in the competition-inducing environment. If the crowding is carried out over a 9-month period, the elevated blood pressure which develops in these mice persists even after they are returned to the isolation cages. Again, increased sympathetic nervous system activity appears involved in the pathogenesis of this hypertension, since increased adrenal levels of catecholamine-forming enzymes are found in the stressed animals. In addition, there are pathological consequences of the hypertension -- larger myocardial weight, myocardial fibrosis and aortic atherosclerosis.

Obviously, the experimental induction of clinically significant hypertension, as done in the Henry, Stephen and Santisteban (1975) studies described above, is not feasible in humans. However, it is possible to take advantage of certain "experiments of nature" to evaluate the possible role of environmental stresses and challenges in the etiology and pathogenesis of hypertension in humans. Perhaps the best example of such an experiment of nature is the finding of Cobb and Rose (1973) that hypertension is significantly more prevalent among air traffic controllers -- whose work is alleged to be more stressful -- than among age- and sex-matched control groups.

Another approach to evaluating the possible role of environmental factors in human hypertension is to expose subjects with and without hypertension (or evidence of a predisposition to hypertension) to various experimental stressors; and if hypertensives or those with a predisposition to develop hypertension show physiologic hyperresponsivity, then it is possible to at least infer that such psychophysiological mechanisms may be involved in the pathogenesis of hypertension. McKegney and Williams (1967) found that patients with hypertension showed more pronounced and prolonged diastolic pressor responses to a personal interview in comparison with normotensive patients. It is possible to argue, however, that such hyperresponsivity is the result rather than a cause of hypertension. Such an argument is not so strongly applied to recent findings of Falkner, Onesti, Angelakos, et al. (1981). They followed 50 adolescents who met criteria for borderline hypertension and found that 28 (56%) progressed to a state of sustained hypertension within a 4-year follow-up period. The adolescents who progressed to essential hypertension were differentiated from those who remained normotensive in several important respects. First, those who progressed to sustained hypertension had a significantly stronger family history of hypertension, recalling the genetic/environmental interaction noted above with regard to animal models of hypertension. Strongly supportive of the participation of psychophysiological mechanisms in hypertension pathogenesis was the finding that those who progressed to sustained hypertension showed significantly greater heart rate and blood pressure increases in response to a mental arithmetic stress than did those who did not progress to sustained hypertension. Coming at this issue from the other side, Hastrup, Light and Obrist(1979) found that among young subjects who showed larger heart rate and blood pressure responses to a reaction time stress, the likelihood of having parents with documented hypertension is higher than among subjects with smaller heart rate and blood pressure responses to the same stress.

In summary, it appears that in some animal models it has been clearly demonstrated that environmental stresses and challenges are capable of resulting in sustained blood pressure elevations, as well as pathological sequellae of such elevations, particularly in animals known to be susceptible to developing hypertension on the basis of genetics or early experience. Among humans, there is extensive evidence that persons with hypertension or a predisposition to develop hypertension show enhanced cardiovascular responses to experimental stressors. Among air traffic controllers, who are presumably exposed to high levels of naturally occurring stressors, the incidence of hypertension is much higher than would be expected in similar age- and sex- matched groups.

Further evidence for the participation of behavioral and psychosocial factors in the pathogenesis of hypertension will now be presented in terms of psychological factors which appear to characterize those who go on to develop hypertension.

Psychological Characteristics Predisposing to Hypertension

A number of early observations that hypertensives are chronically angry but unable to express this anger have been questioned in terms of whether this psychological characteristic of high anger and inability to express it is a result or a cause of the hypertension. Evidence that such a psychological makeup predates the appearance of hypertension -- and, hence, may be a cause rather than a result of hypertension -- is to be found in two recent well-conducted studies. McClelland (1979) reports that suppressed anger -- as indexed by higher need for power than for affiliation and high activity inhibition on the Thematic Aperception Test -- distinguishes young males who go on to develop elevated blood pressure and signs of hypertensive pathology, 20 years after testing, from young males who do not develop hypertension.

In a study analogous to that of Falkner et al. (1979), in which borderline hypertensives who went on to develop sustained hypertension were found to display cardiovascular hyperresponsivity to stress prior to developing sustained hypertension, Julius and Cottier (in press) have reported that young persons with borderline hypertension differ from normotensive controls in terms of several psychological characteristics, as assessed by the Cattell 16 PF Personality Inventory. Compared to normotensive controls, borderline hypertensives scored higher on scales indicating high levels of submission, sensitivity, warmth and sociability. These findings were replicated in a Yugoslavian sample of borderline hypertensives and normotensive controls using a translation of the 16PF. To check whether these personality differences between borderline hypertensives and controls were also reflected in behavior, Julius went on to subject borderline hypertensives and normotensive controls to an opinion change bargaining situation and found borderline hypertensives more likely to both change their opinion and report greater liking for their experimental partner than normotensive control subjects.

To the extent that environmental factors and psychological characteristics play any role in the etiology and pathogenesis of hypertension, it is self-evident that the brain is the transducer of these influences into pathophysiological processes. We shall now consider the evidence documenting the participation of the central nervous system (CNS) in hypertension.

Brain Mechanisms in Hypertension

The clearest evidence documenting the participation of CNS mechanisms in hypertension is to be found in the several animal models of hypertension. This role has already been alluded to above with regard to the evidence that stress-induced hypertension in animals appears associated with increases in sympathetic nervous system activity, an effect which almost certainly must originate in the CNS. More direct evidence is provided by the demonstration that electrolytic destruction of the nucleus tractus solitarii (NTS), where baroreceptor afferents terminate, leads to sustained or fulminant hypertension in several species (Brody, Haywood & Touw, 1980).

In addition, it is now clear that CNS mechanisms are involved in forms of experimental hypertension that were previously thought to be mediated only by peripheral renal and fluid volume mechanisms. In rats given DOCA and salt--a model of hypertension felt initially to occur through renal and fluid volume effects -- Lamprecht, Richardson, Williams and Kopin(1977) found that administration of 6-hydroxydopamine (6-OHDA) -- a neurotoxin specific for adrenergic nerve endings -- into the lateral cerebral ventricle prevented the appearance of hypertension with DOCA-salt administration. In contrast, administration of 6-OHDA into the cisterna magna did not prevent DOCA-salt hypertension. This suggests that catecholaminergic neurones in the hypothalamus and higher centers, but not in the lower brain stem or spinal cord, are essential for the development of DOCA-salt hypertension.

In this same study it was found that if animals in whom hypertension was not prevented by intraventricular 6-OHDA administration prior to starting DOCA-salt were given intraventricular 6-OHDA 4 weeks after the start of DOCA-salt, the developing hypertension was not only arrested but blood pressure levels fell to control levels. In contrast, if the intraventricular 6-OHDA administration was delayed until 7 weeks following the start of DOCA-salt administration, the hypertension was not reversed. These findings are important because not only do they clearly demonstrate that certain rostrally situated catecholaminergic neurones are involved in the pathogenesis of a form of animal hypertension which may approximate at least some hypertension in humans, but they also demonstrate that these catecholaminergic neurones are necessary for the initiation but not for the maintenance of the resultant hypertension.

With regard to CNS mechanisms involved in the pathogenesis of hypertension in humans, two centrally integrated physiological response patterns have been proposed (see Williams, 1981, for review). The first is the classical defense reaction, or fight-flight response, which consists of a beta-adrenergically mediated increase in cardiac output with shunting of that increased output from skin and viscera to skeletal muscle. In addition, there is increased somatomotor activity and increased neuroendocrine response, the latter characterized by increased secretion of catecholamines, cortisol, and possibly prolactin (Williams, Lane, White, Kuhn & Schanberg, 1981). In contrast to the CNS-mediated defense reaction, there appears to be another centrally integrated pattern of response which is associated with attention to sensory inputs and characterized by decreased somatomotor activity and a blood pressure increase mediated not by increased cardiac output as with the defense reaction, but by an alpha-adrenergically mediated increase in total peripheral resistance.

Both these mechanisms could be important from the standpoint of pathogenesis of hypertension. Increased cardiac output has been described among borderline hypertensives (see, e.g., Julius, in press), and heightened activation of the defense reaction is felt by many to be a key mechanism for the development of hypertension. Anderson (1981) and Williams (1981) have both noted, however, that the other pattern, characterized by increased total peripheral resistance and observed most characteristically in association with behavioral demands to pay close attention to environmental stimuli, could also be importantly involved in the pathogenesis of hypertension. Supporting this view is the finding of Julius (in press) that even though treatment of borderline hypertensives with a beta-blocking agent results in a reduced cardiac output, it does not normalize the blood pressure. If patients so treated are then given an alpha-adrenergic blocking agent, 30% will show a normal blood pressure. This finding has led Julius to conclude that in addition to the contribution of increased cardiac output to the elevated blood pressure among patients with borderline hypertension, there is also a contribution from an alpha-adrenergically mediated increase in total peripheral resistance, at least among the 30% of borderline hypertensives whose blood pressure normalizes with the combination of both beta- and alpha-adrenergic blockade.

Even more compelling support for the view that CNS mechanisms resulting in increased total peripheral resistance deserve consideration with regard to pathogenesis of hypertension is the finding of deLeeuw, Kho and Birkenhager (1981) that among male first year medical students who were not aware of their blood pressure level, there was no difference in cardiac output (or cardiac index) between those with clearly normal and those with elevated blood pressures. There was, however, observed a significantly increased total peripheral resistance in the hypertensive group, relative to the normotensive sample. These findings led deLeeuw et al. to conclude that the previously reported elevated cardiac output among borderline hypertensives may have been a result of the fact that most patients in such studies were referred to a hospital and were aware of being hypertensive; in contrast, the medical students in their study were unaware of their blood pressure status, leading to the conclusion that "...in a naive population, cardiac output plays no role in determining the level of resting blood pressure." (deLeeuw et al., 1981, p. 61).

Summary and Conclusions

1. Environmental stresses and challenges have been shown to result in clinically significant hypertension in animal studies, and there is evidence for a similar set of effects in humans.

2. Prospective studies have identified a set of psychological characteristics -- suppressed anger -- which appear to predispose to the development of hypertension.

3. To the extent that environmental stresses and challenges, along with psychological characteristics, are playing a role in the pathogenesis of hypertension, it is likely that CNS mechanisms involving cardiovascular and neuroendocrine responses are mediating the pathophysiological process.

4. The studies and findings leading to the above conclusions provide clues as to the people and processes to which behavioral treatment approaches for hypertension might be directed.

REFERENCES

Anderson, D.E. Inhibitory behavioral stress effects upon blood pressure regulation. In S.M. Weiss, J.A. Herd, & B.H. Fox, (eds.) Perspectives on Behavioral Medicine. New York: Academic Press, 1981.

Brody, M.J., Haywood, J.R. & Touw, K.B. Neural Mechanisms in Hypertension. Annual Review of Physiology, 1980, 42, 441-53.

Cobb, S. & Rose, R.M. Hypertension, peptic ulcer and diabetes in air traffic controllers. Journal of American Medical Association, 1973, 224(4), 489-492.

deLeeuw, P.W., Kho, T.L. & Birkenhager, W.H. Hemodynamic and endocrinologic data in early essential hypertension. In G. Onesti, & K.E. Kim (eds.) Hypertension in the Young and the Old. New York: Grune & Stratton, 1981.

Falkner, B., Onesti, G., Angelakos, E.T., Fernandez, M. &

Lonagman, C. Cardiovascular response to mental stress in normal adolescents with hypertensive parents. Hypertension, 1979, 1, 23.

Five-Year Findings of the Hypertension Detection and Follow-up Program. I. Reductions in mortality of persons with high blood pressure, including mild hypertension, Journal of American Medical Association, 1979, 242, 2562-2571.

Friedman, R. & Iwai, J. Genetic predisposition and stress-induced hypertension. Science, 1979, 193, 161-162.

Hastrup, J.L., Light, K.C. & Obrist, P.A. Relationship of cardiovascular stress response to parental history of hypertension and to sex differences. Presented at Annual Meeting, 1979, Society for Psychophysiological Research, Cincinnati, Ohio.

Henry, J.P., Stephens, P.M. & Santisteban, G.A. A model of psychosocial hypertension showing reversibility and progression of cardiovascular complications. Circulation Research, 1975, 36, 156-164.

Julius, S. & Cottier, C. Behavior and hypertension. In T.M. Dembroski, and T. Schmidt (eds.) Biobehavioral bases of coronary heart disease. Karger, Basel, in press (1982).

Kvetnansky, R. & Mikulaj, L. Adrenal and urinary catecholamines in rats during adaptation to repeated immobilization stress. Endocrinology, 1970, 87, 744-749.

Kvetnansky, R., Weise, V.K. & Kopin, I.J. Elevation of adrenal tyrosine hydroxylase and phenylethanolamine-n-methyl transferase by repeated immobilization of rats. Endocrinology, 1970, 87, 744-749.

Lamprecht, R., St. Richardson, J., Williams, R.B. and Kopin, I.J. 6-Hydroxydopamine destruction of central adrenergic neurons prevents or reverses developing DOCA-salt hypertension in rats. Journal of Neural Transmission, 1977, 40, 149-58.

Lamprecht, F., Williams, R.B. & Kopin, I.J. Serum dopamine-beta-hydroxylase during development of immobilization-induced hypertension. Endocrinology, 1973, 92, 953-956.

McClelland, D. C. Inhibited power motivation and high blood pressure in men. Journal of Abnormal Psychology, 1979, 88(2), 182-190.

McKegney, F.P. & Williams, R.B. Psychological aspects of hypertension: II. The differential influence of interview variables on blood pressure. American Journal of Psychiatry, 1967, 123, 1539-1543.

Rothlin, E., Cerletti, A. & Ammenegger, H. Experimental psychoneurogenic hypertension and its treatment with hypdrogenated ergot alkaloids (hydergine). Acta Medica Scandinavica, 1955, 307-312, 27-35.

Williams, R.B. Behavioral factors in cardiovascular disease: An
     update. In J.W. Hurst, (ed.) Update: The Heart, Vol V.
     New York: McGraw-Hill, 1981.
Williams, R.B., Lane, J.D., White, A.D., Kuhn, C.M. & Schanberg,
     S.M. Type A behavior pattern and neuroendocrine response
     during mental work. Presented at Annual Meeting, American
     Psychosomatic Society, Boston, MA, 1981.

HYPERTENSION FROM THE STANDPOINT OF BEHAVIORAL MEDICINE

David Shapiro

University of California at Los Angeles
Department of Psychiatry
Los Angeles, CA 90024

Hypertension is a major chronic debilitating and life-threatening disorder of contemporary life, involving many complex factors in etiology and treatment. The diseases associated with hypertension account for a large proportion of all deaths of middle-aged individuals. As it involves multiple psychosocial and behavioral processes in its etiology, treatment, and prevention, hypertension is a major topic of research and discussion in behavioral medicine.

Nature and Consequences of Hypertension

Hypertension is defined primarily by levels of blood pressure considered excessive in relation to age and sex norms. Ninety-five per cent of all cases of hypertension are "essential" (idiopathic, primary), that is, without a known cause, and the remainder are secondary to kidney disease, tumors of the adrenal glands, narrowing of the aorta, and primary aldosteronism (Kaplan, 1980). Secondary cases are usually cured by surgery.

The diagnosis of hypertension is not made precisely but rather relates to a gradient of blood pressure levels and of risks of other disorders related to this gradient. Blood pressure levels tend to increase with age, but this progression is not uniform in all societies and does not appear to be a biological necessity. The Framingham Study (Kannel & Sorlie, 1975) documented the role of hypertension in the development of other diseases, based on a study of more than 5,000 people over a period of two decades. Accompanying rises in blood pressure with age over time were parallel increases in cardiovascular and other diseases. For men and women between the ages of 45 and 74, the

age-adjusted risk of various diseases was much greater in
hypertensives (>160/95 mmHg) than in normotensives (<140/90)
The rates for borderline cases (140/90 - 160/95) fell in between.
In coronary disease involving the greatest overall risk with
associated heart attacks and sudden death, compared to
normotensives the risk for hypertensives was 2.5 times greater.
In men, the risk for borderlines was 1.5 times greater. The
ratios were even greater for women.

High blood pressure is not like other diseases arising from
specific infections or obvious physical abnormalities. It is
functional in nature. However, as the disorder progresses over
time, it becomes associated with thickening of the walls of blood
vessels and increased vascular resistance, in addition to cardiac
changes such as left ventricular hypertrophy and reduction in
stroke volume. Drug therapy, the most common treatment approach
to hypertension, is also functional in nature.

Drug Therapy of Hypertension

Drugs are used to control blood pressure through various
pathways: to alter central or peripheral nervous system
processes affecting blood pressure, to deplete sodium and fluid
volume, to dilate blood vessels, or to alter functions of the
renin-angiotensin system. The intended goal of such treatment is
to modify or correct hemodynamic abnormalities as a means of
lowering blood pressure. Whatever corrections are achieved with
drugs typically result on a trial-and-error basis and in
principle are no more rational as therapy than corrections
achieved by non-pharmacologic means.

In controlling blood pressure by whatever means, the goal is
to prevent further progression of blood pressure to higher levels
and to reduce the likelihood of damage to various bodily organs
(heart, brain, blood vessels) adversely affected by sustained
high levels of pressure. Results of major clinical trials
indicate that active drug treatment can significantly reduce the
risk of life-threatening disorders associated with high blood
pressure. These beneficial effects were demonstrated by The VA
Cooperative Study Group on Antihypertensive Agents (1967, 1970)
for severe (115-129 mmHg diastolic) and moderate (104-115mmHg)
hypertension. The most comprehensive study was recently
completed by the Hypertension Detection and Follow-Up Program
Cooperative Group (1979), involving a total of 10,940 people
classified by entry pressure level into mild, moderate, and
severe categories (90-104, 105-114, 115+ mmHg diastolic). The
study compared the effects of a systematic antihypertensive
treatment program (Stepped Care) with referral to community
medical therapy (Referred Care). Five-year mortality for all
causes was 17% lower overall and 20% lower for the mild group

(90-104) in Stepped Care as compared with Referred Care. The mild group constituted about 70% of all cases of hypertension.

Drug treatment, however, is not free of complications. First, drugs are not effective for all patients. Second, particular drugs may provoke complex physiological readjustments which necessitate additional drugs to compensate for these reactions. For example, vasodilator drugs which lower pressure by decreasing peripheral resistance may produce regulatory changes which in turn increase blood pressure by causing fluid retention and increasing cardiac output (Kaplan, 1980). Diuretics and sympathetic nervous system inhibitors are usually added to vasodilators to counter these effects. Third, aside from undesired and sometimes harmful physiological side effects, patients may complain of psychological complications (e.g. drowsiness, sexual impotence, inability to concentrate) or of other undesirable effects of drugs. Fourth, inasmuch as drug therapy of hypertension is still relatively new (20-30 years), possible adverse consequences of long-term drug usage have not as yet been thoroughly assessed. For example, chronic use of diuretics may have adverse effects on glucose and lipid metabolism (Ames & Hill, 1976; Lewis, Petrie, Kohner & Dollery, 1976). It could be said that in general chronic drug use produces inevitable complications related to various compensatory adjustments to the drugs.

As will be discussed later on, behavioral and other non-pharmacological treatment strategies may provide viable alternatives to drug therapy or they may be used in conjunction with drug therapy as a means of providing further control of blood pressure or of reducing drug dosage. Understanding the nature of drug-behavioral interactions is of special importance in the treatment of hypertension, and is a topic requiring much more investigation. Behavioral and non-pharmacological approaches to hypertension may be of particular value for patients who do not wish to comply with drug treatment for whatever reasons.

Complex Causation of Hypertension

Beneficial though it may be, drug therapy for hypertension is primarily symptomatic in nature and does not deal with specific causal processes underlying the disorder. Although the causes of essential hypertension are presumed unknown, this refers to causes of a physical nature that are ordinarily dealt with by the traditional techniques of modern medicine (drugs, surgery). A vast amount of literature has accumulated on various causal factors: genetic constitutional predispositions, diet (salt, fats, caffeine), smoking, body weight, amount of exercise, and stress and psychosocial factors. Many of these factors often figure in the general recommendations of physicians to their patients -- to ingest less salt, cholesterol, and caffeine, to lose weight if overweight, to exercise regularly, to avoid stress, to relax regularly, and the like. Unfortunately, we seem not to be able to identify which factors are operative in subgroups or individual patients.

It is well known that hypertension runs in families, suggesting that genetic factors play a role in predisposing certain individuals to respond to certain environmental stimuli with increases in pressure (see references in Light & Obrist, 1980). In animal research, this interaction has been studied. Rats with a genetic susceptibility to high blood pressure, when exposed to a conflict situation involving programmed shock associated with responses for food, showed large sustained increases in blood pressure (Friedman & Dahl, 1975). It is supposed that repeated elicitations of pressor responses in "pre-hypertensive" individuals lead eventually to sustained hypertension.

Use of salt has been correlated with incidence of hypertension in different cultures, but a high-salt diet can be compatible with low blood pressure. However, a consistent very low salt intake may be incompatible with the development of hypertension (see Henry & Stephens, 1977).

Behavior plays an indirect role in diet and exercise, and behavioral science offers a means of understanding how people acquire food habits and gain weight and of indicating how such habits may possibly be changed. Behavioral science also is important in providing methods for studying side effects and problems in adhering to drug therapy requirements. The difficulty in evaluating the evidence on salt or weight or exercise lies in the fact that each risk factor is closely interlocked with other risk factors. How the various factors interact is not well understood, and it is difficult to draw conclusions about any one factor in isolation from the others.

The issues are compounded when we turn to the large mass of data and writings on the role of cultural and psychosocial factors in hypertension (see Shapiro & Goldstein, 1980). At the social cultural level, the evidence is substantial that increased blood pressure is related to work pressure and occupational stress, community disruption, prolonged illness, social and natural threats and disasters, economic threats and uncertainties, and rapid social change. At the individual behavioral level, demanding situations requiring continuous behavioral adjustments, conflict, or frustrated goal achievement may precipitate increases in blood pressure which may be transformed eventually into sustained hypertension. At the level of individual personality, hypertensives have been characterized as poorly adjusted, submissive, anxious, and inhibited in the expression of anger. The psychosocial sources of hypertension are varied and ubiquitous.

Neurogenic Factors in Hypertension

If behavior has an influence on blood pressure and hypertension, the disorder must be mediated by central and autonomic nervous system mechanisms involved in regulation of the cardiovascular system. Little is known about the exact sequence of steps leading from behavior on the one hand to hypertension on the other, but researchers have speculated about the role of excessive sympathetic nervous system activity and idiopathic high cardiac output states (see Shapiro, Mainardi, & Surwit, 1977). It seems likely that neurogenic hypertension represents a major part of the total population of individuals with high blood pressure, at least as can be judged by the large mass of epidemiological, sociological, and behavioral data previously discussed. In an attempt to differentiate neurogenic from non-neurogenic hypertension, Esler, Julius, Zweifly, Randall, Harburg, Gardiner, and DeQuattro (1977) distinguished two groups of patients with mild essential hypertension, those patients who had relatively elevated levels of plasma renin activity and those with normal levels. The high renin subgroup also had a higher heart rate and an elevated plasma norepinephrine concentration. Through the use of autonomic blocking agents, these investigators established that the high renin group sustained their increased blood pressure by means of overactivity of the sympathetic nervous system. Through psychological test methods, it was learned that as a group patients with mild high renin essential hypertension were controlled, guilt-prone and submissive, had a high level of unexpressed anger, and appeared to sustain their blood pressure by means of overactivity of the sympathetic nervous system. Normal renin patients were not different from normal control subjects in these psychological characteristics. Esler et al. (1977) concluded that the pathogenesis of elevated blood pressure in the high renin mild hypertensive group may involve a behavioral pattern (primarily suppression of anger) which is related to increased sympathetic nervous system activity, but they also suggested that the behavior pattern could follow from increased sympathetic nervous tone rather than the other way around. Their research does not make clear why increased blood pressure is the major outcome in this group of patients (high renin) rather than some other symptom, inasmuch as excessive sympathetic nervous system activity has been believed to be a causal factor in other psychophysiological disorders as well (Shapiro & Katkin, 1980). It is not known whether the heightened sympathetic nervous tone represents a constitutional predisposition in these patients, or whether the sustained and

heightened sympathetic nervous tone results from repeated
elicitations of behaviorally-induced patterns of nervous
reaction.

Many invesitgators believe that fixed essential hypertension
develops as a compensatory process in response to idiopathic
cardiac output states. The process apparently develops in
genetically predisposed individuals who are cardiovascular
hyperreactors under emotional or environmental stress (see
Shapiro et al., 1977). A recent paper supports this view (Light
& Obrist, 1980). Normotensive individuals with similar blood
pressure and heart rate under low stress or relaxation conditions
showed wide and stable differences in their cardiovascular
responses to stressful stimuli. High heart rate reactivity and a
tendency toward occasional elevations in systolic blood pressure
contributed to blood pressure levels observed under stress
conditions. Light and Obrist also demonstrated an association
between high heart rate reactivity to stress and incidence of
hypertension in the subjects' parents.

Alternative viewpoints have also been presented. Kaplan
(1980) believes that the more likely initiating process of
hypertension is the gradual overfilling of the vascular bed
secondary to a defect in the ability of the kidneys to excrete
sodium (likely a genetic trait). Many years of high sodium
intake lead to an expanded body fluid volume and thus to
hypertension eventually. Are there any ways to account for a high
salt intake in some individuals and not others? Is high salt
intake associated with other hypertension-related processes?

Complex Causal Interactions in Hypertension

How the various psychosocial and behavioral factors interact
with each other and with genetic predisposition, dietary factors,
exercise, and other environmental factors is still unknown.
Consider the cultural variability in the rise of blood pressure
with age. Depending on one's theoretical orientations or
convictions, various differences from society to society or group
to group may be explained in genetic, dietary, constitutional, or
psychosocial terms. Correlational studies involving multiple
factors are difficult to interpret. In addition, disentangling
the interactions of factors by statistical analysis is hazardous
at best. Henry and Stephens (1977) discuss as an example a study
showing no relationship between coffee drinking and death due to
coronary disease, a finding that was presumably corrected for the
influence of obesity, lack of exercise, and smoking. Henry and
Stephens point out, however, that a certain type of personality
may be associated with all of these factors. Thus, smoking and
coffee drinking as well as emotional responses all increase
plasma catecholamines, and these may be associated with high

blood pressure. Statistical analysis by itself cannot provide a definitive answer. Rather, Henry and Stephens suggest that experimental studies in controlled animal populations may be the only way to arrive at the critical knowledge.

In reviewing further the complex interactions involved in the variable relationship between blood pressure and age as possibly related to cultural differences and cultural change across different societies, Henry and Stephens (1977) sum up their position as follows:

> The conclusion was that a man living in a stable
> society and well-equipped by his cultural
> background to deal with the familiar world around
> him will not show a rise in blood pressure with
> age. This thesis holds whether he is a modern
> technocrat who became a fighter pilot early in
> life or a Stone Age Bushman who is a skilled
> lumber-gatherer in the Kalahari Desert. However,
> when radical cultural changes disrupt his
> familiar environment with a new set of demands
> for which past acculturation has left him
> unprepared, his social assets are then critical.
> Should they fail to protect him, he will be
> exposed to emotional upheavals and ensuing
> endocrine disturbances that may eventuate in
> cardiovascular disease. (p. 203)

They argue throughout that problems of psychosocial adaptation are importantly connected to hypertension, as well as to coronary heart disease, cancer, and other illnesses.

Task for Behavioral Medicine

The task for behavioral medicine in regard to hypertension (or other complex chronic disabling disorders) is to begin to integrate the accumulated knowledge, concepts, and methods that have been developed by investigators in the various disciplines (behavioral sciences, cardiovascular physiology, neurophysiology, pharmacology, epidemiology, immunology, sociology, genetics) in attacking the issues of etiology, treatment, and prevention of the disorder. These will need to include basic experimental studies in man and other species, and long-range longitudinal studies, as well as research designed to evaluate various approaches to treatment and prevention (pharmacological and non-pharmacological). Simplistic unifactorial or unidisciplinary approaches are not likely to succeed. Particularly critical for behavioral medicine is to increase our understanding of how the central nervous system regulates normal cardiovascular processes and our understanding of the nature of the disregulation (Schwartz, 1979) that results in high blood pressure. That knowledge is critical to the design of rational methods of prevention and treatment.

Behavioral Strategies in Treatment

The remainder of this chapter will focus on behavioral methods that are relevant to the treatment and prevention of essential hypertension. These methods, as in the case of drug therapy, have evolved out of various assumptions and theories and typically on an empirical or trial-and-error basis. Detailed reviews and appraisals of behavioral treatment procedures were presented recently by Agras and Jacob (1979) and Reeves and Victor (in press). Agras and Jacob emphasize the use of relaxation and biofeedback treatment approaches and also consider in detail the prevention of poor adherence to use of antihypertensive drugs. Reeves and Victor review the various methods of behavioral and physiological self-regulation (biofeedback, relaxation, meditation, autogenic training, hypnosis), and they also summarize and discuss research on diet (salt restriction, caloric restriction, caffeine, ethanol) and exercise, and the potential of these latter strategies for controlling blood pressure. These excellent reviews are comprehensive, and it is not necessary here to go over the same ground. Rather, an attempt will be made to outline the major behavioral strategies of prevention and treatment, with a particular focus on integration of behavioral and physiological concepts and methods.

It is clear from an overview of the great variety of non-pharmacological methods that blood pressure can be lowered to varying amounts and in many different ways. For simplicity, the methods that have been examined in research will be grouped into indirect methods and direct methods, the former focused primarily on dietary factors, exercise, and drug adherence, the latter on various methods of relaxation, biofeedback, self-regulation, and reduction of reactivity to stress.

Indirect Methods of Control

Salt Restriction. The role of sodium in hypertension has been debated extensively, primarily on the basis of epidemiological evidence. The effects of salt intake on blood pressure apparently depend on many factors, including the possibility that humans may vary in their genetic sensitivity to salt. Strains of rats have been bred for a hypertensive response to salt loading, and another strain in which this response does not occur (Dahl, 1972; see Reeves & Victor, in press). Several clinical studies indicate that a 50% reduction in salt use (daily intake of 5 grams) can result in a small but reliable lowering of blood pressure (about 10/5 mmHg reduction) (see Reeves & Victor, in press). Thus, even moderate salt reduction may be useful. To my knowledge, individual differences in human hypertensive response to salt loading have not been identified specifically. If there were a reliable means of determining who are likely salt-sensitive hypertensive individuals, if such is the case, then it would greatly aid in selecting salt restriction as a method of treatment for such persons (and also as a method of prevention for "salt-sensitive pre-hypertensives").

Weight Loss. Body weight has generally been correlated with blood pressure level in various studies, and recent studies suggest that weight reduction in obese patients can lower blood pressure independently of salt restriction and that even modest weight loss may be beneficial for hypertensive patients (see Reeves & Victor, in press).

Other Dietary Factors. Heavy caffeine and alcohol use are associated with hypertension, and their control may be useful in the treatment of hypertension (Reeves & Victor, in press). Caffeine has a short-term hypertensive effect, but it is far from certain how chronic usage of this substance or of alcohol causally relates to hypertension.

Exercise. Although there is apparently no relationship between exercise and blood pressure in normotensive individuals, the evidence suggests that exercising is beneficial in lowering blood pressure and in incresing general cardiovascular fitness in individuals with high blood pressure (see Reeves & Victor, in press).

In sum, with respect to the control of dietary factors and exercise, the evidence supports their usefulness in reducing blood pressure in hypertensive patients. In applying any of these particular approaches with an individual patient, the problem is to identify whether or not a particular practice is of significance to the hypertension response of that individual. It seems clear enough that really excessive use of salt, caffeine, or alcohol or excessive body weight is likely detrimental to an individual with high blood pressure (and probably detrimental in general to one's health), but it is less certain whether control of these factors is beneficial for all hypertensive individuals. By the same token, regular exercise is probably beneficial, but primarily for hypertensives who are not in very good physical condition.

Simply knowing about the role of diet and exercise and educating patients about these factors may not be sufficient to alter the lifelong habits involved. To change eating and exercise habits requires a consideration of various individual-

and group-geared methods of education, behavior modification, and self-management (see Pomerleau & Brady, 1979).

In addition, from a research standpoint, the causal interpretation of the effects of any one treatment approach is a difficult issue. In addition to eating less or using less salt or exercising regularly, an individual may be coincidentally altering other aspects of lifestyle and response to stress. These complex interactions will require careful analysis in future studies.

Drug Adherence. The complex issues of drug adherence have been discussed by Agras and Jacob (1979). Given that antihypertensive drugs can effectively reduce blood pressure, and probably without serious side effects when the drugs are properly selected, the problem of drug adherence is of considerable significance. Proper selection of drugs requires careful attention to the particular physiological and psychological make-up of the individual and the particular side effects of the drugs. The poor adherence that is commonly observed in the care of hypertensive patients may be due in no small way to the insufficient monitoring of these reactions on the part of the physician and patient alike. Aside from the problem of undesired side effects, patients may not take their drugs because of the apparent lack of felt symptoms associated with high blood pressure, fear and ignorance about the drugs and how they work and why they are important, and some general bias against drug taking in many patients. Agras and Jacob cited the work of Sackett, Gibson, Taylor, Haynes, Hockett, Roberts, and Johnson (1975), showing that only slightly more than half of all hypertensives take over 80% of their medication. Medication taking and adherence to various treatment procedures can be improved by various means, e.g. maximizing the convenience of medical care and other means of reducing dropout from treatment, enhancing appointment keeping, simplifying dosage regimens, reducing number of pills to be taken per day, more careful consideration of side effects problems, counseling to remedy poor adherence, among others (Agras & Jacob, 1979).

One approach in fostering drug adherence is to engage the patient more directly and actively in the process of establishing the correct choice of drugs and dosage. Typically, physicians prescribe a drug regimen on the basis of knowledge of the pharmacology of the substance, prior experience, clinical reports and other studies, particular preferences, and various indications and counterindications in the patient. The patient is asked to take the drugs for a period of time (2-4 weeks) and then to return for blood pressure evaluation and to report about side effects. The reporting of side effects is not systematic and physicians do not always elicit such reports. With a little instruction and effort, patients can take a more active part in setting their drug therapy by recording their pressure at home 2-3 times daily and also by registering information about side effects, the latter on special checklists or rating scales. With guidance from their physician, patients can then modify either the specific drug used or the drug dosage required to obtain the desired goal of an acceptable level of blood pressure with acceptable or no side effects. This form of taking responsibility for one's own treatment may be especially valuable for certain patients for whom such active involvement fosters further cooperation and adherence. It is as if they have determined their own treatment.

Other factors may come into play in interaction with the presumed effects of this particular approach to drug adherence and pressure control. Thus, self-recording of blood pressure (see below) by itself may be beneficial (Laughlin, Fisher & Sherrard, 1978). In a like vein, for many patients, simply being under medical care may be beneficial. How these influences come about remains for further definitions and investigation. Calling them "placebo" effects does not explain them. Continued observation and long-term follow-up studies may elucidate their persistence and nature.

Finally, further research is in order on the degree to which individuals can sensitize themselves to variations in their own blood pressure, possibly as a means of guiding their own drug treatment, in taking steps to reduce their stress responses, and the like.

Direct Methods of Control

The "direct" rubric groups together various methods of controlling blood pressure derived primarily from psychological and psychophysiological conceptions and methods.

Regulation of Blood Pressure or Critical Cardiovascular Functions. Since the first systematic study on the control of blood pressure in hypertension by means of systolic blood pressure biofeedback training (Benson, Shapiro, Tursky & Schwartz, 1971), many reports have appeared which have followed this strategy, that is, of providing information to patients of ongoing changes in blood pressure with the expectation that the information can be utilized by the patient in reducing blood pressure specifically (see Agras & Jacob, 1979, and Reeves & Shapiro, 1978, for reviews). The results of various clinical studies indicate that this method can be beneficial to patients and that the observed reductions in pressure are relatively specific (Kristt & Engel, 1975).

A study recently completed in my laboratory attempted to compare the effectiveness of biofeedback against an alternative behavioral method, standard drug therapy, and a simple control procedure. (Goldstein, Shapiro, Sambhi & Thananopavarn, Note 1). Patients with mild hypertension were studied twice a week for eight weeks under one of the four following conditions: blood pressure biofeedback, Benson's relaxation response (Benson, Rosner, Marzetta & Klemchuk, 1974), drug treatment, and a control procedure consisting of the home self-monitoring of blood pressure. The groups were composed of nine patients each, primarily males between the ages of 35 and 60 years. Extensive baseline, treatment, and follow-up assessments were obtained of home blood pressure recordings and psychophysiological variables in the laboratory (heart rate, blood pressure, skin conductance, breathing rate, and frontalis muscle tension). Drug treatment was found to be markedly superior to all of the behavioral procedures in the regulation of blood pressure. Biofeedback was more effective than either relaxation or the control procedure but its effects in lowering diastolic blood pressure were comparable to drug treatment. The relaxation response did not prove to be effective.

In general, the biofeedback method of physiological control is an attractive strategy as the treatment is oriented to modification of specific physiological response systems. It is akin to the use of drugs to control blood pressure by blocking or reducing beta-adrenergic activity (e.g. propanalol) or by decreasing peripheral resistance (e.g. vasodilators). Thus, the method can be used to effect specific control of cardiovascular parameters other than systolic blood pressure, e.g. diastolic blood pressure, peripheral blood flow, skin temperature, or heart rate. There is also the possibility of providing feedback for a pattern of simultaneous changes in several cardiovascular functions. For example, Surwit, Shapiro, and Good (1978) gave patients feedback for simultaneous reduction of blood pressure and heart rate, although the outcome of this procedure was not superior to others studied. In a case study, Williams (1975) trained a patient in the simultaneous reduction of heart rate and increase in forearm blood flow, and found significant reductions in systolic and diastolic pressure without compensatory increases in cardiac output. The strategy of tailoring the biofeedback method specifically to the functions that are of unique significance to an individual patient is one which requires further investigation and development. It also requires the ability to determine which particular patients are proper candidates for such an approach.

Regulation of Sympathetic Nervous System Activity. Given the acknowledged significant role of sympathetic nervous system activity in blood pressure regulation and in hypertension, the strategy of control over such activity has been discussed by a number of authors. Direct recording of sympathetic activity in specific neural pathways has advanced considerably in recent years (see Wolf, 1979), but to my knowledge no attempt has been made to have subjects regulate such activity that may be related to blood pressure regulation or pressure reduction.

The strategy of sympathetic nervous system control has been emphasized by Patel (1977), revolving around the technique of skin resistance (a sympathetically-innervated response) biofeedback training (Patel, 1973). In her studies, however, Patel incorporated breathing-relaxation and passive concentration (meditation) procedures along with other methods of stress management and in some cases EMG feedback (see Patel, 1975a, 1975b; Patel & North, 1975). Patel reported substantial reductions in pressure level and medication use in addition to reduced pressor reactions to exercise and cold pressor tests following this multi-modal behavioral treatment, as compared to no changes in control conditions. It is not possible to determine which components of such a program were mainly responsible for the beneficial outcomes reported.

Relaxation. Agras and Jacob (1979) provide a comprehensive review of the use of various forms of relaxation training in the treatment of hypertension. The relaxation approaches have used variations of certain meditative disciplines, paced respiration and breath control, autogenic training, progressive relaxation, hypnosis, and the relaxation response. In some instances, muscle biofeedback training may be utilized to facilitate relaxation. In general, the various methods of relaxation have been shown to be effective to varying degrees either in terms of reduced blood pressure or reduced medication requirements.

Other studies have been devoted to a comparison of different relaxation methods and relaxation against biofeedback. No firm conclusion can be drawn about the relative value of different approaches. •Moreover, Agras and Jacob concluded that "biofeedback of blood pressure has no greater clinical effect than relaxation therapy, thus, the extra investment in blood pressure feedback equipment would seem unwarranted unless more sophisticated and powerful methods are developed" (p. 225).

Tape cassette relaxation would appear to be a promising relaxation approach because it can be easily taught to patients and with some guidance can be used by the patient at home on a daily basis. Taylor, Farquhar, Nelson, and Agras (1977) and Bali (1979) have reported success in utilizing taped-relaxation techniques. In our laboratory, we are currently utilizing tape cassette relaxation with the self-recording of blood pressure by the patient immediately prior to and following the relaxation exercise. In this way, relaxation plus minimum feedback of blood pressure change information indicating that relaxation can indeed reduce pressure are combined. In addition, we ask patients to record their own blood pressure and pulse rate at other times at home during the day (morning on awakening, evening before going to sleep), and at work if possible. Self-monitoring alone has been shown to have positive effects in several studies (see Goldstein, Shapiro, Thananopavarn, Note 1).

Stress Management. An alternative strategy focuses attention directly on the impact of stressful life events and situations on the individual's blood pressure response. The role of cardiovascular reactivity I have already discussed in the development of hypertension. A behavioral strategy focusing on stress reduction may be especially beneficial in borderline or labile cases. Moreover, it is uncertain whether relaxation, which is the most commonly employed behavioral strategy, has a significant effect on stress reduction. Patel's research, previously discussed, suggests this kind of outcome, although she usually incorporated stress management techniques into her treatment. On the other hand, Goldstein et al. (Note 1) reported that various degrees of blood pressure lowering associated with drug and behavioral treatments did not result in differential pressor responses to laboratory stressors (mental arithmetic, cold pressor test).

Various techniques of stress management may be employed either with individuals alone or in groups of patients. These may include the deliberate practice of relaxation under stress stimulation, stress desensitization procedures, learning to discriminate under what conditions blood pressure will tend to increase, and appropriate assertiveness under demanding work and interpersonal situations. Various methods of cognitive behavior modification including behavioral rehearsal, self-statement, and visualization of pleasant experiences may also prove useful. Techniques such as these have been employed effectively in cardiac stress management training (see Suinn & Bloom, 1978). Efforts of this kind with hypertensives are currently underway, but little has been reported in the recent literature. In many respects, programs employing multiple behavioral procedures and a lot of attention to patient motivation and patient adherence appear to produce the most substantial benefits.

Prevention and Assessment of Hypertension

Almost any of the procedures and variables discussed previously in this chapter have a potential application in preventing hypertension. To the extent that we take seriously the role in hypertension of various psychosocial factors, stressful life events, behavior patterns, and certain individual characteristics (such as cardiovascular hyperreactivity, family history of hypertension), then we can make recommendations concerning ameliorative and preventative procedures. The critical need is to develop systematic means of assessing the specific impact of these variables in a given individual or group and of identifying the critical individual and group characteristics. If there are reliable ways of identifying "pre-hypertensives" (e.g. family history, cardiovascular hyperreactivity), then preventative studies aimed at these groups should be undertaken.

The literature on the assessment of hypertension is extensive, and no attempt will be made to review it here. Two broad classes of assessment have concerned behavioral factors. The first involves psychophysiological studies of reactions to stress based on the assumption that hypertensives (and pre-hypertensives) will show abnormal pressor responses to stressful stimuli. The second involves the assessment of psychological characteristics through the use of psychological tests and interviews, research based on the assumption that hypertension is associated with unique personality characteristics which have some specific connection to increased blood pressure. A comprehensive review and critique of the assessment literature and its shortcomings may be found in Goldstein (in press). Assessment studies are often inconclusive because of the failure to identify specific behavioral patterns or complex patterns of behavior, physiological response, and other psychological characteristics. The study by Esler et al. (1977), previously discussed, was able to identify certain psychological characteristics (unexpressed anger) in a subgroup of borderline hypertensive patients (those having high levels of plasma renin activity, cardiac output, and associated physiological responses indicative of high levels of sympathetic nervous system activity). Through such refinements in selection of patients and in the use of multiple behavioral and physiological criteria, further progress can be made in differentiating subgroups of hypertensives along combined physiological and behavioral lines. Such information may permit a more selective approach to the use of behavioral interventions. That is, it may allow us to tailor the available treatment modalities to the characteristics and needs of particular varieties of individual patients. For example, will patients characterized by a _pattern_ of elevated cardiac output and unexpressed anger profit more from techniques oriented to the reduction of cardiac output (or heart rate) _and_ to assertiveness and the appropriate expression of anger as compared with a simple relaxation technique? Will patients with fixed hypertension having elevated peripheral resistance and normal or subnormal cardiac output profit more from biofeedback aimed at reducing diastolic pressure and increasing peripheral blood flow as compared with relaxation or other related therapies?

CONCLUSIONS

This chapter attempted to lay out some of the major issues
of concern to a developing behavioral medicine of hypertension.
These have included psychosocial, behavioral, physiological,
epidemiological, medical, and other research topics. By taking
aim at hypertension as a critical public health problem, I have
given attention to a variety of processes which seem relevant to
the complex puzzle of this disorder. The literature is huge,
although the studies tend to be compartmentalized into one area
or another -- whether about etiology or treatment, whether
physiological or behavioral, whether sociological or
epidemiological. What seems needed now is a continuing effort
toward integration of the diverse facts and theories as well as
of methodologies. Well-planned, interdisciplinary studies are in
order. Through interdisciplinary cooperation and collaboration,
we can hopefully advance our understanding of the disorder and
improve our methods of control and prevention.

One specific recommendation can be made. To foster
collaborative efforts between behavioral and biomedical
scientists, one strategy may be to include behavioral specialists
as regular staff of hypertension clinics in hospitals. Working
alongside physicians and other hospital staff, these behavioral
specialists can participate in diagnostic and other assessments
of patients as well as arranging for various non-pharmacological
and behavioral interventions, assisting in patient education, and
the like. In this way, we can begin to build up a more extensive
base of clinical experience and scientific data, which hopefully
will lead to more sophisticated and definitive research on the
disorder.

REFERENCE NOTE

1. Goldstein, I.B., Shapiro, D., Thananopavarn, C.,
        and Sambhi, M.P. Comparison of drug and
        behavioral treatment of hypertension. Manuscript
        submitted for publicaton, 1981.

REFERENCES

Agras, S. and Jacob, R.G. Hypertension. In O.F. Pomerleau &
    J.P. Brady (eds.) Behavioral medicine: Theory and
    practice. Baltimore: Williams & Wilkins, 1979. Pp.
    205-232.
Bali, L.R. Long-term effect of relaxation on blood pressure and
    anxiety levels of essential hypertensive males: A
    controlled study. Psychosomatic Medicine, 1979, 41,
    637-645.
Benson, H., Beary, J.F., and Carol, M.P. The relaxation

response. Psychiatry, 1974, 37, 37–46.

Benson, H., Shapiro, D., Tursky, B. & Schwartz, G. Decreased
    systolic blood pressure through operant conditioning
    techniques in patients with essential hypertension.
    Science, 1971, 173, 740–742.

Dahl, L.K. Salt and hypertension. American Journal of Clinical
    Nutrition, 1972, 25, 231– .

Esler, M., Julius, S., Zweifler, A., Randall, O., Harburg, E.,
    Gardiner, H. and DeQuattro, V. Mild high–renin essential
    hypertension: Neurogenic human hypertension? New England
    Journal of Medicine, 1977, 296, 405–411.

Friedman, R. and Dahl, L.K. The effect of chronic conflict in
    the blood pressure of rates with a genetic susceptibility
    to experimental hypertension. Psychosomatic Medicine, 1975,
    37, 402–416.

Goldstein, I.B. Assessment of hypertension. In L.A. Bradley &
    C.K. Prokop (eds.) Medical psychology: A new perspective
    New York: Academic Press, in press.

Henry, J.P. and Stephens, P.M. Stress, health, and the social
    environment. A sociobiligic approch to medicine. New
    York: Springer-Verlag, 1977.

Hypertension Detection and Follow–up Cooperative Program.
    Five-year finding of the hypertension detection and
    follow-up program. I. Reduction in mortality of persons
    with high blood pressure, including hypertension.
    Journal of American Medical Association, 1979, 242,
    2562–2571.

Kannel, W.B., and Sorlie, P. Hypertension in Framingham. In O.
    Paul (ed.) Epidemiology and control of hypertension.
    Miami: Symposia Specialists, 1975.

Kaplan, N.M. The control of hypertension: A therapeutic
    breakthrough. American Scientist, 1980, 68, 537–545.

Kristt, D.A. & Engel, B.T. Learned control of blood pressure in
    patients with high blood pressure. Circulation, 1975, 51,
    370–378.

Laughlin, K.D., Fisher, L. and Sherrard, D.J. Blood pressure
    reductions during self-recording of home blood pressure.
    American Heart Journal, 1978, 98, 629–634.

Lewis, P.J., Petrie, A., Kohner, E.M. and Dollery, C.T.
    Deterioration of glucose tolerance in hypertensive patients
    on prolonged diuretic treatment. Lancet, 1976, 1, 564.

Light, K.C. and Obrist, P.A. Cardiovascular reactivity to
    behavioral stress with and without marginally elevated
    casual systolic pressures. Hypertension, 1980, 2, 802–808.

Patel, C.H. Yoga and biofeedback in the management of
    hypertension. Lancet, 1973, 2, 1053–1055.

Patel, C.H. Yoga and biofeedback in the management of "stress"
    in hypertensive patients. Clinical Science and Molecular
    Medicine, 1975, 48, Suppl., 171–174.

Patel, C.H. Biofeedback-aided relaxation in the management of hypertension. Biofeedback and Self-Regulation, 1977, 2, 1-41.

Patel, C.H. and North, W.R.S. Randomized controlled trial of yoga and biofeedback in management of hypertension. Lancet, 1975, 2, 93-95.

Pomerleau, O.F. and Brady, J.P. (eds.) Behavioral medicine: Theory and practice. Baltimore: Williams & Wilkins, 1979.

Reeves, J.L. and Victor, R.G. Behavioral strategies in hypertension. In P.A. Boudewyns & F.J.Keefe (eds.) Behavioral medicine for the primary care physician. Reading, Mass.: Addison-Wesley, in press.

Reeves, J.L. and Shapiro, D. Biofeedback and relaxation in essential hypertension. International Review of Applied Psychology, 1978, 27, 121-135.

Sackett, D.L., Gibson, E.S., Taylor, D.W., Haynes, R.B., Hockett, B.C., Roberts, R.R. and Johnson, A.L. Randomized clinical trial of strategies for improving medication compliance in primary hypertension. Lancet, 1975, 1205-1207.

Schwartz, G.E. Disregulation and systems theory: A biobehavioral framework for biofeedback and behavioral medicine. In N. Birbaumer & H.D. Kimmel (eds.) Biofeedback and Self-Regulation. New Jersey: Erlbaum Associates, 1979.

Shapiro, D. and Goldstein, I.B. Behavioral patterns as they relate to hypertension. In J. Rosenthal (ed.) Clinical pathophysiology of arterial hypertension. Heidelberg: Springer-Verlag, 1980. (in German)

Shapiro, D. and Katkin, E.S. Psychophysiological disorders. In A.E. Kazdin, A.S. Bellack & M. Herson (eds.) New perspectives in abnormal psychology. New York: Oxford University Press, 1980.

Shapiro, D., Mainardi, J.A. and Surwit, R.S. Biofeedback and self-regulation in essential hypertension. In G.E. Schwartz & J. Beatty (eds.) Biofeedback: Theory and Research. New York: Academic Press, 1977. Pp. 313-347.

Suinn, R.M. and Bloom, L.J. Anxiety management training for pattern A behavior. Journal of Behavioral Medicine, 1978, 1, 21-36.

Surwit, R.S., Shapiro, D. and Good, M.I. A comparison of cardiovascular biofeedback, neuromuscular biofeedback, and meditation in the treatment of borderline essential hypertension. Journal of Consulting and Clinical Psychology, 1978, 46, 252-263.

Taylor, C.B., Farquhar, J.W., Nelson, E. and Agras, S. Relaxation therapy and high blood pressure. Archives of General Psychiatry, 1977, 34, 3390-342.

VA Cooperative Study Group on Antihypertensive Agents. Effects of treatment on morbidity in hypertension. Results in patients with diastolic blood pressure averaging 115 through 129 mm Hg. Journal of American Medical Association, 1967, 202, 116-122.

VA Cooperative Study Group on Antihypertensive Agents. Effects
     of treatment on morbidity in hypertension. II: Results in
     patients with diastolic blood pressure averaging 90 through
     114 mm Hg.Journal of American Medical Association, 1970,
     213, 1143-1152.
Williams, R.B. Heart rate and forearm blood flow feedback in the
     treatment of a case of severe essential hypertension.
     Psychophysiology, 1975, 12, 237. (abstract)
Wolf,   S.L.   Microneurography   and   cardiovascular   control.
     Psychophysiology, 1979, 16, 164-170.

CONTROL OF CARDIOVASCULAR REACTIVITY AND THE TREATMENT

OF HYPERTENSION

Andrew Steptoe

St. George's Hospital Medical School
London University
London, England

Training in the voluntary control of cardiovascular
functions is typically carried out in comfortable, quialet
laboratories or clinics. It is presumed that abilities
developed in these settings will generalize to the more
demanding conditions of everyday life, so that patients will
continue to regulate cardiovascular activity when faced with
social and physical challenges. This chapter argues in favor of
an alternative treatment strategy: direct training in the
control of cardiovascular reactivity. This involves the
administration of voluntary control techniques while the patient
is simultaneously exposed to stressors or stimuli that provoke
pressor responses. The rationale for this approach is first
outlined, and the methods of regulating reactivity are then
considered.

CONSTRAINTS ON VOLUNTARY CONTROL

When training is carried out in physically restful
conditions, the reductions of cardiovascular activity produced
during sessions of biofeedback or relaxation tend to be modest.
Mean decreases of 2 to 11 mmHg in systolic pressure have been
recorded in experimental studies with normotensives; the
corresponding within-session reductions of diastolic pressure
range approximately from 3 to 8 mm (Shapiro, Tursky & Schwartz,
1970; Fey & Lindholm, 1975; Elder, Gamble, McAfee & Van Veen,
1979). The effects of biofeedback on heart rate or pulse
transit time have also been undramatic in most investigations,
although more elaborate training strategies may accentuate
responses (McKinney, Geller, Gatchell, Barber, Bothner & Phelps,
1980).

A number of factors may account for this limited success. Firstly, experimental studies with volunteers are necessarily brief, so the time during which subjects can practice a particular technique tends to be short. This argument undoubtedly has some force, since the response patterns of individuals entering the psychophysiological laboratory for the first time are not typical of their subsequent performance. Nevertheless, improvements in voluntary control are not invariably recorded across sessions, so other constraints must be present. A second restriction may lie in the use of normotensive populations. The cardiovascular systems of patients with essential hypertension tend to be more sensitive to environmental stimuli; witness their heightened reactivity to physical and behavioral challenge (Steptoe, 1981). It has been assumed that the responses of normotensives exemplify in miniature trends that are of greater significance in patients. But the arguments for and against the parallel between normotensives and hypertensives are complex. The hemodynamic disturbances in hypertension are multifaceted, so that mechanisms contributing to the maintenance of high blood pressure may not be influential on the normal circulation (Guyton, 1978). Yet on the other hand, similarities have been observed between the patterns of voluntary control in healthy volunteer and patient groups.

The conditions pertaining in the training setting itself may also limit the magnitude of cardiovascular modifications. Since the cardiovascular system is regulated to a level which satisfies metabolic demand, deviations from the balance between supply and demand are countered by multiple servocontrol mechanisms (including baroreceptor and chemoreceptor reflexes, autoregulation and pressure diuresis). The metabolic requirements of the human body at rest are modest, so blood pressure and heart rate may be low even before the introduction of training in voluntary control.

This "physiological floor" argument has frequently been invoked to explain the poor returns from biofeedback and relaxation training in lowering cardiovascular activity (Malcuit & Beaudry, 1980). It suggests that attempts to reduce blood pressure, pulse transit time or heart rate from elevated levels may offer greater scope for behavioral interventions. Increased levels can of course be generated both by behavioral stimuli and physical exercise, but there are several reasons why behavioral provocation is attractive in the case of essential hypertension. Firstly, essential hypertensives respond with substantial increases in heart rate and blood pressure even when the motor concomitants of behavioral coping are small. Thus, circumstances frequently occur in which hemodynamic reactions to behavioral stressors are in excess of metabolic demands

(Steptoe, 1981). Moreover, the generalization of voluntary control outside the laboratory may be facilitated by training in the presence of behavioral stressors. One of the purposes of treatment in the clinic is to develop skills which will then be useful to patients in their everyday lives. The conditions of ordinary life may be mimicked to some extent by laboratory tasks, so that patients can learn to control adverse reactions directly. The strategies emphasized in the treatment setting may then be integrated into the general behavioral repertoire of the individual.

A further reason for studying cardiovascular reactions to behavioral challenges emerges from consideration of the role of disturbed hemodynamic function in the etiology of essential hypertension.

CARDIOVASCULAR REACTIVITY AND EARLY ESSENTIAL HYPERTENSION

The mechanisms of essential hypertension that are most sensitive to behavioral influence have been described in Chapter 10. Here it is pertinent only to highlight the role of transient hemodynamic reactions in the early stages of the disorder. For amongst the characteristics measured in youth, one of the firmest predictors of later dysfunction is a pattern of transient elevations in blood pressure (Julius & Schork, 1971). The risk in such cases is substantially raised, although stable hypertension is not of course an inevitable outcome. It is possible that the repeated challenge of autonomically mediated reactions may precipitate the structural (vascular) and functional (renal and neural reflex) adjustments that serve to maintain blood pressure at a high level.

This sequence of events has been telescoped in animal experiments by using direct electrical stimulation to the central nervous system, or by imposing intense behavioral stressors (Folkow & Rubinstein, 1966; Forsyth, 1969). The pattern is less easy to observe in man, due to the extended time span. It is also probable that not all behavioral stressors provoke hemodynamic adjustments of the appropriate type. Longitudinal studies of essential hypertensives suggest that early stages of the dysfunction are maintained predominantly through increases in cardiac output rather than peripheral resistance. Although this "hyperkinetic" pattern may be due in part to preferential distribution of blood to the cardiopulmonary (central) circulation, increased traffic in the cardiac sympathetic nervous pathways is almost certainly involved (Birkenhager & Schalekamp, 1976).

Consequently, attempts to modify reactivity by voluntary means might usefully focus on conditions in which cardiac sympathetic (β-adrenergic) responses are prominent. This appears particularly true during active coping behavior; active coping can be elicited by challenging environmental stimuli which call for instrumental responses with uncertain outcome (Obrist, 1981). Individual differences in hemodynamic response topography should not be ignored however, since recent investigations suggest that heightened reactivity may be an important predictor of risk, even before increases in tonic blood pressure levels have emerged.

REACTIVITY IN THE PRE-HYPERTENSIVE PROFILE

Although cardiovascular hyperreactivity is characteristic of early hypertension, this does not necessarily imply a causal link; for it is possible that exaggerated reactivity is a consequence of pre-existing pathophysiology. Nevertheless, a wide range of reactivities is typically observed even amongst individuals without elevated blood pressures. Moreover, studies of young people at risk for essential hypertension suggest that hyperreactivity precedes identifiable rises in tonic levels. Falkner and his associates (1979) compared adolescents with and without a family history of hypertension during the performance of mental arithmetic under harassment. The rises in systolic and diastolic pressure were greater in those with a family history plus occasional high pressure readings in their own records (labile group) than in controls; children with positive family histories alone displayed an intermediate response. Interestingly, Hofman et al. (1979) have shown that plasma noradrenaline concentration is also raised in young people with slightly elevated blood pressures on screening.

Exaggerated heart rate reactions to behavioral stressors may also characterize those at risk. Light and Obrist (1980) categorized a cohort of volunteers on the basis of cardiac reactions to a demanding reaction time task known to provoke cardiac sympathetic activation. The blood pressures recorded at home or under resting laboratory conditions were indistinguishable in the low and high heart rate reactors. But during behavioral stress, the high cardiac reactors produced heightened systolic pressure responses as well. At present, the results are not entirely consistent, since some studies have failed to identify these differential reaction patterns (Lawler, 1980). However, Steptoe and Ross (1981) demonstrated that high heart rate responders also show substantial modifications in pulse transit time, so the cardiovascular reaction pattern tends to be coordinated. In addition, cardiovascular hyperreactivity is comparatively specific, and is not associated with exaggerated lability in other autonomic variables. It is possible therefore that the specific dysfunctions of reactivity delineated in recent investigations will help to predict risk populations with increasing accuracy.

METHODS OF REGULATING CARDIOVASCULAR REACTIVITY

Pathophysiological considerations, and the arguments about training conditions outlined earlier, converge on the same conclusion: that it may be valuable to employ voluntary control strategies to regulate the heightened cardiovascular reactivity displayed by patients at early stages of the spiral towards sustained hypertension. Significantly, the relationship between lability and control is also emphasized elsewhere in this volume, by Birbaumer (Chapter 21) in the case of vasomotor function and migraine headache. The argument is given further impetus by recent analyses of the physiological processes mediating the effects of behavioral treatments. Hemodynamic investigations of patients attempting to reduce blood pressure using biofeedback indicate that short-term responses are effected through modifications in cardiac output, while chronic reductions are sustained by decreases in total peripheral resistance (Messerli, Decarvalho, Christie & Frohlich, 1979). Patel (Chapter 4) found that the falls in blood pressure produced by borderline groups undertaking a multifactorial voluntary control regime were associated with decreases in plasma renin activity and aldosterone. Other studies also indicate that relaxation operates through autonomic pathways that include the cardiac sympathetic innervation (Davidson, Winchester, Taylor, Alderman & Ingels, 1979).

The remaining sections of this chapter detail some of the methods that may be useful in controlling cardiovascular reactivity. Techniques found valuable in the control of hemodynamic reactions to physical challenges such as static and dynamic exercise are described elsewhere in this volume (Johnston, Chapter 9). The present discussion is therefore confined to behavioral stressors. Yet even here there is a wide variation, both in the methods employed and in the nature of the behavioral demands that provoke cardiovascular adjustments.

A considerable amount of work has been devoted to the application of instructional techniques such as progressive muscle relaxation or autogenic training in threatening conditions. Psychophysiological studies of systematic desensitization and other behavioral interventions frequently include cardiac indices amongst the dependent variables, but these are assessed primarily for their association with anxiety or fear reduction. However, Connor (1974) showed that the phasic heart rate accelerations following threat of heatburn or loud tones were reliably attenuated in volunteers trained in muscle relaxation. A more paradoxical pattern was identified by

Goleman and Schwartz (1976) in their study of responses to a distressing movie in meditators and non-meditators. Meditators showed larger phasic electrodermal reactions to the threatening incidents in the movie, while also recovering more rapidly. The anticipatory heart rate increases were followed by greater bradycardias from meditators. Similarly, Lehrer and his associates (1980) found that volunteers trained in non-cultic meditation produced larger cardiac decelerations following the presentation of loud tones than those given progressive muscle relaxation.

These studies suggest that meditation may be associated with enhanced psychophysiological recovery following aversive stimulation, although not all reports have been positive (Puente & Beiman, 1980). Unfortunately, they are compromised by the lack of cardiovascular reactivity assessments before training. Selection factors and individual differences in reactivity may therefore account for the results, irrespective of treatment conditions. For even with random assignment, systematic pre-training differences in response magnitude may emerge between the comparatively small groups used in psychophysiological studies. In our laboratory, subjects are routinely allotted to groups according to pre-treatment cardiovascular reactivity, so that comparisons are based on similar cohorts.

These problems have been avoided in the investigations which have employed biofeedback-assisted relaxation as a method of modifying reactivity. However, in many cases a failure of response generalization precludes the application of such techniques to cardiovascular patients. For example, Gatchel and coworkers (1978) explored the use of frontalis EMG biofeedback in controlling reactions during the anticipation of electric shock. The method effectively prevented a rise in frontalis EMG, but had no impact on cardiovascular or electrodermal responses. Likewise, the EMG training procedures examined by Nielson and Holmes (1980) did not reduce skin conductance reactions to a distressing movie. Steptoe and Greer (1980) confirmed this lack of generalization in their study of relaxation assisted by skin conductance feedback; although subjects were able to reduce electrodermal reactions to demanding mental tasks, the concomitant heart rate increases were unaffected.

None of these programes of self-control have been primarily focused on cardiovascular reactivity. Nevertheless, the report by Patel (1975) suggests that such interventions may be valuable to the hypertensive patient. Patients with high blood pressure were administered a physical exercise (step-up) test and the cold pressor before and after training in biofeedback-assisted relaxation or control procedures. The blood pressure reactions to both challenges were reliably attenuated by the relaxation treatment, while recovery back to baseline was also more rapid. For example, the mean systolic pressure rise was reduced from 19.3 to 10.3 mmHg in the treatment group, but increased from 22.3 to 33.8 mmHg in controls.

Direct effects on cardiovascular reactivity may also be facilitated with biofeedback of heart rate, blood pressure or pulse transit time. Studies of heart rate feedback suggest that the technique has promise in this context, although results are not entirely consistent. Sirota and his associates (1974) showed that the cardiac accelerations produced in anticipation of electric shock could be modified with heart rate biofeedback, and the observation has been repeated elsewhere (DeGood & Adams, 1976). More recently however, Malcuit and Beaudry (1980) studied the impact of cardiac feedback on the elevated heart rates generated during short mental arithmetic trials. Biofeedback was no more effective than instructions alone in promoting heart rate reductions under these conditions. On the other hand, a series of experiments conducted by Shapiro and his colleagues indicate that cardiac reactions to the cold pressor test can be reliably modified with heart rate feedback, although instructions also play an important role (Reeves, Shapiro & Cobb, 1979).

All these investigations have been concerned with the generalization of voluntary control to stressful conditions. Training is carried out in undisturbed settings and subjects are then tested for their ability to regulate psychophysiological reactivity under stress. However, the arguments presented earlier suggest that it may be appropriate to train people in the challenging conditions themselves. Furthermore, voluntary control of reactions to active stressors, as opposed to the passive conditions (cold pressor, shock anticipation, etc.) generally studied, may be profitable in the case of essential hypertension.

DIRECT TRAINING IN CONTROL OVER CARDIOVASCULAR REACTIONS

A series of studies is currently in progress, in which methods of training during exposure to active stressors are being developed and compared. A major problem lies in the comparatively rapid habituation of cardiovascular reactions to repeated presentations of behavioral tasks; although large heart rate and blood pressure reactions may be elicited under a variety of conditions, these effects are not sustained. Consequently, it is difficult to distinguish genuine voluntary control from the reductions in pressor response that occur on increased familiarity with task demands. This effect can perhaps best be overcome by employing a number of different behavioral tasks, so that the patient has only limited exposure to each.

Since our clinical data from borderline hypertensives identified through mass screening is not yet complete, strategies for direct training in reactivity control will be outlined using data from healthy volunteers. The effects of relaxation training have been compared with biofeedback of pulse transit time. Pulse transit time is a continuous measure of hemodynamic function associated with arterial pressure, but also with the cardiac pre-ejection period (Steptoe, 1980). It is valuable in the present context, since both these variables are sensitive to active behavioral stressors, responding to increased discharge through cardiac sympathetic pathways. Moreover, pulse transit time accurately monitors phasic cardiovascular adjustments, so that continuous feedback can be provided during periods of extreme lability. The technique has already been used to a limited extent in the management of essential hypertension (Walsh, Dale & Anderson, 1977).

Since visual pulse transit time feedback has been employed, the behavioral tasks are presented auditorily in order to avoid peripheral sensory interference. Figure 1 outlines the results of one study, in which pulse transit time feedback and

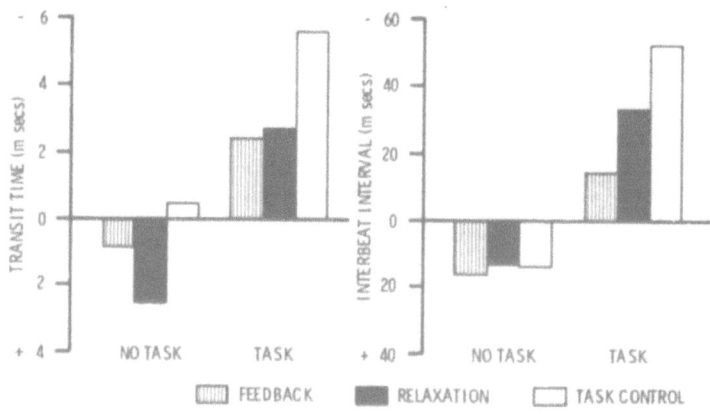

Figure 1. Mean pulse transit time (left) and interbeat
           interval (right) responses during training
           trials with and without simultaneous task
           performance. Data plotted as changes from the
           average baseline, with upward bars representing
           greater cardiovascular activation.

relaxation were administered both under conventional resting conditions, and during task performance (Steptoe & Ross, in press). Tasks included sequences of mental arithmetic and verbal reasoning problems, presented at fast rates. The two methods were compared with a control group that performed the tasks but did not attempt to modify reactivity; this condition controlled for the diminution in reactivity that occurs with habituation to task demands. Pulse transit time and interbeat interval data from four training sessions are summarized in the figure (shorter transit times and interbeat intervals reflect increased cardiovascular activity).

Different patterns were displayed in training trials with and without simultaneous task performance. In no-task/training trials, neither relaxation nor feedback produced impressive reductions of cardiovascular activity; this is consistent with the evidence discussed earlier in the chapter. Effects were more striking during task/training trials. All groups showed pressor reactions, with shortening of transit time and interbeat interval from intertrial basal levels. However, the cardiovascular reactions were attenuated with feedback and relaxation compared with control. In the case of transit time, the two strategies produced similar modifications, while the reduction of heart rate reactions was greater in the feedback condition. It should be noted that even task controls showed decreases in reactivity, since the pre-training responses to tasks averaged -12.8 and -87.3 msec in transit time and interbeat interval respectively. Nevertheless, the size of reductions was magnified with these training strategies.

The arguments presented in earlier sections suggested that cardiovascular hyperreactors may be especially suitable for this approach to management, since their labile systems may place them at high risk for permanent elevation in blood pressure. The association between reactivity and self-control was accordingly explored using correlational analyses. The relationship was assessed by calculating change scores between initial reactions and average transit time and interbeat interval levels in task/training trials; for example, if the initial interbeat interval reaction was -150 msec and averaged modification during training was -100 msec, an individual's change score would be +50 msec. These change scores were then correlated with initial reactions.

High negative correlation coefficients emerged from this analysis, since hyperreactors produced greater reductions in reaction during training. More importantly however, differences in the coefficients of regression or slope, were identified. Both feedback and relaxation groups showed larger regression coefficients than task controls, and the differences were statistically significant (Steptoe & Ross, in press). This means that hyperreactors benefited selectively from training with these techniques, in comparison with their counterparts in the control group. The interventions had little impact on those who do not show heightened cardiovascular reactivity on initial exposure to the stressors. The methods may therefore have particular value for patients who react to behavioral stressors with exaggerated blood pressure and heart rate changes.

CONCLUSIONS

This chapter has described an alternative to the conventional strategy for training in cardiovascular self-control. The technique of helping patients to cope with physiological reactions to behavioral stressors has not yet been evaluated in depth with clinical groups; however, relaxation training and stress management have been applied in analogous fashion as methods for coping with anxiety and subjective distress. It remains to be seen whether reactivity control is a useful adjunct to the behavioral approaches to essential hypertension outlined elsewhere in this volume.

Voluntary control of reactivity may have greater potential in the preventive rather than treatment mode. If individuals displaying excessive reactivity can learn to control this tendency, they may avoid the accumulation of transient pressor reactions that spiral towards sustained hypertension (Steptoe, 1981). This enterprise is difficult to evaluate, owing to the long time span involved, and the fact that only a limited proportion of borderline hypertensives progress to higher levels. Nevertheless, it offers a challenge for which behavioral medicine is uniquely equipped, since few physicians are sanguine about employing antihypertensive drugs under such circumstances.

NOTE

The research described in this chapter was supported by the Medical Research Council, United Kingdom.

REFERENCES

Birkenhager, W.H., & Schalekamp, M.A.D.H. Control mechanisms in essential hypertension. Elsevier, Amsterdam 1976.
Connor, W.H. Effects of brief relaxation training on autonomic response to anxiety-evoking stimuli. Psychophysiology, 1974, 11, 591-599
Davidson, D.M., Winchester, M.A., Taylor, C.B., Alderman, E.A. & Ingels, N.B. Effects of relaxation therapy on

cardiac performance and sympathetic activity in patients with organic heart disease. Psychosomatic Medicine, 1979, 41, 303-309.

DeGood, D.E., & Adams, A.S. Control of cardiac responses under aversive stimulation. Biofeedback and Self-Regulation, 1976, 1, 373-378.

Elder, S.T., Gamble, E.H., McAfee, R.D., & Van Veen, W.J., Conditioned diastolic blood pressure. Physiological Behavior, 1979, 23, 875-880.

Falkner, B., Onesti, G., Angelakos, E.T., Fernandes, M., & Langman, C. Cardiovascular responses to mental stress in normal adolescents with hypertensive parents. Hypertension, 1979, 1, 23-30

Fey, S.G., & Lindholm, E. Systolic blood pressure and heart rate change during three sessions involving biofeedback or no feedback. Psychophysiology, 1975, 12, 513-519

Folkow, B., & Rubinstein, E.H. Cardiovascular effects of acute and chronic stimulation of the hypothalamus defense area in the rat. Acta Physiologie Scandinavia, 1966, 68, 48-57

Forsyth, R.P. Blood pressure responses to long term avoidance schedules. Psychosomatic Medicine, 1969, 31, 300-309

Gatchel, R., Korman, M., Weis, C., Smith, B., & Clarke, L. A multiple response evaluation of EMG biofeedback performance during training and stress-induction conditions. Psychophysiology, 1978, 15, 253-258

Goleman, D.J., & Schwartz, G.E. Meditation as an intervention in stress reactivity. Journal of Consulting & Clinical Psychology, 1976, 44, 456-466

Guyton, A.C., Essential cardiovascular regulation - the control linkages between bodily needs and circulatory function. In: C.J. Dickinson and J. Marks (eds.), Developments in Cardiovascular Medicine. Lancaster: MTP, 1978.

Hofman, A., Boomsma, F., Schalekamp, M.A.D.H., & Valkenburg, H.A. Raised blood pressure and plasma noradrenaline concentrations in teenagers and young adults selected from an open population. British Medical Journal, 1979, 1, 1536-1538.

Julius, S., & Schork, M.A. Borderline hypertension - a critical review. Journal of Chronic Disease, 1971, 23, 723-754.

Lawler, K.A. Cardiovascular and electrodermal response patterns in heart rate reactive individuals during psychological stress. Psychophysiology, 1980, 17, 464-470

Lehrer, P.M., Schoicket, S., Carrington, P., & Woolfolk, R.L. Psychophysiological and cognitive responses to stressful stimuli in subjects practicing progressive relaxation and clinically standardized meditation Behavior Research Therapy, 1980, 13, 293-303.

Light, K.C., & Obrist, P.A. Cardiovascular reactivity
    to behavioral stress in young males with and without
    marginally elevated casual systolic pressure. Hypertension,
    1980, 2, 802-807.
Malcuit, G., & Beaudry, J.Voluntary heart rate lowering
    following   a   cardiovascular   arousing   task.   Biological
    Psychology, 1980, 10, 201
McKinney, M.E., Geller, D., Gatchell, R.J., Barber, G., Bothner,
    J. & Phelps, M.E. The production and generalization
    of large magnitude heart rate deceleration by contingently
    faded biofeedback. Biofeedback & Self-Regulation, 1980, 5,
    407-416
Messerli, F.H., Decarvalho, J.G.R., Christie, B., & Frohlich,
    E.D.Systemic hemodynamic effects of biofeedback in
    borderline hypertension. Clinical Science, 1979, 57, 437s.
Nielson, D.H., & Holmes, D.S. Effectiveness of EMG biofeedback
    training for controlling arousal in subsequent stressful
    situations. Biofeedback & Self-Regulation, 1980, 5, 235-248
Obrist, P.A. Cardiovascular Psychophysiology. New York: Plenum,
    Press, 1981.
Patel, C. Yoga and biofeedback in the management of "stress" in
    hypertensive patients. Clinical Science, 1975, 48, 141-154
Puente, A. & Beiman, I. The effects of behavior therapy,
    self-relaxation, and TM on cardiovascular stress responses.
    Journal of Clinical Psychology, 1980, 36, 291-295
Reeves, J.L., Shapiro, D., & Cobb, L.F.Relative
    influences of heart rate biofeedback and instructional set
    in the perception of cold pressor pain. In N. Birbaumer
    and H. D. Kimmel, (eds.) Biofeedback and Self-Regulation.
    Hillsdale, N.J.: LEA, 1979.
Shapiro, D., Tursky, B., & Schwartz, G.E. Control of blood
    pressure in man by operant conditioning. Circulation
    Research. 26-27, Suppl. 1: 27-32 (1970).
Sirota, A.E., Schwartz, G.E., & Shapiro, D. Voluntary control
    of human heart rate: effect on reaction to aversive
    stimulation. Journal of. Abnormal Psychology, 1974, 83,
    261-267.
Steptoe, A. Blood pressure. In: Techniques in psychophysiology,
    P. Venables and I. Martin, (eds.), Wiley, Chichester
    1980.
Steptoe, A. Psychological factors in cardiovascular disorders.
    London: Academic Press, 1981.
Steptoe, A. & Greer, K. Relaxation and skin conductance
    feedback in the control of rections to cognitive tasks.
    Biological Psychology, 1981, 10, 127-138
Steptoe, A. & Ross, A. Psychophysiological reactivity and the
    prediction   of   cardiovascular   disorders.   Journal   of
    Psychosomatic Research, 1981, 25, 23-31

Steptoe, A. & Ross, A. Voluntary control of cardiovascular
    reactions   to   demanding   tasks.   Biofeedback   &
    Self-Regulation, in press.
Walsh, P., Dale, A., & Anderson, D.E. Comparison of
    biofeedback of pulse wave velocity and progressive
    relaxation in essential hypertensives. Perceptual Motor
    Skills, 1977, 44, 839.

RELAXATION TREATMENT FOR HYPERTENSION

Guido L.R. Godaert

University of Utrecht
Department of Clinical Psychology
3512 JK Utrecht, The Netherlands

PREVIEW

One purpose of this conference is to bring various disciplines together. Thus for those scientists who are not yet too familiar with the behavioral sciences a number of basic data are included. For those who are acquainted with this discipline, a number of important areas for future research and practice of relaxation therapy with hypertensive patients are outlined. Since several good reviews have appeared (Seer, 1979; Agras & Jacob, 1979; Vaitl, this volume), the literature is not reviewed extensively. Rather, a few investigations are mentioned by way of illustration. Throughout the different sections -- the description of results, the section on the active ingredients of the techniques, methodologic considerations and trends for the future -- a central idea is that relaxation training is not an isolated procedure. The presence and clinical attention of the trainer, together with other factors such as lifestyle and adherence to medication or diet, are important for the results. The individual characteristics of the patient must also be considered. It is strongly suggested that future work should take into account data from psychophysiological investigations, in the search for criteria that will predict which patients will benefit from a particular relaxation program.

RELAXATION AND HYPERTENSION

Techniques of relaxation form one category of behavioral intervention for a variety of problems including anxiety, stress and psychosomatic complaints. Their purpose is to get the

responses of the patient who faces problematic situations under
control. Most attention has been paid to somatic and autonomic
responses; attempts have been made to lower muscle tension,
cardiac frequency and respiration amongst other variables.
Benson, Beary and Carol (1974) state that various techniques can
evoke the so-called relaxation response. This response, termed
the "trophotropic response" by others, is an integrated
hypothalamic pattern consisting of lowered sympathetic and
perhaps enhanced parasympathetic nervous system activity. As
far as physiological changes are concerned, we may see a reduced
frequency of respiration and pulse rate, a lowering of blood
pressure and muscular tension, and an increase in skin
resistance, as well as a synchronization of EEG activity. This
forms the counterpart of, and a protection against, the
ergotropic fight-or-flight response. According to Benson et al.
(1974), four elements are needed to produce this relaxation
response: (1) a mental device, or constant stimulus, the
purpose of which is to shift attention away from logical,
externally orientated thoughts; (2) a passive, receptive
attitude, disregarding distracting thoughts; (3) a lowered
muscular tension; (4) quiet surroundings.

These elements are found in various relaxation techniques,
and they are said to evoke the relaxation response in similar
ways. Davidson and Schwartz (1976) however have pointed out
that somatic, cognitive and attentional effects can be evoked to
different extents by the various techniques. We will return to
this issue later. Here it is sufficient to say that this
concept offers a general basis for the use of relaxation
techniques in the treatment of essential hypertension. For as
has been pointed out (Benson, Kotch & Crasweller, 1979; Julius,
1977), overactivation of the sympathetic nervous system, leading
to a loss of balance with the parasympathetic, plays an
important role in essential hypertension. We will therefore
briefly review the use of relaxation in essential hypertension,
distinguishing the different techniques. The effectiveness of
the procedures in terms of blood pressure response is discussed
by Vaitl in Chapter 14.

TECHNIQUES AND EFFECTS

Progressive Muscle Relaxation

Progressive muscle relaxation (P.R.) was originally
developed by Jacobson (1939). Different groups of muscles are
separately tensed and relaxed. The attention of the trainee is
directed towards feeling the difference between tension and
relaxation, bringing about proprioceptive differentiation.
Jacobson's original technique demands a long training period,

but nowadays shorter variants are used. Patients are taught the technique in about six sessions, and are then advised to practice by themselves.

Thus far, some ten studies have been done in which P.R. was used as a treatment for hypertension. Here, we concentrate on two of them, in order to illustrate the general design in this type of research. The comparison between these two studies also underlines the importance of the setting in which training takes place. Shoemaker and Tasto (1975) compared P.R. and biofeedback in three groups of hypertensives, matched on diastolic blood pressure. One group was offered taped P.R. instructions for six sessions. The trainees of the biofeedback group were shown their systolic and diastolic pressure written on a polygraph every ninety seconds. The blood pressure of the control group was measured for an equivalent number of sessions. The systolic blood pressure of the P.R. group decreased within sessions by an average of 6.8 mmHg, the diastolic by 7.6 mmHg. These reductions were significant, while in the biofeedback group only the decrease in diastolic pressure was significant. There was no decrease at all in the control group. Only the blood pressure of the P.R. group fell significantly from the first session (systolic 141, diastolic 91) to the last (systolic 131, diastolic 84). This proves that the P.R. technique is important, and that the decreases are not only due to attentional factors, or adaptation to the situation.

The second study was carried out by Brauer, Horlick, Nelson, Farquhar and Agras, (1979). One group received ten weeks of individually guided relaxation training, while a second group were trained mainly by way of audio cassettes for use at home. The third group received non-specific individual psychotherapy. Immediately after the training there was no difference between the groups. After six months however, the group who underwent therapist-conducted training showed significantly lower systolic and diastolic blood pressure than the two other groups. The authors conclude that the technique is certainly of importance: the group who had non-specific psychotherapy received as much attention as those with individual training, but they did not show any lasting fall of blood pressure. But in view of the results from the taped relaxation group, the technique in itself was not enough. The authors conclude: "thus it appears that both intensive therapist contact without relaxation techniques and relaxation techniques without regular therapist contact are ineffective in lowering blood pressure." The characteristics of the therapist also turned out to be important: one of the three therapists produced better results with his non-specific therapy patients than with his relaxation therapy patients. Thus, while Shoemaker and Tasto (1975) make it clear that technique is of

importance, the investigation of Brauer et al. draws attention
to the fact that effects are also modulated by other elements.

Autogenic Training (A.T.)

Autogenic training (Schultz & Luthe, 1969) consists of six
basic exercises. Central to these is concentration on a
physical feeling. Successive exercises focus on feelings of
heaviness, warmth and complete relaxation; this is done by way
of auto-suggestive sentences like: "my left arm feels heavy".
Other exercises concern the heart ("my heart beats quietly and
evenly") and respiration. The published reports on the use of
A.T. in essential hypertension are summarized by Vaitl (Chapter
14). In our own research (Godaert & Schreurs, see Reference
Note) we have used a form of A.T., asking clients to concentrate
on a feeling of warmth, heaviness, complete relaxation and quiet
respiration. Group A were given general information on
hypertension, followed by six weekly training sessions. They
practiced twice daily at home with the aid of an audio cassette,
and then discussed their experiences in the group. During these
group meetings more information was given and discussions were
held about the role played by salt, overweight, and stress in
hypertension; individual experiences were reported but there was
no formal group psychotherapy. Every patient received a
portable device to measure blood pressure twice a day. Group B
received the same treatment program, but without the extended
information and discussion of other topics; so A.T. was the main
subject of conversation. During the treatment period a control
group C came together weekly to have their blood pressure taken,
but were given no training. It was however suggested to them
that regular measurement could be of help in lowering the blood
pressure.

Six weeks after training, both groups A and B had
significantly lower systolic pressures compared to the
pre-treatment levels (from 141 to 135, and from 149 to 133 mmHg
respectively); the diastolic pressures of group B fell from 101
mmHg to 95 mmHg. After a follow-up of 18 months, six patients
from groups A and B and eight from group C could be reassessed.
Group B still showed significantly lower systolic pressures (for
this n=6 from 148 mmHg to 135 mmHg). The diastolic had not
risen (from 95 mmHg at the six-week follow-up to 96 mmHg during
a follow-up after a year) but it was no longer significantly
different from pre-treatment. After 18 months the control group
had the same blood pressure as on the first measurement
(systolic 143, diastolic 103). These results suggest that the
treatment program caused the blood pressure responses; in the
case of group B the systolic decrease was maintained even after
18 months. In group B, emphasis was placed on relaxation
training only; therefore the extra information and discussion in

group A was presumably of no help. Perhaps better results could be obtained when trainees are taught how to use the technique in specific situations where tensions usually build up (Suinn, 1975; Bloom & Cantrell, 1978). It is important that further research is undertaken with this form of relaxation training.

Meditation and Yoga

Various investigators use the form of Transcendental Meditation (TM) developed by Benson (1975). During every expiration you think of the number "one" or pronounce it subvocally; if distracted you simply go back to counting. By concentrating on this number (or another repetitive stimulus) conceptual thinking and worrying about problems is avoided, so attention is focused on relaxation. Despite the extensive use of this and other meditation procedures, results have been modest and equivocal (see Vaitl's chapter). Furthermore, the study by Seer and Raeburn (1980) suggests that mental repetition is not essential, and that regular interruption of daily activities may be sufficient to produce blood pressure responses.

Yogic breathing exercises have also been used with some success to treat essential hypertensives in uncontrolled studies (Datey, Desmukh, Dalvi, & Vinekar, 1969). The magnitude of effects suggests that such procedures might fruitfully be explored in more detail. Combined treatment packages have been extensively developed by Patel, whose work is described elsewhere in this volume. Two points should be made here however. Firstly, there is general agreement that this multimodal program is effective for lowering blood pressure (Reeves & Victor, in press). However, a methodological problem remains in that no evaluation of the various components of training has been carried out. The second important result was reported by Patel (1975a). Patients underwent a series of stress tests, and it turned out that the treated patients showed significantly shorter recovery times following blood pressure reactions to the tasks. This should be pursued in future research, in view of recent ideas on the importance of controlling blood pressure responses to stress in essential hypertensives (see Chapter 12 by Steptoe).

General Conclusion: Effects of Relaxation

The reported research has not as a whole been perfect in terms of the methodological considerations which will be outlined in the next sections. Despite this, there have been sufficient controlled studies to be able to say that relaxation techniques can be valuable when treating hypertension. Various authors stress that relaxation should be seen as an addition to pharmacological treatment and that medication should not be stopped too early (Agras & Jacob, 1979). We agree that relaxation cannot be used in isolation. Besides medication, however, there are many other factors (salt, weight, alcohol, way of life ...) of importance. We will return to this point later on.

ACTIVE INGREDIENTS

It is unlikely that all the elements of a treatment are equally important. Some elements could perhaps be left out for the sake of cheaper and more efficient intervention. Besides this, scientific curiosity has also stimulated a number of investigators to trace the "active ingredients" of treatment. Agras and Jacob (1979) list a number of components that can be found in all relaxation techniques to a greater or lesser degree. They suggest that the role of muscular tension is unclear, because of the equivocal results in the literature. According to them, the need for mental focusing is also unsupported. Seer (1979) shares this opinion. Task awareness and expectancy are not sufficient to lower blood pressure in view of the negative results from credible non-specific control treatments, but maybe they are necessary conditions. Regular training is perhaps the most important factor. In the study conducted by Brady, Luborsky and Kron, (1974) blood pressure rose after regular practice was discontinued; however, only a small number of patients were involved. In our own research (Godaert & Schreurs, see Reference Note) briefly mentioned earlier, we found that patients in the treatment groups maintained lower systolic pressures after an 18-month follow-up. Yet on inquiry it appeared that very few still practiced the relaxation exercises, and none did so regularly. Thus it is even uncertain that continued practice is absolutely necessary. There is a growing realization that the personal attention given by the therapist during training is significant (Brauer et al, 1979; Redmond, Gaylor, McDonald, & Shapiro, 1974). Patel too has stressed the importance of the doctor-patient relationship in her treatments. A further aspect which has been almost completely neglected in investigating relaxation techniques is the effect on other variables. We have observed that patients can become more alert to diet and the use of salt, and may adjust their way of life quite markedly; these aspects are often the hardest to quantify.

METHODOLOGY

I will first discuss a number of factors in experimental control and secondly consider the criteria for therapeutic success. It is of course essential to have a sound baseline and adequate control groups. Without a sufficiently long baseline, habituation to the measuring situation can be confounded with training effects. It is very difficult to give a clear recommendation for the frequency and duration of baseline determinations, since no systematic research has been carried out. But Seer (1979) recommends measurement on four occasions over a four-week period.

Control groups are used to distinguish between the specific and non-specific effects of a treatment program. Clinical attention, and the prospect that something will be done about the problem, may lower blood pressure considerably. A careful selection of patients is also required with a multifactorial problem such as hypertension. If experimental groups are not homogeneous, it is difficult to interpret the results of the investigation. This poses two problems. Clinically, it is often difficult to find enough patients from which to select. Moreover, it is by no means clear what the relevant selection criteria are. Age, duration of hypertension, the blood pressure level, and the use of medication are all important. It is conceivable, however, that medications with different working mechanisms show different interactions with behavioral training procedures. Taking this into account causes many more problems.

The duration of training and spacing of sessions are also important considerations. It takes some time to integrate a relaxation technique into the patient's daily routine; there must be an opportunity to practice for some time and then return to the trainer with questions and experiences. Spreading six and eight sessions over eight weeks gave satisfactory results in our investigations.

It was pointed out earlier that changes in diet can have a drastic influence on blood pressure. It is therefore important to register any such "side effects" of the training. Perhaps this is even more true for control of medication intake. Casual observations have taught us that patients follow their medication regimen more rigorously during the course of behavioral training. Such changes in drug compliance or adherence can be of great significance to the investigation.

The effects of a relaxation technique cannot be assessed solely by the magnitude of pressure reductions obtained in the training situation. Patients must maintain control over their blood pressure after the training session as well. A recent study (Agras, Taylor, Kraemer, Allen, & Schneider, 1980) shows promising results in this respect. Control outside the training site (clinic, consulting room) should be shown as well. Ambulatory measurements are valuable for this purpose; however, these devices are very expensive and problematic. Technological developments, including improvements in the portability of

apparatus and the use of other indices of the blood pressure
such as pulse transit time (Wesseling, DeWit, Snoeck, Weber,
Hyndman, Nijland, & Van der Hoeven, 1978), may lead to a
breakthrough in due course. But for the present it is
worthwhile having patients take their measurements at home or at
work. During our investigations, it appeared that the average
reductions of blood pressure achieved by patients at home
correlated moderately with our own control measurements outside
the training situation.

TRENDS FOR FUTURE INVESTIGATION

Three main areas of interest will be outlined, since
developments in these aspects will be of great importance to the
application of relaxation techniques and other behavioral
approaches to hypertension. On practical grounds, the search
for criteria that will predict which patients will benefit is a
major priority. The level of blood pressure and heart rate in
borderline hypertensives predict for later development of
sustained hypertension, so this group might qualify for
preventive treatment (Julius, 1979). The size of the elevation
in the blood pressure is also a good predictor of the expected
reduction during training (Agras & Jacob, 1979). Another
prognostic indicator may be the reactivity of blood pressure and
other cardiovascular responses (Steptoe, 1979). Obrist and his
colleagues (1979) have emphasized the importance of specifying
what reactivity refers to. Experimental tasks that demand and
allow an active response from the subject -- such as a choice
reaction time task -- evoke cardiovascular reactions that are
primarily mediated by beta-adrenergic sympathetic nervous system
activity. Such reactions occur more often in young students who
are normotensive themselves but one or both of whose parents
have hypertension, than in controls (Falkner, Onesti, Angelakas,
Fernandes & Langman 1979). From epidemiologic research it
appears that the chance of developing hypertension is
considerably higher when one or both parents have hypertension
(Feinleib, 1979). Thus when we combine these data, it seems
that there may be a connection between reactivity and the risks
of developing hypertension. However, this possibility must be
treated cautiously at present, since only prospective research
can provide definitive answers (see chapter by Steptoe, this
volume).

A second aspect concerns the development of treatment
programs directed towards several components of the problem
simultaneously. This is a logical development in view of the
multifactorial nature of hypertension. Patel has already
demonstrated important reductions of blood pressure with the
help of a combined biofeedback-relaxation-information package.
Efforts should be directed at standardizing methods and making

them usable for other researchers, so that replications can be undertaken in other treatment centers. Perhaps the characteristics of the individual patient can also be taken into account at this stage. It is clear that the starting point for psychophysiological disregulation, resulting in elevated blood pressure, can differ widely between individuals. Schwartz (1977), for instance, has distinguished four stages at which disregulation can take place. Although much of this is hypothetical at present, it ties in with clinical experience. It is important to determine where the most important disturbances are located in an individual, so as to adjust treatment appropriately (see chapter by Johnston, this volume). Thus, on the level of environmental demands, the specific conditions under which the blood pressure elevation - or behavior related to it -- occur, have to be identified. An interesting development in this respect is the application of coping mechanisms to cardiovascular problems (Pittner & Houston, 1980). Relaxation can also be considered as a method of coping with the psychophysiological disregulation of hypertension. This may not be the optimal treatment for an individual patient however. Different forms of relaxation may be appropriate, perhaps with selective emphasis on muscular or cognitive relaxation (Davidson & Schwartz, 1976).

The final development concerns the link between treatment and psychophysiological research. The importance of blood pressure reactivity in prediction and evaluation has already been emphasized. It seems equally essential to investigate the influence of neurogenic (sympathetic) activation on other components of cardiovascular regulation. Zanchetti has already done important work of this type on the neurogenic influences upon kidney functioning (Zanchetti, Stella, Bocelli & Mancia, 1979); in view of the central part the kidney plays in the blood pressure regulation (Guyton, Coleman, Cowley, Manning, Norman, & Ferguson, 1974), the effects of relaxation techniques will also have to be investigated.

CONCLUSIONS

The tone of this review on the contribution of relaxation techniques to treatment of hypertension has been generally positive. This is due both to our evaluation of the published literature, and consideration of the relationships between the psychophysiological characteristics of high blood pressure on the one hand and of relaxation techniques on the other. However, it is still necessary to integrate relaxation techniques into a comprehensive treatment approach, and to be flexible on the level of the individual patient; only when this is done will relaxation techniques yield their maximum profits.

REFERENCE NOTE

Godaert, G. and Schreurs, P., manuscript in preparation,
    Group-instruction relaxation treatment for essential
    hypertension.

REFERENCES

Agras, S. & Jacob, R. Hypertension. In O. Pomerleau & J. Brady,
    (eds.) Behavioral Medicine. Baltimore: Williams and
    Wilkins, 1979.
Agras, W. S., Taylor, C. B., Kraemer, H. C., Allen, R. A., &
    Schneider, D. S. Relaxation training: Twenty-four hour
    blood pressure reduction. Archives of General Psychiatry,
    1980, 37, 859-863.
Benson, H. The relaxation response. New York: Morrow, 1975.
Benson, H., Beary, J. F., & Carol, M. P. The relaxation
    response Psychiatry, 1974, 37, 37-46.
Benson, H., Kotch, J. & Crasweller, K. D. Stress and
    hypertension: Interrelations and Management. In G. Onesti &
    A. N. Brest (eds), Hypertension Mechanisms, Diagnosis and
    Treatment. Philadelphia: Davis Cy., 1979.
Bloom, L., & Cantrell, D. Anxiety management training for
    essential hypertension in pregnancy. Behavior Therapy,
    1978, 9, 377-382.
Brady, J. P., Luborsky, L., & Kron, R. E. Blood pressure
    reduction in patients with essential hypertension through
    metronome conditioned relaxation. Behavior Therapy, 1974,
    5, 203-209.
Brauer, A. P., Horlick, L., Nelson, E., Farquhar, J. W., &
    Agras, W. S. Relaxation therapy for essential
    hypertensives: A Veterans Administration Outpatient Study.
    Journal of Behavioral Medicine, 1979, 2 (1), 21-29.
Datey, K. K., Desmukh, S. N., Dalvi, C. P., & Vinekar, S. L.
    Shavasan: a yogic exercise in the management of
    hypertension. Angiography, 1969, 20, 325-333.
Davidson, R. J., & Schwartz, G. E. The psychobiology of
    relaxation and related states: a multiprocess theory.
    In D. I. Mostowsky (Ed.), Behavior Control and
    Modification of Physiological Activity. Englewood Cliff,
    N.J.: Prentice-Hall, 1976.
Falkner, B., Onesti, G., Angelakos, E. T., Fernandes, M., &
    Langman, C. Cardiovascular response to mental stress in
    normal adolescents with hypertensive patients.
    Hypertension, 1979, 1,(1).
Feinleib, M. Genetics and familial aggregation of blood
    pressure. In Hypertension: Determinants, Complications and
    Interventions. New York: Grune and Stratton, 1979.
Guyton, A. C., Coleman, T. G., Cowley, A. W., Manning, R.D.,
    Norman, R.A., & Ferguson, J.D. A systems analysis approach

to understanding long range arterial blood pressure control and hypertension. Circulation Research, 1974, 35, 159–176.

Jacobson, E. Variation of blood pressure with skeletal tension and relaxation. Annals of Internal Medicine, 1939, 12, 1194–1212.

Julius, S. Borderline Hypertension: Significance and Management. In G. Onesti & A. N. Brest, (Eds.) Hypertension: Mechanisms, Diagnosis, and Treatment. 1979.

Patel, C. Yoga and biofeedback in the management of "stress" in hypertension patients. Clinical Science and Molecular Medicine, 1975, 48 (suppl.), 171–174.

Patel, C. Biofeedback-aided relaxations and meditation in the management of hypertension. Biofeedback & Self-Regulation, 1977, 2, 1–41.

Pittner, M. S., & Houston, B. K. Response to stress, cognitive strategies, and Type A behavior pattern. Journal of Personality & Social Psychology, 1980, 39, 147–157.

Redmond, D. P., Gaylor, M. S., McDonald, R. M., & Shapiro, A. P. Blood pressure and heart rate response to verbal instructions and relaxation in hypertension. Psychosomatic Medicine, 1974, 36, 285.

Reeves, J., & Victor, R. Behavioral strategies in hypertension. In P. A. Boudewyns and F. J. Keefe, (eds.) Behavioral Medicine in General Medical Practice. Menlo Park, CA: Addison Wesley, in press.

Schultz, J.H. and Luthe, W. Autogenic therapy: I. Autogenic methods. New York: Grune and Stratton, 1969.

Schwartz, G. E. Psychosomatic disorders and biofeedback: a psychobiological model of disregulation. In Psychopathology: Experimental Models. San Francisco: Freeman and Cy, 1977.

Seer, P. Psychological control of essential hypertension: A review of the literature and methodological critique. Psychology Bulletin, 1979, 86, 1015–1043.

Seer, P., & Raeburn, J. M. Meditation of training and essential hypertension: a methodological study. Journal of Behavioral Medicine, 1980, 3, 59–71.

Shoemaker, J.E., & Tasto, D.L. The effects of muscle relaxation on blood pressure of essential hypertensives. Behavior Research & Therapy, 1975, 13, 29–43.

Steptoe, A. Cardiovascular reactivity and its management with psychological techniques, In D.J. Oborne, M.M. Gruneberg, and D.R. Eiser, (eds.) Research in psychology and medicine. London: Academic Press, 1979.

Steptoe, A., & Ross. Psychophysiological reactivity and the prediction of cardiovascular disorders. Journal of Psychosomatic Research, 1981, 25, 23.

Suinn, R. Anxiety management training for general anxiety. In R. Suinn & G. R. Weigel (eds.) The Innovative Psychological Therapies. Harper and Row, 1975.

Wesseling, K.M., DeWit, B., Snoeck, B., Weber, J.A.P., Hyndman, B.W., Nijland, R., & Van de Hoevan, G.M.A. An implementation of the Penaz Method for measuring arterial blood pressure in the finger and first results of an evaluation. Progress Reports 6, Utrecht: Inst. Med. Phys. 1978.

Zanchetti, A., Stella, A., Bacelli, G., & Mancia, G. Neural influences on kidney function in the pathogenesis of aurerial hypertension. In G. Onesti, and C.R. Klimt, (eds.) Hypertension: Determinants, Complications, and Interventions. New York: Grune and Stratton, 1979.

THE EFFECTIVENESS AND COST-BENEFITS OF BEHAVIORAL METHODS IN THE TREATMENT OF ESSENTIAL HYPERTENSION

Dieter Vaitl and Harald Lachnit

Department of Psychology
University of Giessen
Giessen, Free Republic of Germany

Hypertension is a disease which has challenged scientific and medical efforts in this century in the same way in which infectious diseases have done in the past century (Weiner, 1979). Only an interdisciplinary approach can advance our understanding of this life-threatening risk factor, and may provide adequate methods for its control and prevention. Shapiro (chapter   in this volume) has described the numerous efforts undertaken by medical and behavioral scientists to achieve this goal. The behavioral medicine approach is primarily focused on the multi-modal investigation and management of essential hypertension. Although this strategy appears to be a promising step forward, the heterogeneity of the hypertensive process itself is still one of the most important factors impeding the integration of the diverse facts, theories, and methodologies which have been elaborated by various disciplines up to now. Prior to speculating about the potential usefulness of the behavioral medicine approach to hypertension, it must be decided if and to what extent behavioral methods contribute to reducing high blood pressure (BP).

From the great variety of non-pharmacological methods employed for high blood pressure control (for a review see Agras & Jacob, 1979), one group of behavioral treatments will be considered with respect to its clinical effectiveness--the relaxation techniques. These techniques are widely used in psychotherapeutic settings with the goal of improving patients' individual strategies for coping with stress-arousing situations. None of these methods was originally developed for blood pressure control. Very recently, however, they have been applied as additional tools in the management of essential hypertension, under the assumption that essential hypertension is a stress-related disorder. Whenever relaxation techniques are discussed as a potential treatment of high blood pressure, it must be kept in mind that this global assumption lacks solid empirical evidence (for a discussion cf. Shapiro, Benson, Chobanian, Herd, Juliu, Kaoplan, Lazarus, Ostfeld & Syme, 1979). The justification for the use of relaxation techniques in the treatment of hypertension can therefore only be sustained pragmatically. According to this standpoint all techniques are legitimate if they can contribute to a reduction of high blood pressure within the range of acceptable cost-benefit relations.

BIOFEEDBACK CONTROL OF HYPERTENSION

Various methods have been applied to inform patients either directly about changes in blood pressure or indirectly about other physiological processes associated with cardiovascular responses (muscle tone, heart rate, pulse wave velocity, electrodermal and vasomotor activity). There are numerous technical reports available on biofeedback devices for blood pressure control (Tursky, 1974; Steptoe, Smulyan & Gribbin, 1976; Miller & Dworkin, 1977; Shapiro, Greenstadt, Lane & Rubinstein, 1981). For the interested reader, comprehensive introductions to the theory and practice of biofeedback can be found in the following works (Legewie & Nusselt, 1975; Schwartz & Beatty, 1977; Yates, 1980). The current literature on feedback-aided blood pressure control which has been reviewed very extensively permits an up-to-date evaluation of the general usefulness of these methods in the field of behavioral medicine (Frumkin, Natha, Prout & Cohen, 1978; Seer, 1979; Yates, 1980; Vaitl, 1981).

Although it has been shown that the blood pressure of normotensives and hypertensives can be brought under stimulus control by means of direct blood pressure feedback to some extent, in the majority of studies the clinical effectiveness has not been sufficiently demonstrated (see also Shapiro's chapter in this volume). The positive results found in single case reports (i.e. anecdotal case report, systematic case study, single-subject experiment) have not been replicated when a controlled group outcome design was applied (for reviews cf. Seer, 1979; Vaitl, 1981).

Apart from efforts to treat hypertension with direct blood pressure feedback, there have been several attempts to use other forms of biofeedback. The assumption underlying the indirect approach is that high blood pressure may be lowered by reducing either sympathetic arousal or the level of muscle tension. Sedlacek and coworkers (1978) employed EMG and temperature feedback methods on 30 hypertensives and found a reduction in blood pressure from 144/95 mmHg to 130/83 mmHg. These results proved to be relatively stable, as shown by a follow-up four months after the end of training (136/85 mmHg). Love and his associates (1974) treated 40 hypertensives with a combination of frontal EMG feedback and other relaxation techniques. For the 23 patients who were available for an eight-month follow-up, both systolic and diastolic blood pressures were further reduced.

However, in other studies using EMG feedback alone, no corresponding drop in blood pressure could be found (Frankel, Patel, Horowtiz, Friedwald & Gaardner, 1978; Surwit, Shapiro & Good, 1978; Blanchard, 1979). Patel and her group chose a multi-modal treatment approach in which EMG and GSR feedback served to assist other relaxation techniques (i.e. breathing exercises, yoga). The positive results reported by this group are impressive (see chapter by Patel in this volume). She has shown that the combination of relaxation training with feedback is more effective in reducing blood pressure than mere instructions to relax passively without feedback (Patel & North, 1975). However, the question which remains unanswered is whether active relaxation training alone is as effective as active relaxation training aided by various forms of feedback.

The following conclusions may be drawn from the present literature:
  a. Despite encouraging results of single-case studies, the methodological differences and deficiencies of controlled group studies do not permit the firm conclusion that direct feedback of blood pressure changes is an adequate tool for treating hypertension (for a discussion of the methodological flaws see Frumkin et al., 1978; Seer, 1979).
  b. It is still questionable whether the blood pressure reductions produced in the biofeedback laboratory may be generalized to daily life situations.
  c. Feedback-aided bodily relaxation appears to be an adequate strategy in the treatment of hypertension. The form of feedback, however, which may serve as a helpful adjunct has not yet been established empirically.
  d. It may be possible that the permanent control of feedback signals, as usually required in a biofeedback setting, may counteract bodily relaxation.

Despite these preliminary conclusions, biofeedback techniques contain some positive aspects when employed in the control of hypertension which are worth considering. In general, direct blood pressure feedback or other forms of feedback used in blood pressure control provide patients with information on the arousing effects of individual mental and somatic activities. By means of feedback they can learn to

inhibit these processes and to find appropriate alternatives.
Moreoever, patients can be taught that there is a close
relationship between specific behavioral stimuli and
cardiovascular responses. Thus, biofeedback may be introduced
as the first step of a multi-modal treatment program in order to
convince patients to pay particular attention to such arousing
internal or external stimuli.

RELAXATION TECHNIQUES

Compared to biofeedback, relaxation studies are
characterized by a higher quality of outcome control (Seer,
1979)--the samples of treated hypertensives are larger, changes
in blood pressure are reported more often for periods outside
the clinical setting, and the follow-up periods are usually
longer. Therefore, the depressor effects of these methods can
be assessed more accurately. On the other hand, the range of
the reductions in blood pressure obtained between the individual
studies is considerable. Reductions vary between maximal
effects of 37/23 mmHg (Datey, Deshmukh & Vinekar, 1969) and
minimal effects of only 7/4 mmHg (Benson, Rosner, Marzetta &
Klemchuck, 1974 a,b).

None of the relaxation techniques employed was used in its
original form and in its entirety in the treatment of
hypertension. Instead, modifications of the techniques
(simplifications, standardizations, etc.) are always offered.
In the following section the reductions in blood pressure which
have been attained with the individual methods shall be briefly
described (see also Godaert's chapter in this volume).

Hypnosis

Investigations aimed at altering blood pressure using
hypnosis either alone or in combination with other procedures
are sparse. Deabler, Fidel and Elder (1973) found that
medicated and unmedicated hypertensives treated by hypnosis
combined with skeletal-muscular relaxation lowered their blood
pressure after eight sessions over a four-day period. Control
subjects did not change their blood pressure. No data are
reported which indicate whether any between-session or long-term
decreases in tonic blood pressure levels had occurred. Friedman
and Taub (1977) attempted to explore the clinical effectiveness
of hypnosis, direct blood pressure feedback, and combinations of
both. They found that in a one-month follow-up, hypnosis alone
was more effective than biofeedback training alone. The
hypnosis group showed a blood pressure reduction of 11/7 mmHg
from baseline (143/93 mmHg) to follow-up. The combination of
both techniques was as ineffective as simply taking repeated
measurements. Reductions in blood pressure by means of hypnosis

were achieved immediately after training had begun, and were probably stabilized by daily practice at home after these patients had learned the technique of "autohypnosis." In comparison to the other groups, this group was highly susceptible to hypnotic suggestions (according to the Stanford Hypnotic Susceptibility Scale, Form A).

Autogenic Training

The principles of autogenic training are outlined by Godaert (Chapter 12). A beneficial influence on labile hypertension was first reported by Luthe (1963). In a single-group outcome study Klumbies and Eberhardt (1966) were able to show that after a training period of four months, the blood pressure of 26 essential hypertensives (pre-treatment BP: 165/100 mmHg) was lowered to an average level of 130/80 mmHg. Patients were required to practice only two of the six standard exercises ("heaviness" and "warmth"). The authors maintained that reductions in blood pressure were mainly brought about by the general bodily relaxation and emotional calmness. In terms of compliance it is worth noting that only 26 out of 89 patients who began this training adhered to the treatment. Another study failed to show any significant depressor effect of autogenic exercises when combined with biofeedback and progressive relaxation (Frankel et al., 1978).

Lantzsch and Drunkenmoller (1975) examined the hemodynamic changes occurring during the first two autogenic exercises ("heaviness" and "warmth") in ten hypertensives. As in normotensives, they found a reduced cardiac output which was partially accompanied by an increase in the total peripheral resistance. The vasodilatation of renal and splanchnic arteries provoked by a decreased sympathetic outflow apparently elicited this compensatory mechanism.

Meditation

The term "meditation" includes various forms of mental exercise derived from esoteric disciplines and tailored to the western culture. They were primarily devised to develop non-specific benefits for lifestyle and "well- being." For experimental and clinical purposes, Benson and his coworkers have devised a simplified and standardized meditative technique based on transcendental meditation and zen, specifically focusing on relaxation (Benson, Beary & Carol, 1974c).

Single group studies have been carried out by several authors (Benson et al., 1974a, b; Blackwell, Hensenson, Bloomfield, Magenheim, Gartside, Nidich, Robinson & Zigler, 1976; Pollack, Weber, Case & Laragh, 1977). They indicate that

daily practice of meditation or similar exercises may produce a
reduction of blood pressure in a small number of hypertensives.
However, the changes in blood pressure are minimal and the
absence of controls limits the contribution of these studies.

Surwit et al. (1978) compared the depressor effects of the
relaxation response methods developed by Benson with two
different biofeedback methods (EMG feedback and a combination of
blood pressure and heart rate feedback) on a group of 24
borderline hypertensives. None of the three groups showed
significant reductions in blood pressure in the follow-up (six
weeks and one year after the end of training). The authors
attributed the lack of success to the low initial values of
blood pressure during baseline periods.

The investigation of Seer and Raeburn (1980) is considered
one of the best controlled group outcome studies. They compared
the reduction in blood pressure of three different
groups--meditation group, relaxation group (= placebo controls),
and waiting list control group. Forty-one unmedicated essential
hypertensives were randomly assigned to one of the above
treatments. After a training period of 13 weeks the blood
pressure of the meditation group was reduced by 5/7 mmHg
(baseline: 152/103 mmHg), whereas the relaxation group reduced
its blood pressure by 3/8 mmHg (baseline: 147/100 mmHg); on the
other hand, the blood pressure rose for the waiting list
controls by 2/3 mmHg (baseline: 150/102 mmHg). During the
follow-up (12 weeks after the end of training) the blood
pressure was 2/7 mmHg lower than baseline for the meditation
group, whereas it was 8/12 mmHg lower for the relaxation group.

The slight differences in diastolic blood pressure between
these two groups is hardly surprising. A plausible explanation
may be due to the fact that meditating does not essentially
differ from just sitting in a relaxed fashion and interrupting
the daily activities. This is probably true for all relaxation
techniques which can be practice at home.

Yoga

From the diverse yoga techniques, breathing exercises are
mainly employed in the treatment of hypertension (e.g.
regulation of breathing pattern, abdominal breathing,
"Shavasan").

Datey et al. (1969) trained 47 hypertensives (essential,
renal, and arteriosclerotic) in the technique of "Shavasan"
(posture control, slow rhythmic diaphragmatic breathing,
regulation of pauses before and after inspiration, paying
attention to the nostrils). Successful reduction of

hypertension was achieved in 52% of the sample studied. In some cases the consumption of antihypertensive drugs could be markedly reduced. Ten unmedicated hypertensives were able to reduce their blood pressure from 184/109 mmHg (pre-treatment) to 147/86 mmHg after treatment. A second group consisted of 22 patients whose blood pressure was pharmacologically maintained at an average level of 137/86 mmHg. After the Shavasan training the dosage of antihypertensive drugs could be reduced by 32% for 13 patients without a consequent increase in blood pressure. For a third group of patients who blood pressure could not be sufficiently controlled by drugs a decline in drug usage and in mean blood pressure could be achieved.

Using a similar technique of breathing control, Bali (1979) was able to demonstrate that the blood pressure of nine hypertensives was reduced compared to nine matched controls. The blood pressure control acquired by regular practice of these techniques was maintained over a 12-month follow-up (pre-treatment: 149/97 mmHg; follow-up: 137/88 mmHg).

Patel and coworkers have also incorporated yoga exercises into their treatment package. Their results are described in detail in Chapter 3.

From these reports, we may conclude that combined methods procedures bring about greater decreases in blood pressure than just resting, attention, no treatment, or instruction just to relax (Patel & North, 1975). However, it is impossible to differentiate the relative effectiveness of feedback, meditation, and relaxation as part of the "treatment package." Another central component of the combined treatment is most likely the way in which patients are taught to integrate the learned relaxation response into their daily activities.

Progressive Relaxation

Jacobson was one of the first researchers who referred to the relation between muscle relaxation and blood pressure reduction. Shoemaker and Tasto (1975) examined this relation using a small sample of hypertensives. The observed reductions in blood pressure (7/8 mmHg) after two weeks of progressive relaxation exercises were greater than those achieved by both a group treated by direct blood pressure feedback and a waiting list control group. Taylor and coworkers (1977) compared the blood pressure reducing effects of progressive relaxation with drug and psychotherapeutical treatments. After termination of the training period, significant reductions in blood pressure could only be found for the relaxation group (14/5 mmHg). No difference in the blood pressure changes remained between the groups during the follow-up.

Summary

     A review of the literature on the various methods of bodily
relaxation shows that blood pressure reduction has been reported
for patients with borderline and/or mild hypertension (Stage I
and II).   When carefully examined according to the same
standards used for evaluating clinical pharmacological studies
of drugs for hypertension, the relaxation studies are mostly of
Phase I type (i.e. short-term treatment effects in a small
number of subjects).   Controlled trials involving a comparison
with established effective agents (Phase II) are rare (Shapiro
et al., 1978).   Furthermore, most of the clinical outcome
studies available up until the present have methodological
faults, many of which can be considered quite serious.   These
deficiencies   include   patient   selection,   concurrent
pharmacological treatment, baseline assessment of blood
pressure, control groups, length of training, follow-up period,
home practice, and lifestyle changes (for details see Seer,
1979; Vaitl, 1981).

     Despite the dearth of sufficiently controlled studies, it
has been shown that relaxation techniques in general are
superior to mere placebo therapy in reducing blood pressure
(Jacob, Karemer & Agras, 1977).   Since it is not yet possible to
differentiate which form of relaxation is most valuable in
controlling blood pressure, one can consider them as a class of
different procedures primarily aimed at producing the same
reaction pattern--the relaxation response.   If one agrees with
this general view, one principal question remains--Are the
depressor effects of relaxation techniques as a form of
behavioral treatment comparable to the pharmacologically
produced changes in blood pressure in Stage I and II
hypertensives?

COMPARISON OF BEHAVIORAL (BIOFEEDBACK/RELAXATION) AND DRUG
TREATMENT

     To clarify this question, data were gathered from published
reports.   The depressor effects on diastolic blood pressure
(DBP) only were considered, as this is of greater clinical
significance for determining the risk profile of a given
population.   Thus, the changes in DBP, as reported by
biofeedback and relaxation studies, were contrasted with the
changes in DBP achieved by the multicenter high blood pressure
drug intervention programs for mild hypertension (DBP $\leq$ 110
mmHg).   For this purpose, separate linear regression analyses
were made for both samples of studies to determine the mean
changes in DBP in relation to the pre-treatment blood pressure
levels.   Furthermore, the slopes of the regression lines
obtained for biofeedback/relaxation and drug treatment were

statistically compared with the ideal regression ($y' = 90-x$), which represents blood pressure reductions resulting in DBP of 90 mmHg.

Criteria for the selection of biofeedback/relaxation studies were: a) more than six subjects studied; b) more than three blood pressure reading before treatment entry (average DBP 90-110 mmHg); c) follow-up period of at least one month; d) information available on concurrent drug treatments; e) absolute rather than relative change scores in blood pressure.

According to these criteria, data on 19 groups treated by biofeedback/relaxation were selected (Klumbies & Eberhardt, 1966; two groups of Datey et al., 1969; Patel, 1973 a,b; Benson et al., 1974 a,b; two groups of Patel & North, 1975; Blackwell et al., 1976; Taylor, Farquhar, Nelson & Agras, 1977; Pollack, 1977; three groups of Friedman & Taub, 1977; Sedlacek & Cohen, 1978; Bali, 1979; two groups of Seer & Raeburn, 1980).

Criteria for selection of multicenter antihypertensive drug studies were: a) more than three blood pressure readings before treatment entry; b) controlled regimen with antihypertensive drugs; c) documentation of cohorts with mild hypertension (DBP 110 mmHg); d) documentation of changes in DBP during the first year of intervention.

Twenty-one sets of data from the following drug studies were included: the Hypertension Detection and Follow-up Program (1979) (selected subgroups: Stratum I (DBP 90-104 mmHg), Stratum II (DBP 105-114 mmHg), Stepped Care, Referred Care, sex, race); the Australian National Blood Pressure Study (Reader, 1979) (selected subgroups: entering DBP 94, 99, 104, and 111 mmHg); the Coe, Norton, Oparil, Tatai & Pullman, study (1977) (selected subgroups: entering DBP 110 mmHg, treatment designed by a computer program or by physician); the Smith study (1977) (selected subgroup: active drug).

The mean changes in DBP were: 9.0 mmHg for biofeedback/relaxation (pre-treatment DBP = 97.9 mmHg) and 8.7 mmHg for drug intervention (pre-treatment DBP = 101.8 mmHg), respectively. The results of the linear regression analyses are presented in Figure 1. The regression lines for the DBP reduction in both samples of studies are parallel (relaxation: $y' = 65.6 - 0.76x$; drugs: $y' = 50.2 - 0.58x$) and do not differ significantly from each other. Thus, for DBP the degree of pressure reduction likely to result from treatment can be predicted on the basis of the initial blood pressure both for relaxation and drugs. However, the two treatments are not equivalent when compared with the ideal regression line ($y' = 90-x$).

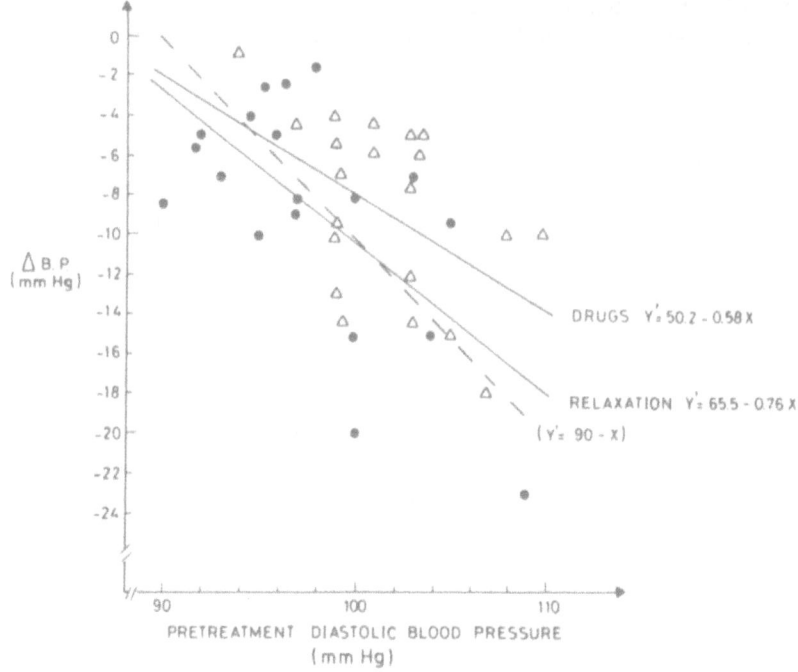

Figure 1.   Changes in diastolic blood pressure during bio-
feedback/relaxation (●) and drug treatment (▲)
as reported by clinical outcome studies.   Each
point represents the relationship between the
reduction in blood pressure during treatment
(BP mmHg) and pretreatment blood pressure

            - - - - = ideal regression line (y' = 90-x)
                      (for explanation see text)

The drug treatment differs significantly from it (t=1.81; df=19,
p≤ .05, one-tailed), whereas the biofeedback/relaxation techniques
do not (t = 1.08; df = 17, p > 05, one tailed).

     According to these results, the depressor effects on DBP
produced by biofeedback and/or relaxation techniques appear
fairly promising.   They can at least be assumed to be adequate
procedures for reducing mild hypertension when compared to the
mean changes in DBP achieved by drug intervention.   However,
this preliminary finding cannot be considered to be proof of
their   general   effectiveness.   The   reasons   limiting   the
generalizability of the present result are as follows:
     a.   The relaxation-induced changes in DBP are predominantly
short-term effects.   The follow-up data range from one month to
one year (mean = three months).   In contrast, drug intervention
studies   as   used   for   the   above   statistical   analysis   include
one-year follow-up periods.

b. The most important factor concerns the duration of blood pressure reduction. Usually, antihypertensive drug treatments adequately administered result in long-term blood pressure changes. Compared to these changes in blood pressure levels, the blood pressure responses induced by relaxation techniques or biofeedback are most likely restricted to training and/or to periods in which blood pressure is measured. It can therefore be argued that the observed blood pressure changes are merely conditioned cardiovascular responses without any beneficial impl4act on blood pressure level over longer time periods.

c. Since patients were additionally under medication in several biofeedback/relaxation studies, the marked falls in blood pressure may also have been produced by the combination of both treatments.

d. Patients participating in a relaxation study are often highly motivated to control their blood pressure by alternative methods. Thus, the observed reduction in DBP may also have been brought about by non-specific changes in patients' lifestyles.

e. In a statistical comparison based only on those subgroups of drug studies in which the patients' adherence to drug regimen were adequately controlled (e.g. in Stepped Care groups of the Hypertension Detection and Follow-up Program), the relaxation techniques would certainly turn out to be inferior to the drug treatment.

CONCLUSIONS

For these reasons, the contributions of biofeedback and relaxation to the behavioral management of mild hypertension are limited, and they should not be recommended as alternatives to pharmacological intervention. However, it seems reasonable to use these techniques as adjuncts to medication, due to their potential blood pressure reducing effects. In contrast to biofeedback, which requires a sophisticated technology, the relaxation techniques are easy to apply in a clinical setting and may help to ameliorate emotional stress in general. Furthermore, no contraindications or side effects have been reported for hypertensive patients.

A central component of all relaxation techniques consists of regular practice. Only when this is maintained can one expect positive effects (cf. Klumbies & Eberhardt, 1966; Elder & Eustis, 1975; Patel & North, 1975). This probably accounts for the fact that the blood pressure feedback methods are relatively ineffective--patients are not able to practice at home, as they are dependent on the apparatus.

Discussions of relaxation methods must end with the question--What type of relaxation technique is able to motivate

the patient to regular training? Patients' compliance can be
viewed as a focal point of therapeutic consideration in the same
way in which it is a crucial factor for pharmacological blood
pressure treatment. Only when the compliance to drug and
relaxation regimen has been established is it possible to
discuss whether or to what extent drug treatments for mild
hypertension can be replaced by relaxation techniques.

REFERENCES

Agras, S., & Jacob, R.G. Hypertension. In O.F. Pomerleau & J.P.
    Brady, (eds.) Behavioral medicine: Theory and Practice.
    Baltimore: Williams & Wilkins, 1979.
Bali, L.R. Long-term effects of relaxation on blood pressure
    and anxiety levels of essential hypertensive males: A
    controlled study.Psychosomatic Medicine, 1979, 41, 637-645.
Benson, H., Rosner, B.A., Marzetta, B.R., & Klemchuck, H.M.
    Decreased blood pressure in pharmacologically treated
    hypertensive patients who regularly elicited the relaxation
    response. Lancet, 1974a, 1, 289-291.
Benson, H., Rosner, B.A., Marzetta, B.R., & Klemchuck, H.M.
    Decreased blood pressure in borderlfine hypertensive
    subjects who practiced meditation. Journal of Chronic
    Disease, 1974b, 27, 163-169.
Benson, H., Beary, J.F., & Carol, M.P. The relaxation response.
    Psychiatry, 1974c, 37, 37-46.
Blackwell, B., Hensenson, I., Bloomfield, S., Magenheim, H.,
    Gartside, P., Nidich, S., Robinson, A., & Zigler, R.
    Transcendental meditation in hypertension. Individual
    response patterns. Lancet, 1976, 1, 223.
Blanchard, E.B., Miller, S.T., Abel, G.T., Haynes, M.R., &
    Wicker, R. Evaluation of biofeedback in the treatment of
    borderline essential hypertension. Journal of Applied
    Behavior Analysis, 1979, 12, 99-109.
Coe, I.L., Norton, E., Oparil, S., Tatai, A., & Pullman, T.N.
    Treatment of hypertension by computer and physician - A
    prospective controlled study. Journal of Chronic Disease,
    1977, 30, 81-92.
Datey, K.K., Deshmukh, S.N., & Vinekar, S.L. "Shavasan": A
    yogic exercise in the management of hypertension.
    Angiology, 1969, 20, 325-333.
Deabler, H.L., Fidel, E., & Elder, S.T. The use of relaxation
    and hypnosis in lowering high blood pressure. American
    Journal Clinical and Experimental Hypnosis, 1973, 16,
    75-83.
Elder, S.T., & Eustis, N.K. Instrumental blood pressure
    conditioning in outpatient hypertensives. Behavioral
    Research and Therapy, 1975, 13, 185-188.
Frankel, B.L., Patel, D.J., Horowitz, D., Friedwald, W.T., &
    Gaardner, K.R. Treatment of hypertension with biofeedback

and relaxation techniques. Psychosomatic Medicine, 1978, 40, 276-293.

Friedman, H., & Taub, H.A. The use of hypnosis and biofeedback procedures for essential hypertension. The International Journal of Clinical and Experimental Hypnosis, 1977, 25, 335-347.

Frumkin, K., Nathan, R.J., Prout, M.F., & Cohen, M.C. Nonpharmacologic control of essential hypertension in man: A critical review of experimental literature. Psychosomatic Medicine, 1978, 40, 294-320.

Hypertension Detection and Follow-up Program Cooperative Group Five-year findings of the hypertension detection and follow-up program. II. Mortality by race, sex, and age. Journal of American Medical Association, 1979, 242, 2572-2577.

Klumblies, G., & Eberhardt, G. Results of autogenic training in the treatment of hypertension. In J. J. Thor (ed.) IVth World Congress of Psychiatry (Madrid, September 1966, International Congress Series No. 117) Amsterdam: Excerpta Medica Foundation, 1966.

Jacob, R.C., Karemer, H.C., & Agras, W.S. Relaxation therapy in the treatment of hypertension. Archives of General Psychiatry, 1977, 34, 1417-1427.

Lantzsch, W., & Drunkenmoller, C. Kreislaufanalytische unterschungen bei patieenten mit essentieller hypertonie wahrend der ersten und zweiten standardubung des autogenen trainings. Psychiatria Clinica, 1975, 8, 223-228.

Legewie, H., & Nusselt, L. Biofeedback-therapie. Lernmethoden in der Psychosomatik, Neurologie und Rehabilitation. Munchen: Urban, & Schwarzenberg, 1975.

Love, W.A., Montgomery, D.D., & Moeller, T.A. Working paper 1. Unpublished manscript, Fort Lauderdale, Fla: Nova University, 1974.

Luthe, W. Autogenic Therapy. New York: Grune & Stratton, 1969.

Miller, N.E., & Dworkin, B. Critical issues in therapeutic applications of biofeedback. In G.E. Schwartz & J. Beatty (eds.) Biofeedback: Theory and Research. New York: Academic Press, 1977.

Patel, C. Yoga and biofeedback in the management of hypertension. Lancet, 1973a, 2, 1053-1055.

Patel, C. Yoga and biofeedback in hypertension. Lancet, 1973b, 2, 1327.

Patel, C. & North, W.R.S. Randomized controlled trial of yoga and biofeedback in the management of hypertension. Lancet, 1975, 2, 93-95.

Pollack, A.A., Weber, MA., Case, D.B., & Laragh, J.H. Limitations of transcendental meditation in the treatment of essential hypertension. Lancet, 1977, 1, 71-73.

Reader, R. Therapeutic trials in mild hypertension ongoing

throughout the world. In H.M. Perry & W.M. Smith (eds.)
Mild hypertension: To treat or not to treat. New York: The
New York Academy of Science, 1978,.

Schwartz, G.E., & Beatty, J. Biofeedback. Theory and Research.
New York: Academic Press, 1977.

Sedlacek, K., Cohen, J., & Boxhill, C. Comparison between
biofeedback and relaxation response in the treatment of
essential hypertension. Proceedings of the 8th Meeting of
the Biofeedback Society of America. Alberquerque, New
Mexico, 1978.

Seer, P. Psychological control of essential hypertension:
Review of the literature and methodological critique.
Psychological Bulletin, 1979, 86, 1015-1043.

Seer, P., & Raeburn, J.M. Meditation training and essential
hypertension: A methodological study. Journal of Behavioral
Medicine, 1980, 1, 59-71.

Shapiro, A.P., Schwartz, G.E., Redmond, D.P., Ferguson, D.C.E.,
& Weiss, S.M. Non-pharmacological treatment of
hypertension. In H.M. Perry & W.M. Smith (eds.) Mild
Hypertension: To treat or not to treat. New York: The New
York Academy of Sciences, 1978.

Shapiro, A.P., Benson, H., Chobanian, A.V., Herd, J.A., Julius,
S., Kaplan, N., Lazarus, R.S., Ostfeld, A.M., & Syme, L.
The role of stress in hypertension. Journal of Human
Stress, 1979, 5: 7-25.

Shapiro, D., Greenstadt, L., Lane, J.D., & Rubinstein, E.
Tracking- cuff system for beat-to-beat recording of blood
pressure. Psychophysiology, 1981, 18, 129-136.

Shoemaker, J.E., & Tasto, D.L. The effects of muscle relaxation
on blood pressure of essential hypertensives. Behavioral
Research and Therapy, 1975, 13: 29-43.

Smith, W.M. Treatment of mild hypertension: Results of a
10-year interventional trial. Circulation, 1977, 40, 1-98.

Steptoe, A., Smulyan, H., & Gribbin, B. Pulse wave velocity and
blood pressure change: Calibration and applications.
Psychophysiology, 1976, 13, 488-493.

Surwit, R., Shapiro, D., & Good, M.I. Comparison of
cardiovascular biofeedback, neuromuscular feedback, and
meditation in the treatment of borderline hypertension.
Journal of Consulting and Clinical Psychology, 1978, 46,
252-263.

Taylor, C.B., Farquhar, J.W., Nelson, E., & Agras, W.S. The
effects of relaxation therapy on blood pressure of
essential hypertensives. Archives of General Psychiatry,
1977, 34, 339-342.

Tursky, B. The indirect recording of human blood pressure. In
P.A. Obrist, A.H. Black, J. Brener, & L.V. Dicara (eds.)
Cardiovascular psychophysiology. Chicago: Aldine, 1974.

Vaitl, D. Kontrolle der essentiellen Hypertonie durch
Entspannungs-techniken (Control of essential hypertension

by   relaxation   techniques),   In   D.   Vaitl   (ed.)
Psychologisch-medizinische   Aspekte   der   essentiellen
Hypertonie (Psychological and medical aspects of essential
hypertension). Berlin, Heidelberg, New York: Springer, in
press.

Weiner, H. Psychobiology of essential hypertension. New York:
    Elsevier, 1979.

Yates, A.J.   Biofeedback and the modification of behavior. New
    York and London: Plenum Press, 1980.

EMG FEEDBACK IN NEUROMUSCULAR CONTROL

John V. Basmajian

Department of Medicine
McMaster University & Chedoke-McMaster Rehabilitation
Hamilton, Ontario, Canada

In the rehabilitation or retraining of neurologically handicapped patients, electromyographic (EMG) feedback combined with standard therapies represents a widespread and exciting concept: given instant and continuous electronic displays of their internal physiologic events (by means of meters, banks of lights, and various auditing devices), human beings can be taught to manipulate those otherwise unsensed events voluntarily. As it is now being practiced, biofeedback is a scientific technique rather than a separate science, but the basic concept has stimulated the beginnings of a probable revolution in medicine -- behavioral medicine. Although behavioral medicine emphasizes the treatment of a host of behavioral disturbances which account for more than half of all symptom-complexes seen by physicians, patients with physical handicaps of varying degrees also should share in its general benefits as it matures. However, this chapter is about one part of behavioral medicine: EMG feedback -- also referred to as "myofeedback," "EMG biofeedback," "neuromyometry," "sensory integration," "audiovisual neuromuscular re-education," "artificial proprioception," (my now-abandoned first attempt at nomenclature), etc.

By concentrating on increasing or decreasing the electronic signals indicating the level of the striated muscle activity, a person can alter the level of contractions in the body which are not normally felt or sensed in any way. By the end of the 1960's this ability, which was at first a scientific curiosity, began to show some of its earliest practical applications. Three main scientific sources flowed together to form the broad stream that is modern biofeedback: EEG, cardiovascular phenomena, and EMG. The last of these arose from diagnostic and

201

research work in electromyography (EMG), which dates back to the 1940's.

Clinical electromyographers from the earliest days were quite aware of the considerable help they derived from the instant feedback of the myoelectric signals. Even in the infancy of electromyography we used the sound of motor unit potentials to grade the desired strengths of contractions and often recruited the help of the patient. No one thought to give this phenomenon a name. Certainly it was not used as an intensive therapeutic tool. Only Mims (1956) and Marinacci and Horande (1960) made brief mention of the clinical possibilities in 1956 and 1960, respectively. Quite separately, my colleagues and I were exploring the possibilities in the late 1950's and early 1960's of employing feedback signals to train exquisite controls in normal muscles of handicapped persons to substitute for lost limbs and to augment the strength of weakened parts of the body (Marinacci & Horande, 1960; Basmajian, 1963; Basmajian, 1967; Basmajian, 1972). This ultimately came to be called "single motor unit training" and "artificial proprioception"-- hardly as catchy as "biofeedback" (Figures 1 and 2). Our concerns were two- fold: first, to determine and define the normal mechanisms of motor control in all parts of the body and, second, to develop methods and improved devices for treating neurologically and orthopedically handicapped patients. By the early 1960s we were part of the world-wide movement for the development and use of electrically activated artificial limbs, and the need for scientific understanding was of paramount importance. In the midst of these studies, we found that when our subjects were provided with instant visual and acoustic feedback of the EMG signals arising from invisible and unfelt contractions of their muscles, they could learn to perform elaborate tricks with the tiniest units of muscle -- the motor units.

Because motor units are supplied by a single motor neuron, obviously we were training conscious control of individual motor cells in the spinal cord then considered by most neurologists to be impossible. Equally important, our subjects could put a single cell through elaborate tricks while completely inhibiting the activity of the surrounding cells, they consciously relaxed all the muscle fibers in a muscle (or even a whole limb) while activating the target motor unit "in isolation."

Not only can human subjects "fire" single motor nerve cells with an active suppression or inhibition of neighbors, but they can also produce deliberate changes in the rate of firing. Most persons do this if they are provided with aural or visual cues from their muscles by means of electromyography. Following the implantation of special fine-wire electrodes in any voluntary

Figure 1.   Volunteer undergoing single motor unit training
in laboratory.

skeletal muscle, a subject needs to be given only general
instruction. He is asked to make contractions of the muscle
under study while listening to and seeing the motor unit
potentials on the monitors. A period of up to fifteen minutes
is sufficient to familiarize him with the response of the
apparatus to a range of movements and postures. Subjects are
invariably amazed at the responsiveness of the loudspeaker and
cathode-ray tube to their slightest efforts, and they accept
these as a new form of internal body awareness without
difficulty. It is not necessary for subjects to have any
knowledge of electromyography. After receiving a general
explanation they need only to concentrate their attention on the
obvious response of the electromyograph. With encouragement and
guidance, even the most naive subject is soon able to maintain
various levels of activity in a muscle on the sensory basis
provided by the monitors. Indeed, most of the procedures he
carries out involve such gentle contractions that he is only
aware of them through the apparatus. Following a period of
orientation, the subject can be put through a series of tests of
many hours.

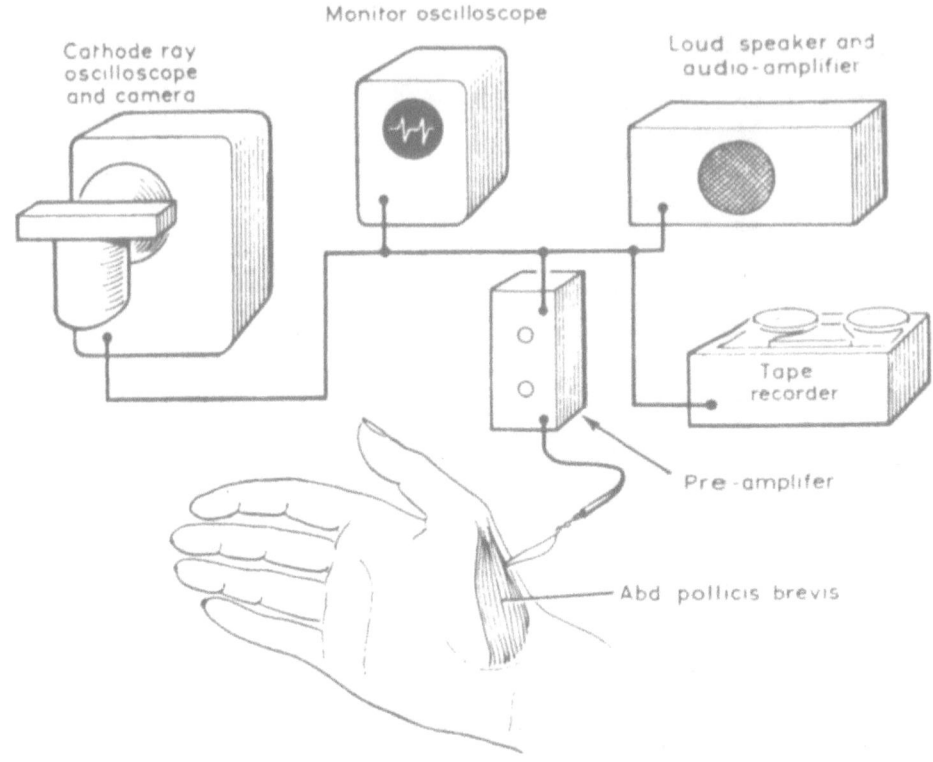

Figure 2. Diagram of set-up in Figure 1

Many subjects then can be trained at greater lengths in any special skills revealed in the earlier part of their testing (for example, either an especially fine control of, or an ability to manipulate a single unit). Finally, the best performers can be tested on their ability to maintain the activity of specific motor-unit potentials in the absence of either one or both of the visual and auditory feedbacks. That is, the monitors can be turned off while the subject tries to maintain or recall a well-learned unit without the artificial "proprioception" (i.e., internal body awareness) provided earlier. About one person in ten can learn this skill in a few hours.

The problem of what happens to the synergistic muscles at the "hold" positionor during movements of a limb was previously unknown.   The   level   of   activity   now   appears   to   be individualistic. Active inhibition of synergists is learned only after training of the motor unit in the prime mover is well established. A subject who focuses his attention on feedback from a single motor unit in one muscle finds that the surrounding muscles become progressively relaxed to the point of complete silence when isolation of the single motor unit is complete. Only such motor units in a limb as are needed to maintain its particular posture are still active. The process of "active inhibition" probably is the more significant part of motor training.

Throughout the 1960's and 1970's many investigations of a highly technical nature have been reported around the world, giving EMG biofeedback a solid foundation. Nevertheless, not more than 2% of rehabilitation personnel have hands-on experience with it. Of course the same might be said for psychologists when it comes to the clinical use of biofeedback. The technique is also spreading to other areas of neuromuscular training, e.g., physical education. Here, too, much research is needed.

Psychophysiological Mechanisms

Is EMG feedback training based on volition or is it operant conditioning similar to that shown for experimental animals by B.F. Skinner? The evidence from various groups is contradictory and reflects their commitment to or rejection of the operant conditioning paradigm. Certainly it is related to it. Conditioning can clearly be employed in modifying electromyographic responses.

In the 1960's, the second source of biofeedback came from operant conditioning experiments in animal psychology. The conditioning of cardiovascular responses in rats by Neal Miller (1968) and his colleagues at the Rockefeller University clearly showed that functions controlled by the autonomic nervous system could be influenced by operant conditioning. Soon the relationship of this work to our own in striated muscle control became clear and the applications to clinical problems became widespread. Obviously, the autonomic nervous system was not autonomous at all -- automatic and semi-automatic would be more precise. Within months of the first reports, studies on patients with various cardiovascular disturbances were under way. In EMG feedback, in spite of its clearer mechanisms, the number of clinicians "plunging in" was smaller except in psychotherapy where it began to flourish.

Muscle Relaxation Therapy

The main spinoff of my early EMG biofeedback studies undoubtedly has been relaxation therapy. It stimulated thousands of clinicians on several continents -- mostly psychologists and psychiatrists -- to apply biofeedback to the relief of various symptoms of stress. Tension headache, chronic back problems and anxiety are prime targets, and the literature on their management with biofeedback relaxation is expanding rapidly. The main problem in this area is confusion about "placebo effects," always a bugbear in psychosomatic medicine. Nevertheless, many patients have received substantial benefit when all earlier treatments proved ineffective. Relaxation training for the treatment of psychosomatic ailments actually predates biofeedback, going back a half-century to Jacobson's (1929) Progressive Relaxation and Schultz's (1959) Autogenic Training.

EMG Feedback in Rehabilitation Medicine

Although EMG feedback is the dominant technique in rehabilitation, a number of other approaches to feedback therapy are quietly gaining recognition: limb load monitors, head-position monitors, and electrogoniometric feedback.

EMG Feedback in Stroke Therapy

In the 1970's a number of medical research groups soon began reporting the efficacy of training the motor functions of a substantial proportion of previously "untreatable" patients (Johnson & Garton, 1973; Basmajian et at.,1975; Basmajian et al., 1977; Brudny et al.,1976). No miracles are wrought, but patients are able to discover and use within themselves motor pathways that apparently have survived the injury and have lain dormant. Other disorders of movement and posture are now also proving to be treatable.

In rehabilitating stroke patients with biofeedback, three major symptom complexes have been targets: footdrop (with or without spasticity), shoulder subluxation and reduced hand function. The treatment of footdrop is emphasized here because it has had the widest application around the world (Figure 3).

Many patients with and without short leg braces were included in a series of treatments conducted at Emory University in Atlanta (Basmajian et al., 1977). Twenty-five of the patients had been treated before biofeedback training with a short leg brace, which appeared to be reasonably efficient in most cases. All but one of these patients (who was seventy) were from thirty-one to sixty-four years old, with a wide spread among the various ages. The shortest duration since the stroke was three months, but almost all the patients had had their strokes many months or several years before biofeedback treatment, so the condition of their footdrop had stabilized. Of these twenty-five patients, sixteen were able to discard their short leg brace entirely following three to twenty-five sessions (mean, 16.6 sessions). Each biofeedback rehabilitation training session lasted approximately one-half hour. The remaining nine patients showed little or no improvement, sometimes for obvious reasons, such as poor motivation, severe spasticity, intercurrent illnesses, and early discontinuance of treatment (e.g., only three or four treatments in four of these nine patients). Some of the patients were even able to discard

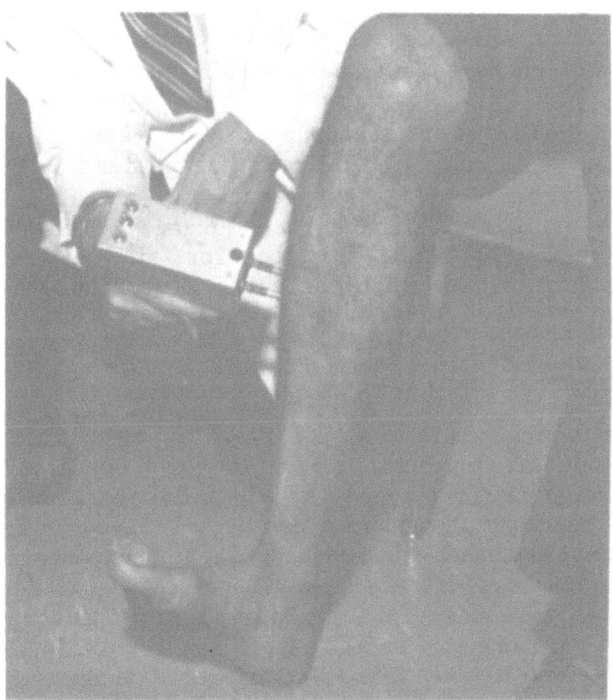

Figure 3. Training a stroke patient to dorsiflex the ankle
          in response to an acoustic and visual signal
          from a portable miniature EMG feedback trainer.
          At this point, the patient is ready for gait
          training with stuck-on electrodes leading the
          portable device hanging from the belt at the
          waist.

their canes for activities of daily living. Several required
their short leg braces intermittently when on their feet for
long periods of time.

Fourteen patients with footdrop had reasonably good
function at the ankle and had not been treated with braces. The
aim of treatment was to produce sharp gains in function. After
three to seventeen sessions in biofeedback rehabilitation
training, ankle function failed to improve in only two patients,
while six had moderate to excellent improvement of strength and
range of motion, which greatly improved their gait.

The patient's age apparently was not directly related to
the effectiveness of biofeedback. Patients in both the thirty
and the sixty year old age groups were among those who discarded
short leg braces. The proportion of men and women in whom
treatment was successful was the same as in the general
population studied. Neither failure nor success of treatment
seemed related to the duration of footdrop. Failure occurred in
patients with either recent or late stroke, while treatment was
successful in patients who had had footdrop for periods ranging
from three months to six and one-half years.

Biofeedback for footdrop in the stroke patient is of
undisputable value, especially if the patient is not confined to
a chair and has mobility. Particularly dramatic results appear
in those patients who have been forced for months or years to
wear a short leg brace for footdrop; in the above series of
treatments involving such patients, almost two-thirds succeeded
in discarding their braces.

Motor Skills Learning

The neural pathways involved in this marked neuromotor
improvement are unclear. There are two possibilities: either
new pathways are developed (highly unlikely), or old persisting
cerebral and spinal pathways can be mobilized by introducing the
artificial feedback loop. The latter explanation is highly
probable; the artificial internal awareness provided by acoustic
and visual responses to a peripheral motor act appears to be a
powerful reinforcer. Undoubtedly, new forms of cognition at the
cortical level are also recruited. What is said about
retraining paralyzed muscles is also apparently true of
voluntary inhibition of spastic muscles.

In recent years, the process of motor learning and control
has received increasing attention. Children are born with a
high level of anarchy in their motor control. As they mature,
the overactivity disappears and is absent in healthy adults. It
reappears in adults under psychological stress, but people can

be trained to inhibit it to varying degrees. In patients with diseases and injuries of the central nervous system, the normal inhibition pattern is lacking; then mass responses from local stimulation of the motor nerve cells in the spinal cord result in an exaggerated mass response described as spasticity.

The inhibition patterning would seem to come in part from obscure processes in diffuse centers of the cerebral cortex. Since inhibition is a central feature, one must consider the possibility that brain stem centers (and perhaps the cerebellum) are critically important in the imprinting of the learning. It is simplistic to consider a schema where an impulse is striated at a tiny area of the cerebral cortex and is then passed directly along a facilitatory path to a desired set of spinal motor cells. The motor learning process probably employs a nerve network, with the "main" pathway for motor activation being almost a small part of the whole.

Stroke patients who succeed in inhibiting marked peripheral spasticity apparently use surviving pathways that increase the inhibition of overactive spinal centers. Using an "override mechanism" they must be succeeding in damping even the influence of the powerful reflexes otherwise unrestrained. In any case, our patients are able to move one muscle while inhibiting the usual hyperactivity.

Relaxation therapy too has a major application, both targeted and generalized, in managing stroke patients who obviously are under great emotional stress. Functional improvement can be gained in stroke patients with general biofeedback and deep relaxation, just as psychosomatic ailments can be improved in the neurologically intact patient.

Shoulder Subluxation

The treatment of shoulder subluxation by biofeedback seems related to the above, and yet it also employs a mechanism that is quite different. In this case patients are trained to mobilize an area by the usual biofeedback technique, which in turn results in the restoration of a passive (but effective) function of a joint -- the locking mechanism of the shoulder joint (Basmajian, 1979). Much superior to the usual treatment of subluxation by the use of slings, this treatment technique relies on an understanding of normal anatomy and simple biomechanics. When the glenoid cavity is restored to its normal position from its downward rotation, the superior part of the shoulder joint capsule (especially the coracohumeral ligament) draws the humeral head upward.

Hand Function Retraining

     Much of the problem in hand function is not simply the
obvious paralysis; muscle spasticity is equally disabling.
Combining inhibition training with neuromotor retraining of the
weak hand and forearm muscles seems a logical approach -- at
present the only approach. It is being thoroughly investigated
in many clinics. Particularly encouraging has been my
experience both at Emory University (until 1977) and our
clinical results and research at McMaster University. Although
our data are not in a final reportable state, it is possible to
say that earlier dismal forecasts for general upper limb
recovery beyond the start of the "plateau period" must be
revised. With EMG feedback, enormous progress is achieved by
some patients and modest but useful results in most of them has
replaced our uniformly poor results previous to inaugurating
biofeedback-augmented therapy. There is, however, one clearly
defined limit: fine manipulative function appears to be not
restorable when it is chronically absent due to a middle
cerebral artery lesion. These research studies and similar but
independent ones at Queen's University in Kingston, Ontario
should be published within a year or two. Meanwhile many
clinics are experiencing substantial improvements (beyond
placebo effects) in upper limb recovery when
biofeedback-augmented therapy is introduced.

Other Rehabilitation Applications of EMG Feedback Spasmodic Torticollis and Related Conditions

     Since Cleeland's first report (Cleeland, 1973) of the
successful use of behavioral techniques including biofeedback, a
substantial number of reports have confirmed the substantial
successes attributable to EMG feedback (Korein et al., 1976;
Cleeland, 1979). Other dyskinesias and dystonias also have been
treated successfully (Cleeland, 1979). Dr. Brudny's chapter
deals with this issue in detail.

Spinal Cord Injuries

     Only a few single case reports have appeared on the use of
EMG feedback for paraplegia. Small successes are possible if
the lesion is (or proves to be) incomplete. The retraining
process is long and tedious, but our eleven patients worked with
for many months at Emory University generally believed the
effort was worthwhile.

Cerebral Palsy

     EMG feedback in different areas of the body has had limited
application with no major controlled studies. Sporadic general

reports have appeared with limited validation. The best work has been with position monitors (see below).

Force, Position and Joint-Angle Detectors

A number of special devices have been described that feed back information on body position or movement. In cerebral palsy patients, both head position monitors (Harris, Spelman & Hymer, 1974; Woolridge & Russel, 1976) and foot-placement and pressure switches (Moore & Byers, 1968; Warren & Lehmann, 1975) are undergoing testing. The results have been encouraging.

Electrogoniometric monitoring of movements at various joints is receiving increasing application (Brown & Nahai, 1979; Brown, et al., 1979). Although no controlled research studies have yet been reported, this mode of feedback provides added motivation and guidance to patients with neural, musculotendinous and articular limitations.

SUMMARY AND CONCLUSION

Clinical EMG biofeedback in rehabilitation has now gained a firm place in the treatment of upper motor neuron lesions particularly in retraining muscles and inducing relaxation in spastic muscles of stroke patients. In cerebral palsy and musculoskeletal disturbances additional feedback transducers (electrogoniometers, pressure-sensitive and position-sensing devices, etc.) are gaining wider use. Spasmodic torticollis has proved to be particularly suitable for behavioral methods of treatment including EMG feedback.

REFERENCES

Basmajian, J.V. Control and training of individual motor units. Science, 1963, 141, 440-441.
Basmajian, J.V. Control of individual motor units. American Journal of Physical Medicine, 1967, 46, 480-486.
Basmajian, J.V. Electromyography comes of age. Science, 1972, 197, 603-609.
Basmajian, J.V. Biofeedback--Principles and Practice for Clinicians. Baltimore: Williams & Wilkins Co., 1979.
Basmajian, J.V. Muscle Alive: Their Functions Revealed by Electromyography. (4th Edition). Baltimore: Williams & Wilkins Co., 1979.
Basmajian, J.V., Kukulka, C.G., Narayan, M.G. & Takabe, K. Biofeedback treatment of foot-drop after stroke compared with standard rehabilitation technique. Archives of Physical Medicine and Rehabilitation, 1975, 56, 231-236.
Basmajian, J.V., Regenos, E.M. & Baker, M.P. Rehabilitating

stroke patients with biofeedback. Geriatrics, 1977, 32, 85-88.

Brown, B. Recognition of states of consciousness through association with alpha activity represented by a light signal. Psychophysiology, 1970, 6, 442-452.

Brown, D.M., DeBacher, G. & Basmajian, J.V. Feedback goniometers for hand rehabilitation. American Journal of Occupational Therapy, 1979, 33, 358-463.

Brown, D.M. & Nahai, F. Biofeedback strategies of the occupational therapist in total hand rehabilitation. In J.V. Basmajian (ed.) Biofeedback--Principles and Practice for Clinicians. Baltimore: Williams & Wilkins Co., 1979.

Brucker, B.S. & Ince, L.P. Biofeedback as an experimental treatment for postural hypotension in a patient with a spinal cord lesion. Archives of Physical Medicine and Rehabilitation, 1977, 58, 49-53.

Brudny, J., Korein, J., Grynbaum, B.B., Friedman, L.W., Weinstein, S., Sachs-Frankel, G. & Belandres, P.V. EMG feedback therapy: Review of 114 patients. Archives of Physical Medicine and Rehabilitaion, 1976, 57, 55-61.

Cleeland, C.S. Behavioral techniques in the modification of spasmodic torticollis, Neurology, 1973, 23, 1241-1247.

Cleeland, C.S. Biofeedback and other behavioral techniques in the treatment of disorders of voluntary movement In J.V. Basmajian, (ed.) Biofeedback--Principles and Practice for Clinicians. Baltimore: Williams & Wilkins, 1979.

Craik, R. & Wannstedt, F. The limb load monitor. In R. Herman (ed.) Proceedings of Conference on Devices and Systems for the Disabled. Philadelphia: 1975.

Green, E. & Green, A. General and specific applications of thermal biofeedback. In J. V. Basmajian (ed.) Biofeedback--Principles and Practice for Clinicians. Baltimore: Williams & Wilkins, 1979.

Harris, F.A., Spelman, F.A. & Hymer, J.W. Electronic sensory aids as treatment for cerebral-palsied children. Physical Therapy, 1974, 54, 354-365.

Jacobson, E. Progressive relaxation. Chicago: University of Chicago Press, 1929.

Johnson, H.E. & Garton, W.H. Muscle re-education in hemiplegia by use of electromyographic devices. Archives of Physical Medicine and Rehabilitation, 1973, 54, 320-323.

Korein, J., Brudny, J., Grynbaum, B., Sachs-Frankel, G., Weisinger, M. & Levidow, L. Sensory feedback therapy of spasmodic torticollis. In Eldrige and S. Fahn (eds.) Advances in dystonia: Results in the treatment of 55 patients. New York: I.R. Raven Press, 1976.

Marinacci, A.A. & Horande, M. Electromyogram in neuromuscular re-education, Bulletin of Los Angeles Neurological Society, 1960, 25, 57-71.

Miller, N.E. & Banuazizi, A. Instrumental learning by curarized rats of a specific visceral response, intestinal or cardiac. Journal of Comparative Physiology and Psychology, 1968, 64, 1-8.

Mims, W.H. Electromyography in clinical practice. Southern Medical Journal, 1956, 49, 804-806.

Moore, A.J. & Byers, J.L. A miniaturized load cell for lower extremity amputees, Archives of Physical Medicine and Rehabilitation, 1968, 57, 294-296.

Schultz, J.H. & Luthe, W. Autogenic training: A psychophysiologic approach to psychotherapy. New York: Grune & Stratton, 1959.

Warren, C.G. & Lehmann, J.F. Auditory feedback from limb and crutch used to control weight bearing during ambulation (abstract). Archives of Physical Medicine and Rehabilitation, 1975, 56, 567.

Woolridge, C.P. & Russel, G. Head positioning training with the cerebral palsied child: An application of biofeedback techniques. Archives of Physical Medicine and Rehabilitation, 1976, 57, 407-414. .

EMG FEEDBACK IN NEUROMUSCULAR REHABILITATION OF SPASMODIC

TORTICOLLIS: THERAPEUTIC ELECTROMYOGRAPHY

Joseph Brudny

New York University Medical Center
Department of Rehabilitation Medicine
New York, New York

Introduction

Researchers in behavioral medicine not unlike in neurosciences strive to achieve understanding of how the nervous system accepts, interprets, integrates and translates neuronal code signals into thoughts, ideas and concepts resulting in human behavior. While pursuing this goal, one takes it for granted that the complex and coordinated interaction of muscles mediating behavior is constant, repetitive and predictable. Lord Adrian (1935) observed that, "The chief function of the nervous system is to send messages to the muscles which will make the body more effective."

In this context, spasmodic torticollis (ST) is a fascinating model of central nervous system (CNS) dysfunction in which control over effectors of behavior has become abnormal, and messages sent to the muscles create a most distorting and senseless movement of the head and neck, resistant to the patient's attempts at its correction. Considering that ST is also resistant to scores of various therapies, but does often respond to behavioral treatment as described in this chapter, the selection of this CNS disorder for presentation and discussion at this symposium is most appropriate.

The clinical manifestations of ST are bizarre and often grotesque (Dandy, 1930; Patterson & Little, 1943; Podivinsky, 1968). In view of the nature of these manifestations, their occurrence only during certain actions and their increased incidence during mental stress, many physicians still believe that ST is purely psychogenic and is the product of a disturbed

mind. From a psychoanalytic point of view, the stiff neck was
considered a symbol of an erect phallus (Abse, 1966), and the
constant position of head turned to the side was interpreted as
symbolic of rejection (Whiles, 1940). The fact that brain
autopsies in patients with ST failed to show histological
abnormalities (Tarlov, 1970) headed further argument for the
psychogenic origin of ST.

It is only within the last decade that understanding and
consensus on the etiology and nature of ST has taken place, not
to a small degree as a result of two international conferences;
on Dystonia (Eldridge & Fahn, eds., 1976) and on the Basal
Ganglia (Yahr, ed., 1976), both held in New York city. During
these conferences, my colleagues and I had an opportunity to
present our form of behavioral treatment of ST (Korein, Brudny,
et al., 1976; Korein & Brudny, 1976) which evolved from clinical
research of this entity first begun by me in 1971 at New York
University Medical Center (Brudny, Grynbaum & Korein, 1974).

ST is considered to be one of the non-hereditary focal
dystonic disorders of adult-onset and of similar pathophysiology
with the generalized torsion dystonia (Marsden, 1976; Couch,
1976). Dystonia is best knwon for its hereditary form (dystonia
musculorum deformans), its early-age onset, and the progressive
course of illness (Eldridge, 1970). The pathology of dystonia
is considered to result from disturbance in processing and
modulation of neuronal signals in basal ganglia (Zeman, 1970)
that serves as ramp function generator for slow, smooth
voluntary movement (Kornhuber, 1974). The underlying
distrubance is biochemical in nature and is due to an existing
imbalance in neuronal transmittal agents relating to the
extra-pyramidal system (Fahn, 1976).

The agent most implciated in dystonia is probably
gamma-amino-butyric acid (GABA). Its main action is inhibitory
in nature, both pre-a dn post-synaptically. The GABA-ergic
system can be envisioned as setting the gain on the sensitivity
of sensory receptors and serving as part of the final common
switching mechanism regulating the release of preprogrammed
neural circuits (Roberts, 1976). The turning on of patterned
activity is most probably due to the excitatory release of
dopamine or other similar agents.

With disturbed inhibition and processing of neuronal
signals, certain thalamo-cortical feedback loops of the complex
servosystem of voluntary movement transmit defective information
causing excessive activity of the motor cortex with the
resulting manifestations of various dystonic dyskinesias.

In the past ten years, I have had the opportunity to interview, examine, study, follow-up and/or treat over 250 patients with various focal dystonic disorders. My colleagues and I systematically studied patients in Bellevue Hospital Center and in the Sensory Feedback Therapy Clinic at the ICD Rehabilitation and Research Center. In addition, a large number of patients were similarly studied and treated in my private practice. Based on this wealth of clinical material, a fairly typical picture of ST can be described.

The onset of ST usually occurs in the fifth decade of life, is of a progressive nature and leads to inability to control the head position within weeks or months. To minimize the oscillatory movements resulting from attempted correction, many patients eventually assume a fixed head position. The position varies from head turned to one side (torticollis) to head turned and tilted with chin facing upwards (retro-torticollis). In cases where there is not turn, the head may only be acutely extended (retrocollis) or acutely flexed (antecollis). The displacement of the head and its position are the result of dystonic manifestations occurring in various muscles of the neck. These muscles are primarily, but not exclusively, the sternocleidomastoid (SCM), the trapezius, the splenius and, less frequently, the levator of the scapula.

Oscillation may or may not be present at rest but invariably occurs during attempted movement. The shoulder to which the chin is pointed is usually elevated to various degrees. Pain may be absent or present, either localized or of radiating quality, and of various severity. Scoliosis of the cervico-dorsal spine is frequent and cervical arthropathy is common.

The greatest relief is usually obtained when lying down or supporting the head against the backrest of a chair. Even lightly touching the chin or cheek with the hand aids in controlling head position for a short time (geste antagonistique). All dyskinetic movements disappear completely during sleep. Alcohol in small amounts may be of temporary help. Recently, the interaction between GABA and ethanol has been described, ethanol apparently potentiating GABA-elicited synaptic inhibition (Tran, Snyder, Major & Hawley, 1981).

Exacerbation of symptoms usually occurs upon standing and walking, especially on streets with heavy traffic. Fatigue, exertion and emotional upset or stress are of an aggravating nature. Just prior to and during menstruation symptoms are usually aggravated. Manual work requiring concentration may be helpful at times. The course is usually progressive with a certain degree of stabilization achieved after five to ten years of illness.

As mentioned before, while not hereditary, ST of adult-onset apparently occurs in patients with a genetic predisposition for a deficit in neuronal transmittal agents (most probably GABA). Such a deficit may go unnoticed through the first four or five decades of life until various triggering factors may precipitate the onset and the symptomatology of the disease. From review of our data, it is obvious that emotional factors may act as a precipitant but the disease is not of psychological origin. Caucasians, especially with light color eyes, are much more affected by this disorder. Statistical review of 153 Caucasian patients points to the relationship between decreased melanin metabolism and a genetic predisposition to focal dystonic manifestations (Korein, 1980). On theoretical grounds one may suppose that there is a link between the tyrosine-dopamine pathways and tyrosine-melanin pathways and their interaction with GABA. Exposure to psychotropic drugs is quite often the precipitating factor as some may cause a functional excess of dopamine (Baldessarini & Kline, 1976). Trauma directly to the neck or head cannot be ruled out as a precipitating factor. Many patients show far advanced degenerative osteoarthritis of the cervical spine with considerable encroachment upon the intervertebral foramina. In these patients, the pain associated with certain positions is radiating and is, no doubt, of radicular nature, as opposed to the localized pain so typical of constant dystonic powerful contraction of a larger muscle. Whatever the precipitating factors are (trauma associated with pain, psychotropic drugs, infection, emotional or exertional factors) they all may act as a precipitant, but to produce the clinical manifestations of ST one must assume an underlying genetic predisposition for imbalance in neuronal transmittal agents rendering the basal ganglia incapable of processing and modulating increased input in a physiological manner.

Therapies of ST range from physical methods such as heat, cold, massage, traction, chiropractic manipulations, electric stimulation, acupuncture, sphenopalatine ganglion block, collars and braces to surgical procedures such as rhizotomies, section of the muscles, neck fusion, labyrinthian suppression by means of "iontophoresis" and cryothalamectomies. Psychiatric and psychoanalytic management are still common. Pharmacotherapy is the treatment most often prescribed by neurologists, while surgical intervention is becoming less common. The rationale in prescribing various medications is to affect the anatomic connections between certain neurons and neuronal cell assemblies in order to therapeutically alter the effects of these synapses, whether they are excitatory or inhibitory in nature. The beneficial response naturally depends on the nature of the neurotransmitters in each of the synaptic connnections. At present, information of the specific neurotransmitter-neuron relationship is available for only a minute number of billions of cells in the cerebrum (Fahn, 1976). Consequently, the clinical results of pharmacotherapy are less than desirable. In a thorough review of 42 papers describing pharmacotherapy of ST, Lal (1977) pointed out that pertinent information is scarce, samples described are too small to be statistically significant, there is a lack of objective evaluation to response and an absence of adequate followup. Furthermore, these potent pharmacological agents by themselves may induce ST (Ayd, 1961) or may cause other unexpected and often irreversible side effects known as tardive dyskinesia (Baldessarini & Tarsy, 1976).

Behavioral modification of ST has been reported to be successful in some patients without using any instrumental feedback (Surwit, 1980). Cleeland (1973) employed a combination of EMG feedback and aversive electric shock in treatment of ten patients with ST and reported success in some of them.

My interest in behavioral treatment of ST stemmed from having originated in 1971 a clinical research study consisting of training patients with various CNS insults to achieve voluntary control over their dysfunctional muscles (Brudny, Korein, Levidow, Grynbaum, Lieberman & Friedman, 1974). Convinced that conventional physical therapy approaches did not provide enough information for motor retraining in these patients, I selected the oscilloscopic displays of electromyographic (EMG) activity as a substitute sensory modality capable of supplying sufficient pertinent information concerning the disposition of intended movement. Such visual information, paralleling the kinesthetic data (change in muscle length and the rate of such change) could aid the patient in quantification of muscle activity and in "recalibration" of central motor programs and motor output.

Historically there was anecdotal evidence that feedback of EMG displays has therapeutic value in the treatment of hemiparesis (Marinacci & Horande, 1960; Andrews, 1964). The diagnostic, "raw" EMG displays used in these studies did not meet the criteria for optimal electromyographic learning (Rubow & Smith, 1971). To be meaningful, EMG displays must be immediate, continuous, pertinent, and capable of reflecting the spatio-temporal changes in activity of a muscle at any point in time. With all vagueness eliminated, the precision of such information should be like an algorithm, when it comes to learning and forming motor patterns in the human brain. It is for these reasons that in the course of our research, a system known as the EMG Bioconditioner (manufactured by Hyperion, Inc., a subsidiary of Cordis Corp., Miami, Florida) was developed, tested extensively and recently made available commercially.

Methods

The Bioconditioner is a two-channel electromyographic device consisting of a microprocessor controller assembly and video monitor. Application of computer technology for control and video display of EMG signals offers an advance in terms of analysis and therapy. The system provides for on-line, real-time computation of EMG activity and permits the measurement and recording of digitally integrated muscle potentials in units of microvolt-seconds on two channels simultaneously (Brudny, Weisinger, & Silverman, 1976). The concept of the microvolt-second is derived from computation of the raw EMG. The recording of microvolt-seconds as an indication of altered muscle activity during movement provides a means of representing and verifying the status of motor control.

The therapeutic potential of this system is based on the generally recognized premise that, when transduced by surface electrodes and integrated digitally during short time periods, the EMG yields an adequate representation of ongoing muscle activity, that is, muscle contraction (or its length), and the rate of changes of such contraction (DeVries, 1960; Gans & Noordergraaf, 1975). Controlled by a microprocessor, such integrated EMG (iEMG) appears instantaneously in the form of traces on the video monitor. Displacement of the trace line along the vertical axis represents the magnitude of activity generated in the muscle, while the rate of change of such activity is reflected by vertical displacements over time (horizontal axis). Therefore such a visual display constitutes a continuous representation of the spatio-temporal events occurring in a muscle during attempted or ongoing movement. Further, during an action's materialization, it provides instantaneously information regarding the occurrence as well as outcome of an intended response. The system has the capacity for storage of various discriminative traces derived from EMG recorded during movement, combined with capabilities for retrieval of such traces to serve as performance models for the patient to replicate in future trials.

The analysis of patient's iEMG response helps to establish specific training goals. In normal subjects the iEMG patterns generated in response to command or volition are quite constant, reproducible, and reflect response activity within appropriate spatial and temporal dimensions. The coordinated nature of facilitation and inhibition of agonist and antagonist activities are readily apparent (Figure 1A).

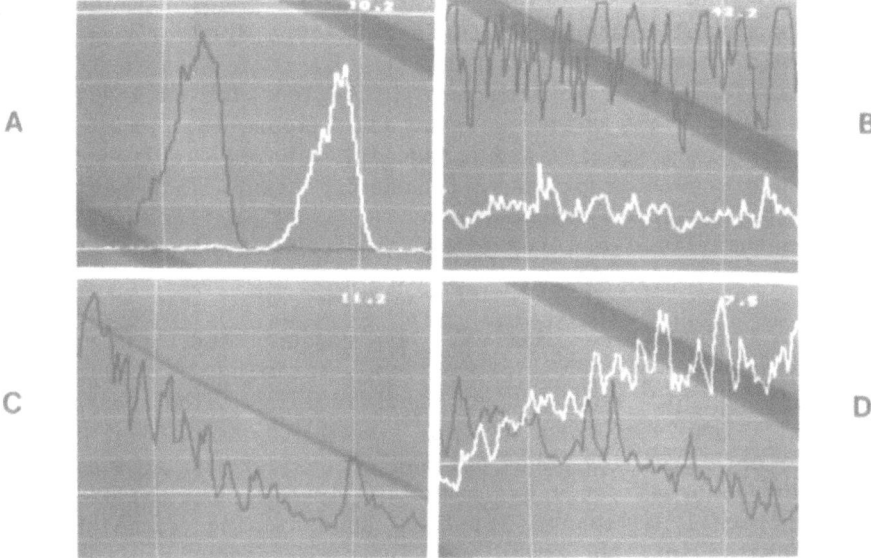

Figure 1.   A.   iEMG displays reflecting activities of both
                 sternocleidomastoid (SCM) muscles at rest
                 and during turn of the head to either side
                 in a normal subject.
            B.   Black trace reflects uncontrollable activity
                 of a dystonic right SCM muscle during the
                 attempt at correction of head position by
                 left SCM muscle (white trace)
            C.   Progressive voluntary reduction of spasmodic
                 activity toward the visually established
                 goal (horizontal white line)
            D.   Training of simultaneous desired alteration
                 of iEMG responses in spasmodic left
                 splenius (black trace) and right SCM
                 (white trace).

     In contrast, iEMG patterns derived from monitoring the
muscles of a patient with ST reveal typical abnormalities of
response. The visual display readily confirms the clinical
impression that although movement can be initiated by the
patient, the spatial and temporal iEMG parameters are not of a
dimension appropriate to fostering functional utility.

Disused muscles display a temporal delay in iEMG reponse together with an inability to reach and sustain steady levels of output needed for function. On the other hand, patients whose muscles reveal dystonic features reflect a loss of spatial and temporal control characterized by an excessive and uncontrollable rise in iEMG level, coupled with an inability to reduce such levels quickly, if at all (Figure 1B). Furthermore, when monitoring agonist and antagonist muscles simultaneously, a pattern of co-contraction is frequently seen. This pattern may completely restrict movement or, if present, movement production may be shortlived and functionally irrelevant due to patient's fatigue.

Following the clinical evaluation and analysis of abnormalities in iEMG production, training procedures are established for each patient individually. The therapy, in general, involves focusing on events in individual primary movers and training the patient to alter volitionally any existing abnormal activity in muscles involved in the production of functional movement. Operant conditioning techniques are incorporated into the training. The simultaneous monitoring and training of agonist and antagonist muscles forms the cornerstone of treatment (Brudny, Korein, Grynbaum, Friedmann, Weinstein, Sachs-Frankel & Belandres, 1976). While observing visually displayed iEMG response which has been produced volitionally upon the therapist's request, the patient, with the therapist's guidance, can easily evaluate otherwise covert and often inappropriate muscle activity. The patient can then alter such inappropriate iEMG patterns towards more normal ones by generating iEMG responses on the video screen that approximate or match visual models displayed thereon. In this way it becomes possible to determine immediately any existing gap between actual performance and intended goal. These visual models take the form of a horizontal threshold line whose vertical positioning on the screen can be adjusted. Their respective positions determine the voltage level of the desired iEMG response of the patient. Should the patient succeed in producing the desired alteration of iEMG response which matches or exceeds the horizontal target line on the screen, a tone is emitted by the Bioconditioner which serves as an additional informational as well as reinforcing event. As a result, with this shaping technique, increasingly greater iEMG responses, as well as an extended ability to sustain such response levels, is often achieved in disused muscles. In spasmodic muscles, volitional reduction of excessive iEMG values follows the successive approximations of patient's response towards horizontal threshold line, which is progressively lowered by the therapist. Such decrease of values of patient's iEMG leads

ultimately toward achieving and maintaining more functional
levels of response (Figure 1C).

The horizontal target line display allows simultaneous
monitoring of agonist and antagonist muscles on separate
channels, combining continuous suppression training of spasmodic
activity with strengthening of disused muscles when needed
(Figure 1D). In time, such a training paradigm allows complete
antagonist inhibition below a threshold level determined and set
by the therapist during the intended agonist activation (Figure
2A).

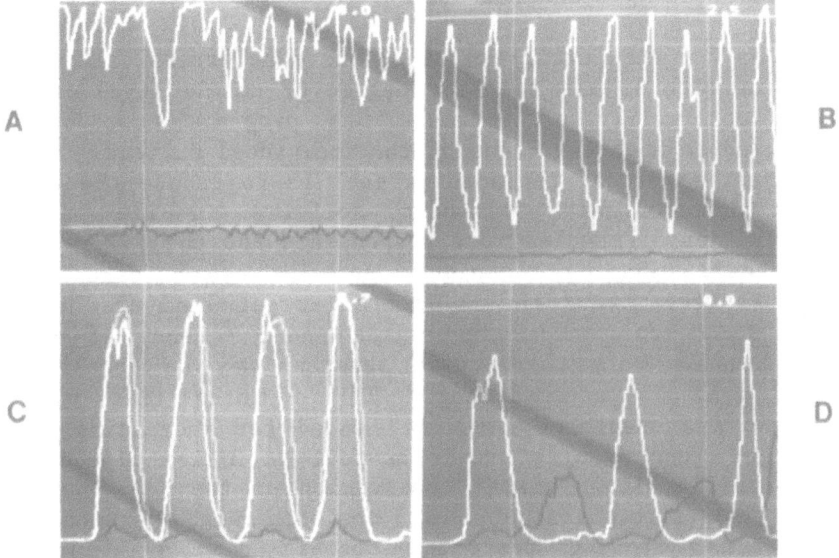

Figure 2: A. Sustained voluntary contraction of agonist
               muscle (white trace) while the spasmodic
               activity of antagonist is practically
               abolished (black trace).
           B. Isometric, brief contractions of left SCM
               muscle, with head in neutral position and
               spasmodic activity of right SCM abolished.
           C. Artifically created trace samples matched
               by patient's iEMG response of unaffected
               left SCM muscle (white trace), with
               spasmodic activity of the right SCM muscle
               kept controlled.
           D. iEMG displays reflecting activities of both
               SCM muscles in an ST patient who learned
               successfully to control head movements at
               rest and during head turning movements
               (blace trace reflecting the previously
               spasmodic muscle)

In more advanced stages training is also carried out by having the patient produce iEMG patterns from monitored muscle that match video screen displayed sample traces derived from a previously successful performance and retrieved from computer memory. For such a performance to sample training paradigm, the Bioconditioner also provides an extensive repertoire of artifically designed variable sample traces. Selection of a model is made by the therapist depending on the stage of the patient's response to training. When the response trace produced by the patient (activity of the monitored muscle) approximates or matches the selected sample trace, a tone is produced by the instrument, again serving as an informational and reinforcing event (Figure 2C).

As for the actual training, one must stress that not every patient with ST has the same degree of muscle involvement, the same abnormality of movement and programs must be custom tailored to meet the needs of the individual patient. The first and most important therapeutic goal is to reduce the excessive and uncontrollable spasmodic activity in the muscle or combination of muscles most responsible for the displacement of the head. Shaping techniques such as escape paradigm, with gradual lowering of the horizontal target line are helpful in most cases to reduce the degree and frequency of spasmodic occurrence. Many therapeutic sessions may be required until this goal is accomplished. It is noted that with each consecutive treatment the time needed for recreating the last session's end results grows shorter. A mirror placed next to the monitor allows visual observation of the head position as well as correlation of such position with particular iEMG patterns. Storing the oscilloscopic display reflecting a successful performance and presenting it in the beginning of the patient's next session as a sample to match is of considerable value. After the amplitude of contractions and the frequency of their occurrence is reduced and a sustained state of activity can be accomplished, one may attempt gradual mobilization of the head through the range of motion. This task is crucial for functional movement of the head during activities of daily living and this is usually accomplished by selecting a muscle unaffected by the dystonic phenomenon and assigning to it the primary role in initiating and executing motion. Through progressive, step-by-step, increase in voluntary contraction of a selected muscle, head mobilization takes place. At times, progress may be quite prompt. In most of the cases, an attempt at turning the head will again trigger off a spasmodic reaction but it usually lasts a short time and is of much less extensive amplitude. Two simultaneous functional tasks, increasing the contraction of one muscle and sustaining the relaxed state of the other, present a true challenge not only to the patient but also to the physician or therapist in charge of the session.

One must provide immediately the proper guide to the patient based on kinesiologic analysis of movement, so that the requirements of the selected training paradigm can be fulfilled. Should failure occur, its cause must be immediately identified. The iEMG traces reflecting the mechanism of failure are recalled from the computer memory, displayed, analyzed, and another attempt at a more successful trial is carried out. In successive sessions it becomes easier and easier to increase the amplitude and the frequency of voluntarily induced contractions (including isometric) of the muscle trained for increase in tonicity. This muscle will assume the primary responsibility for maintaining the head position without triggering off the undesired spasmodic reaction elsewhere (Figure 2B). Turning the head to the right or left respectively with return to neutral position is usually carried out at this stage, mainly by the muscle trained for increase in tonicity, while the dystonic one is activated only partially during movement (Figure 2D).

The patient's discrimination of the involuntary spasmodic occurrences becomes quite acute and the time needed to reduce it grows progressively shorter. When fully successful, the patient can easily and equally rotate the head to either side, flex and extend. Nevertheless, for a long time, he must maintain constant awareness of control over two opposing functional goals -- one being constant suppression of the activity of the dystonic muscle, and the other the gradually modulated increase and decrease of activity in the trained muscle. Patients who successfully learn such a technique of suppression of the spasmodic activity may maintain the head in neutral position for long time periods. For the purposes of simplification, the above described technique refers to two individual muscles (agonist and antagonist). In fact, the number of muscles necessitating retraining is greater in most cases. Four or more muscles are often involved and in order to learn functional control of head position, a step-by-step training of individual muscles is necessary. Voluntary and functional control of head movement is accomplished when the ability to carry out the combined components in a coordinated manner is attained. This ability, to become internalized, and to persist without the benefit of EMG feedback requires extensive time and effort on the part of the patient and the professional in charge. It is not unusual that a period of up to two years is needed for complete normalization of movement to take place. At this stage, there is no need for maintaining awareness and movements become fairly patterned and carried out automatically.

Some patients who learn adequate control over head movement and position while sitting can exercise this learned control while standing and walking. However, many patients are unable to combine the increased complexity of motor tasks involved in mobility with previously learned control in the sitting position only. Such patients require extended training during ambulation. I find that, in my private practice, the use of EMG telemetry will provide the patient with needed mobility while he is receiving constant information regarding his response to increased motor tasks. Miniaturized frequency-modulated EMG transmitters (Midgard Electronics, MXM-100) interfaced with the Bioconditioner are used for this purpose. With two video monitors placed 20 to 30 feet apart, the patient can walk to and fro at different speeds assuming various intended head positions while being constantly monitored for his iEMG activity. To achieve automatic control of head position during mobility the patient may require monthly reinforcement training for additional periods of one or two years.

One group of 80 patients treated and followed for up to four years will serve as a representative sample of the entire patient population (Korein & Brudny, 1976). Among these patients with focal dystonic manifestations, 69 had ST, although 24 of them also had other dyskinetic movements (15 patients had orofacial and oromandibular dyskinesias, one had blepharospasm, and eight patients had focal dystonic involvement of the upper extremity). The remaining 11 patients had focal aspect of their dystonic syndrome primarily in the extremities and facial and mandibular muscles. The median age of onset of illness was 48 years, and the medial duration of illness was four and one-half years (range of duration was from one to 35 years). Although focal dystonic syndrome is not considered hereditary, 12.5 percent of patients had a positive family history of various dyskinesias. Perinatal factors were implicated in five percent of patients. It is also of interest that all patients were Caucasians and 60 percent had blue, grey or green eyes, while the incidence of these colors in unaffected controlled Caucasians is known to be 35 percent.

Approximately 50 percent had psychotherapy, which helped some of them to "live with their illness" but had little effect on altering the movement disorder. Thirty-three percent of patients had acupuncture and/or "iontophoresis". Ten percent had various surgical procedures and 100 percent had various pharmacotherapeutic agents, the response to which was in some patients initially fair, then invariably poor, and in many patients eventually caused worsening of their condition or appearance of new dyskinetic features. The most effective drug reported generally was valium (diazepam). Forty-four percent of patients reported that small amounts of alcohol were of aid. Within three months prior to onset, infection was present in ten percent and trauma to head and neck occurred up to three months prior to the onset of the illness in 13 percent of the patients. Spontaneous partial remission with recurrence after 20 years was noted in one patient. Measurement of quantifiable features of behavior before, during and after therapeutic intervention was collected from most of these patients. Some features were quite apparent and easy to measure. The most pertinent were position of the head with degrees of tilt, turn and vertical displacement; oscillation of the head, its degree and rate; elevation of the shoulders; maneuvers applied for greater control of the head while being interviewed; presence or absence of pain; effect of posture (lying down, sitting down, standing and walking); fatigue and exertion; emotional upset; and effects of alcohol, menstruation and manual work.

All the above variables were included in the baseline neurological and psychiatric examination. The state of the muscles involved was described in detail (hypertrophy, atrophy). Range of motion of the head was determined by standardized measures. Two to four channels of iEMG were used for collecting and storing the information on a magnetic tape for analysis, presently in progress. Recordings were taken from the patients while performing several maneuvers of the head or the involved extremities. A videotape recording and/or still photographs were also taken with the patient performing the same maneuvers. All or some of these were repeated at periodic time intervals during the course of treatment. The iEMG measures in microvolt-seconds were obtained initially before and after the course of treatment. Presently systematic iEMG data collection with long baselines preceding treatment is done in selected cases, using single case methodology as described in documenting response to therapeutic electromyography in hemiplegia (Brudny, in press), dystonia (Bird & Cataldo, 1978) and ST (Martin, in press). Figure 3 illustrates such single case documentation of response.

Most of the 80 patients treated in the research phase of our study were seen in the clinic from three to five times a week and the duration of each session was 45 minutes.

Results

After completion of a twelve-week course of treatment, criteria for improvement in relation to primary goals were as follows:

0 - No significant response even with biofeedback.

1+ - Significant motor responses with feedback but very little carry-over without feedback and very little significant improvement in activities of daily living. Localized neck pain was occasionally relieved.

2+ - Response sustained after feedback was withdrawn. Patient had significant carry-over and could control head

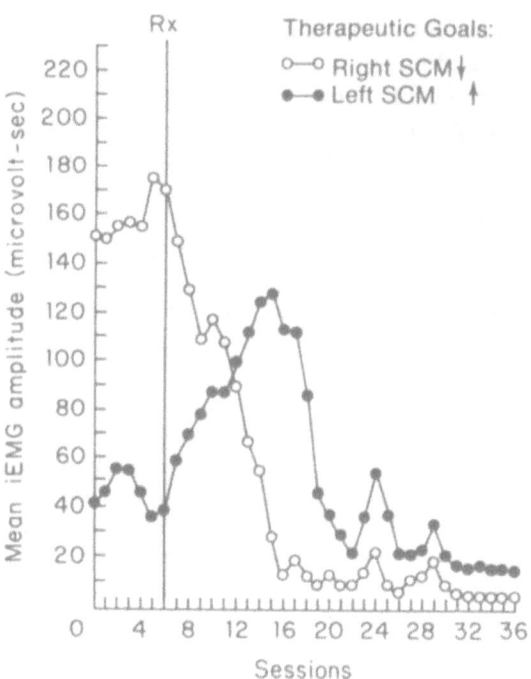

Figure 3: 65-year-old female patient (S.B) with spasmodic
torticollis to the left, five year duration.
Six baseline and thirty therapeutic sessions.
The patient is attempting to assume and to
maintain neutral head position. During the
baseline sessions, the patient is requested
"to do her best," in achieving neutral head
position. The iEMG values are monitored and
recorded but the patient is receiving no feedback information. Each session's data point
represents the average of the peak of iEMG of
six responses from each muscle. With introduction of EMG feedback, the therapeutic goals are
achieved over 30 sessions. During the last six
sessions, the patient assumed neutral head
position at each attempt. Note the relative
elevation of iEMG values in the left SCM trained
for voluntary increase in tonicity.

movements without feedback for days.   Significant change of
activities of daily living occurred, but emotional stress,
fatigue, and exertion still caused loss or decrease of control
of head movement.

3+ - Patient could maintain control of head movement
for extended periods (months) without feedback, and major
changes were seen in range of motion, control of oscillation and
activities of daily living. The patient rarely had episodes of

spasm, but these could be rapidly attenuated when they occurred. Patients no longer used "gestures" or "tricks" to control head movement.

The criteria for improvement were based on response sustained after withdrawal of EMG feedback therapy. Only patients who could control their movement disorders during activities of daily living for days (2+) and months (3+) were considered successes. When there was no change (0) or the response to feedback was present only during therapeutic sessions (1+), the patients were considered therapeutic failures. Of the 80 patients treated, 35 patients or 44 percent were classified as failures and 45 patients or 56 percent were considered therapeutic successes. Along with learning control over previously spasmodic muscles, all these patients demonstrated significant changes in activities of daily living including in some return to work or school, driving a car, resuming social contacts, dancing, swimming, dating, marrying, etc.

Subsequently, within six months regression occurred in nine patients while 36 patients, or 45 percent, maintained their gains during the follow-up of up to four years. We consider this a significant rate of improvement in view of the long duration of their illness and failure to respond to various forms of previous therapy.

Discussion

Certain aspects of therapeutic electromyography need elaboration before presenting the rationale for such training. First and foremost is the sensory plasticity that is rendering the human brain capable of interpreting and utilizing different sensory cues according to need (Granit, 1972). Such sensory plasticity is fundamental to intersensory translation of visual and kinesthetic information (Connolly & Jones, 1970) and to sensory substitution, e.g. the tactile vision substitution system in the blind (Bach-y-Rita, 1972). Second, vision seems to be the dominant modality in strengthening the "perceptual trace" during the development of motor skills (Scott, Kelso & Stelmach, 1976). Three known pathways connect the visual and motor corteces and participate in guiding body and limb movement; the development, therefore, of therapeutic techniques that facilitate the links between the visual input and the motor output seems to be a logical approach. Third, the operant shaping of the motor response necessitates motivational drive for achieving success. The motivational system is essential to the operation of brain algorithms that constitute the basis for information processing, solving problems, and forming motor programs. Finally, awareness is essential in learning control of individual muscles and is a capability that exists apart from the one required in patterning gross movements (Granit, 1972).

Our technique provides a cognitive link between the generated iEMG response and a particular visual model. Such a production to sample paradigm uniquely underlies the human capacity for language learning. Hence, the concept of "iEMG language" suggests itself as it pertains to the patient's capacity to respond to visual information and with its aid to relearn more adequate motor control. It has been said that language gives purpose to sensation and action to movement (Penfield & Roberts, 1976). In this sense the patient, whose cerebral processing of sensory feedback from affected muscles is defective and who thus is unable to quantify and control his motor response, can rely during retraining on visually displayed "iEMG language" signals as a sensory modality substitute. In patients who respond to therapeutic electromyography, these signals apparently can be "understood" by the CNS and translated in terms of kinesthetic information. Such translation of muscular response to a "visual language" which in turn can mediate sensory motor performance in the CNS, involves visually modelled traces which the patient attempts to replicate in training. To match a displayed sample trace during training the patient is required to formulate the precise spatio-temporal patterns of muscle activity. These learned motor strategies or motor patterns can apparently be stored in neural circuits as well as retrieved, as evidenced by the patient's ability to exercise equally precise motor control following withdrawal of feedback.

The most crucial factor is the patient's ability to suppress voluntarily the undesired activity in spasmodic muscles and then incorporate their altered responses into functional movement when feedback is withdrawn. Therefore, one can argue that decoding of visual iEMG signals during training initiates a self-regulatory process implementing a negative feedback system which attempts to correct any discrepancy between desired and actual performance. Over time, such cortical learning will establish a gating mechanism which successfully inhibits the excessive neuronal inputs to the motor cortex. As for the neural substrate underlying such gating and the resulting motor control, one can only speculate that it most probably originates with the thalamo-cortical circuits or possibly even within the basal ganglia itself.

The concept of cerbral gating mechanisms altering other CNS events is not new (Andersen & Sears, 1964). The gate control theory of pain (Melzack & Wall, 1965) describes a system whereby pain perception may be altered through a negative feedback loop of descending fibers from the cortex. Sterman and Bowersox

(1980) also discuss a negative feedback loop mechanism of a thalamic inhibitory gate in patients who have successfully learned to control their epileptic seizures as the result of EEG feedback training. The integrity of this gate mechanism seems to be essential for the filtering of incoming somatosensory signals and the control of corresponding motor responses. If indeed disturbed thalamo-cortical gating circuits can be restored to normal function by means of appropriate behavioral methods (Sterman, 1980), then one can also envision occurrence of a similar gating mechanism in patients with ST and other dystonic manifestations who respond to EMG feedback and who retain motor control following withdrawal of feedback.

Differing pathways may mediate training, and feedforward mechanisms may play a role with use of operant conditioning (Goldberger, 1974). Konorski (1967) assumes that during operant conditioning, "associative learning" takes place, based to a degree on perceptual images that trigger motor action. We speculate that the newly established visual imagery or the "iEMG language" may also assume such a role. In fact, many of our patients reported that initiation and carrying out of a movement is aided by imagery of oscilloscopic display, tracked during successful therapeutic sessions. It seems that memory of "iEMG language" patterns established during retraining enables many patients to inhibit undesired spasmodic activity and assist them in facilitating selective activity in muscles involved in a desired motor task after training has ceased.

All of the above is, of course, of a hypothetical nature. As research in this unique CNS disorder continues, other cerebral mechanisms may explain the process of self-regulation achieved by behavioral intervention such as described in this chapter. Nevertheless, the results obtained in the presented study demonstrate that therapeutic electromyography is a safe and effective form of behavioral treatment of an otherwise hardly tractable disease, and represents a significant advance in neurological rehabilitation of patients with ST. Therapeutic EMG should be the treatment of choice before prescribing treatment that is less predictable in outcome and more apt to result in complications.

With the recent introduction of a small home-type Bioconditioner module that can be interfaced with home TV monitor for display, many patients with ST and other disorders of voluntary movement may become, in the future, treatable at home. The physician no doubt will be the teacher and the initiator of the program, but the interested and motivated individual will finally have the opportunity to reach out for self-help in a manner that reflects our state of knowledge and our state of technology.

The research described in this chapter was made possible by
support from chairmen and faculty members of the Departments of
Rehabilitation Medicine, Neurology and Neurosurgery of New York
University Medical Center, by financial support from the ICD
Rehabilitation and Research Center, and by the essential
contributions and participation of my colleagues Drs. J. Korein
and B.B. Grynbaum.

REFERENCES

Abse, D.W. Hysteria and related disorders. Bristol: Wright,
    1966.
Adrian, E.D. The mechanism of nervous action. Philadelphia:
    University of Pennsylvania Press, 1935.
Andersen, P. and Sears, T.A. The role of inhibition in the
    phasing of spontaneous thalamo-cortical discharge. Journal
    of Physiology (London), 1964, 173, 459-480.
Andrews, J.M Neuromuscular re-education of hemiplegic with aid
    of electromyograph. Archives of Physical Medicine and
    Rehabilitation, 1964, 45, 530-532.
Ayd, F.J. A survey of drug induced exptrapyramidal reactions.
    Journal of the American Medical Association, 1961, 175,
    1054-1060.
Bach-y-Rita, P. Brain mechanisms in sensory substitution. New
    York: Academic Press, 1972.
Baldessarini, R.J. and Tarsy, D. Mechanisms underlying tardive
    dyskinesia. IN M.D. Yahr (ed.) The basal ganglia, New
    York: Raven Press, 1976.
Bird, B.L. and Cataldo, M.F. Experimental analysis of EMG
    feedback in treating dystonia. Annals of Neurology, 1978,
    3, 310-315.
Brudney, J. Biofeedback in chronic neurological cases. IN L.
    White and B. Tursky (eds.) Clinical biofeedback: Efficacy
    and mechanisms. New York: Gilford Press (in press).
Brudny, J., Grynbaum, B.B.and Korein, J. Spasmodic torticollis:
    Treatment by feedback display of the EMG. Archives of
    Physical Medicine and Rehabilitation, 1974, 55, 403-408.
Brudny, J., Korein, J., Grynbaum, B.B., Friedmann, L.W.,
    Weinstein, S., Sachs-Frankel, G. and Belandres, P.V. EMG
    feedback therapy: Review of treatment of 114 patients.
    Archives of Physical Medicine and Rehabilitation, 1976,
    57, 55-61.
Brudny, J., Korein, J., Levidow, L., Grynbaum, B.B., Lieberman,
    A. and Friedmann, L.W. Sensory feedback therapy as a
    modality of treatment in central nervous system disorders
    of voluntary movement. Neurology, 1974, 24, 925-932.
Brudny, J., Weisinger, M. and Silverman, G. Single system for
    displaying EMG activity designed for therapy, documentation

of results and analysis of research. In R. Foulds and R. Lund (eds.), 1976 Conference on systems and devices for the disabled. Boston: Biomedical Engineering Center, 1976.

Cleeland, C.S. Behavioral technics in modification of spasmodic torticollis. Neurology, 1973, 23, 1241-1247.

Connolly, K. and Jones, B. Developmental study of afferent-reafferent integration. British Journal of Psychology, 1970, 61, 259-266.

Couch, J.R. Dystonia and tremor in spasmodic torticollis. IN R. Eldridge and S. Fahn (eds.) Advances in Neurology, New York: Raven Press, 1976.

Dandy, W.E. The brain. New York: Harper and Ros, 1969.

DeVries, H.A. Efficiency of electrical activity as a physiological measure of the functional state of muscle tissue. American Journal of Physical Medicine, 1968, 47, 10-22.

Eldridge, R. The torsion dystonias: Literature Review and Genetic and Clinical Studies. Neurology, 1970, 20, 1-78.

Eldridge, R. and Fahn, S. (eds.) Advances in Neurology. Dystonia. New York: Raven Press, 1976.

Fahn, S. Biochemistry of the basal ganglia. In R. Eldridge and S. Fahn (eds.) Advances in Neurology. Dystonia. New York: Raven Press, 1976.

Gans, B.M. and Noordergraaf, A. Voluntary skeletal muscles: A unifying theory on the relationship of their electrical and mechanical activities. Archives of Physical Medicine and Rehabilitation, 1975, 56, 194-199.

Goldberger, M.R. Recovery of movement after CNS lesions in monkeys. In D.G. Stein, J.J. Rosen and N. Bputters (eds.) Plasticity and recovery of function in the central nervous system. New York: Academic Press, 1974.

Granit, R. Constant errors in the execution and appreciation of movement. Brain, 1972, 95, 649-660.

Konorski, J. The integrative activity of the brain. Chicago: The University of CHicago Press, 1967.

Korein, J. Iris pigmentation (melanin) in segmental dystonic syndromes including torticollis. Annals of Neurology, 1980, 8, 118.

Korein, J. and Brudny, J. Integrated EMG feedback in the management of spasmodic torticollis and focal dystonia: A prospective study of 80 patients. In M.D. Yahr (ed.) The basal ganglia. New York: Raven Press, 1976.

Korein, J., Brudny, J., Grynbaum, B.B., Sachs-Frankel, J., Weisinger, M. and Levidow, L. Sensory feedback therapy of spasmodic torticollis and dystonia: Results in treatment of 55 patients. In R. Eldridge and S. Fahn (eds.) Advances in Neurology. Dystonia. New York: Raven Press, 1976.

Lal, S. Pathophysiology and pharmacotherapy of spasmodi torticollis: A Review. Canadian Journal of Neurological Sciences, 1979, 6, 427-435.

234 J. BRUDNY

Marinacci, A.A. and Horande, M. Electromyogram in neuromuscular
    re-education. Bulletin of teh Los Angeles Neurological
    Society, 1960, 25, 57-71.
Marsden, C.D. The problem of adult-onset idiopathic torsion
    dystonia and other isolated dyskinesias in adult life. IN
    R. Eldrige & S. Fahn (eds.) Advances in Neurology Dystonia.
    New York: Raven Press, 1976.
Martin, P.R. Spasmodic torticollis: Investigation and treatment
    using EMG feedback training. Behavior Therapy, in press.
Melzack, R. and Wall, P.D. Pain mechanisms: New theory.
    Science, 1965, 150, 971-979.
Patterson, R.M. and Little, S.C. Spasmodic torticollis.
    Journal of Nervous and Mental Diseases, 1943, 98, 571-599.
Penfield, W. and Roberts, L. Speech and brain mechanisms. New
    York: Atheneum, 1976.
Podivinsky, F. Torticollis. IN P. J. Vinken and G.W. Bruyn
    (eds.) Handbook of Clinical Neurology, Vol. 6: Diseases of
    the basal ganglia. Amsterdam: North-Holland, 1968.
Roberts, R. Some thoughts about GABA and the basal ganglia. In
    M.D. Yahr, (ed.) The basal ganglia. New York: Raven
    Press, 1976.
Rubow, R.T. and Smith, K.V. Feedback parameters of
    electromyographic learning. American Journal of Physical
    Medicine, 1971, 50, 115-131.
Scott Kelso, J.A. and Stelmach, G.E. Central and peripheral
    mechanisms in motor control In G.E. Stelmach (ed.) Motor
    control, issues and trends. New York: Academic Press, pp.
    1-40, 1976.
Sterman, M.D. EEG biofeedback in the treatment of epilepsy:
    An overview circa 1980. In L. White and B. Tursky (eds.)
    Clinical Biofeedback: Efficacy and mechanisms. New York:
    Guilford Press, in press.
Sterman, M.B. and Bowersox, S.S. Sensorimotor EEG rhythmic
    activity: A functional gate mechanisms. Waking and
    sleeping. (in press).
Surwit, R. Routable discussion. In L. White and B. Turksy
    (eds.) Clinical biofeedback: Efficacy and mechanisms. New
    York: Guilford Press (in press).
Tralov, E. On the problem of pathology of spasmodic torticollis
    in man. Journal of Neurology, Neurosurgery and Psychaitry,
    1970, 33, 457-463.
Trans, V.T., Snyder, S.H., Major, L.F. and Hawley, R.J. GABA
    receptors are increased in brains of alcoholics. Annals of
    Neurology, 1981, 9, 289-291.
Yahr, M.D. The basal ganglia. New York: Raven Press, 1976.
Zeman, W. Pathology of the torsion dystonias. Neurology, 1970,
    20, 79-88.

BEHAVIORAL TREATMENT OF FECAL INCONTINENCE

Bernard T. Engel

Baltimore City Hospital
Baltimore, Maryland

Introduction

Fecal incontinence is said to occur when there is evidence of soiling or staining on the clothing. Soiling means the presence of stool on the garment, staining means discoloration. Although fecal incontinence is normal in very young children, regular soiling beyond the age of four years usually is pathognomonic. It is important to recognize that incontinence is a physical sign, and that there are a number of clinical states which can include incontinence as one of the presenting features. Table 1 is a list of some of the disorders which can include incontinence as one of their sequelae.

Although the incidence of incontinence in the general population is about 1/1000 (Milne, 1976), incontinence is more common in young children or the elderly (Schuster, 1977). About 40% of all children born with spina bifida are incontinent (Lorber, 1971). However, most cases of incontinence in children cannot be attributed to any organic cause (Liebman, 1979). Likewise, in the elderly the cause of incontinence often is unclear although the course usually is progressive. Because of the social stigma associated with incontinence, its impact goes far beyond the inconvenience it imposes upon its victim or his caretaker. For example, Welbourn (1975) notes that incontinence often imposes an insurmountable obstacle to school placement of children; Evans, Hickman, and Carter (1974) report that none of 13 fecally incontinent, adult patients with histories of meningomyelocele whom they studied had married, whereas 13 of 63 patients with urinary incontinence had married; Evans et al., (1974) also noted that only four of the 13 fecally incontinent

235

Table 1.   Some Disorders Associated
with Fecal Incontinence

1. Congenital

    a.   Hirschsprung's disease

    b.   Spina bifida

2. Endocrine

    a.   Hyperparathyroidism

    b.   Hypothyroidism

    c.   Diabetes mellitus

3. Iatrogenic

    a.   Ano-rectal surgery

    b.   Pharmacotherapy--anti-cholinergic drugs such as phenothiazine

         compounds or anti-Parkinsonism agents.

4. Other

    a.   Scleroderma

    b.   Multiple sclerosis

    c.   Chronic lead poisoning

    d.   Memory deficit

patients (31%) had ever worked, whereas 49 of the 65 urinary incontinent patients (78%) had worked; Milne (1976) notes that incontinence (urinary or fecal) is a prominent reason for institutionalizing elderly patients; and Willington (1976a) notes that hospital care of incontinent elderly patients consumes an excessive amount of staff time.

Incontinence can occur because the patient is unable to retain stool--e.g., following extensive ano-rectal surgery; or in association with chronic, severe retention of stool. The apparent paradox of incontinence coupled with severe constipation can be explained by a consideration of the physiology of the rectum. Stool retention results from sustained hypertonicity of the internal anal sphincter or of the distal rectum. As a result of the hypertonicity, there is no progression of stool to the anus, and there is an accumulation of stool in the rectum. Because the rectum has a large volume capacity, it will expand to retain stool. In such patients radiological examination will reveal a large, fecal mass and a greatly distended rectum which is characterized as megacolon. Normally, rectal distension is associated with a sense of fullness and discomfort and leads to defecation. However, as the rectum becomes fuller, the stretch receptors which sense rectal fullness become habituated or their limit of sensation is passed so that further increases in fullness no longer are appreciated. However, it is noteworthy that during periods of severe constipation such as that which might be present after three or four weeks without a bowel movement, patients often report sensations of extreme discomfort and irritability (Schuster, 1977). Eventually, the impaction becomes so great that the internal anal sphincter is incapable of maintaining a sufficient degree of tonus to retain the fecal matter. This coupled with seepage of liquid stool around the impaction leads to loss of liquid or solid stool.

The Physiology of Continence

In the normal subject distension resulting from appearance of stool in the rectum leads to relaxation of the internal anal sphincter and contraction of the external anal sphincter. The internal sphincter is a smooth, annular muscle which is innervated by autonomic fibers found in the pelvic and hypogastric nerves; the external sphincter is a striated muscle which is innervated by somatic fibers found in the pudendal nerve (Schuster, 1968). The internal sphincter response is characterized by a slow relaxation phase and a prolonged recovery phase so that the normal response period is about 20 sec, whereas the external sphincter contraction is brief and recovery is rapid. It is possible to simulate this process by using a balloon (Figure 1) which is briefly distended to simulate rectal distension, and by using recording balloons placed at the sphincters to monitor their actions (Schuster, Hookman & Hendrick, 1965; Engel, 1978). The response of the internal sphincter appears to be a reflex; however, the contraction of the external sphincter, which once was regarded as a reflex, probably is a highly overlearned voluntary response since it can be readily suppressed in the normal subject (Whitehead, Orr, Engel, et al., in press).

Behavioral treatment of fecal incontinence

Because of the presence of social as well as medical factors in patients with fecal incontinence, this disorder must be understood as a psychophysiological disorder (Engel, in press-a). Table 2 presents a list of five different clinical diagnoses which are associated with fecal incontinence. Each of these classes of patients has shown to be responsive to behavioral treatment. In the ensuing sections, each of these diagnoses and treatments will be reviewed.

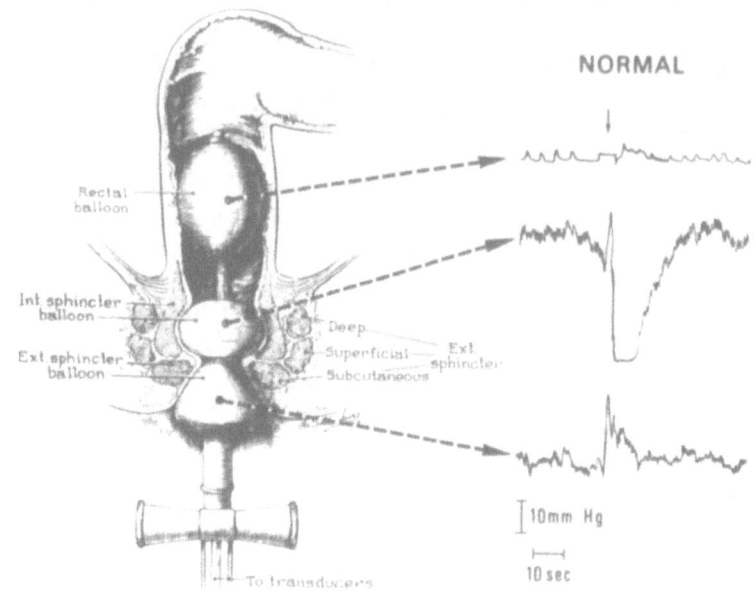

Figure 1.   Illustration of the retro-sphincter responses.
Distention of the rectal balloon leads to
relaxation of the internal (Int.) anal sphincter and contraction of the external (Ext.) anal
sphincter.

1.   Overflow incontinence

As already indicated, overflow incontinence occurs secondary
to stool retention. This problem can be present in persons of any
age; however, it is most prominent in children and the elderly
(Schuster, 1977). Frequently, there is no known cause (Lieberman,
1979). Young (1973) described a behavioral strategy for dealing
with this problem which appears to be effective, and with some
variations probably is the most widely used method. He proposed a
combination of classical conditioning, operant conditioning and
medical management. It is well known that most people experience
an urge to defecate shortly after a meal. Young (1973) has
exploited this phenomenon (which may or may not be a reflex), and
he has shown that if young children with histories of chronic
fecal retention are trained to toilet within about one-half hour
after breakfast, 15/19 (79%) can be trained to become continent in
an average period of five months, and will remain continent on
follow-up (average period = 29 months). Rovetto (1979) has

modified this procedure to treat a series of eight adult women who
were chronically constipated.

Olness, McParland, and Piper (1980) recently reported on the
use of biofeedback involving rectal manometry to treat a series of
50 children. Tenof the children had received rectal surgery for
imperforate anus and the remaining 40 had chronic constipation of
unknown etiology. The method they used was a variation on that
originally described by Engel, Nikoomanesh, and Schuster (1974).
Since this method will be discussed in detail below, only the
Olness et al. (1980) method will be described here. In their
procedure, Olness et al. (1980) educated the children concerning
the nature of their problems and showed them how their ano-rectal
responses differed from normal. Children who could emit anorectal
responses were rewarded for making these responses regularly
following threshold levels of rectal distension ("discrimination
training"), whereas children who could not emit any ano-rectal
responses were rewarded for trying various procedures which were
effective ("response shaping"). Following training, Olness et al.
(1980) reported that 39/40 (98%) of the constipated children and
8/10 (80%) of the imperforate anus group all had regular bowel
movements and 62% of the improved constipated patients and 75% of
the improved imperforate anus patients did not soil. Since all
the children in their series had failed to respond to traditional
therapeutic interventions, these data are very impressive.

2. Memory deficit

Most young children who are diagnosed as mentally defective
and elderly patients who are diagnosed as senile are incontinent.

Table 2. Behavioral Strategies in the
Treatment of Fecal Incontinence

Primary Diagnosis	Behavioral Intervention
1. Overflow incontinence	Classical conditioning and contingency management
2. Memory deficit	Contingency management
3. Anal sphincter dysfunction	Biofeedback: discrimination training
4. Peripheral nerve dysfunction	Biofeedback: discrimination training
5. Spina bifida	Biofeedback: response shaping

Typically, these patients are doubly incontinent. In most of these patients ano-rectal anatomy and physiology is normal. Clinically, these patients present problems in management secondary to their cognitive defects which include poor motivation, confusion or the use of soiling behavior to obtain attention. Foxx and Azrin (1973) have shown that mentally retarded children are very responsive to habit training procedures in which contingencies are placed upon correct performance. Several authors have suggested that these procedures also may be effective in the elderly (Grosicki, 1968; Willington, 1976b); however, there are relatively few data in this age group. Because the incontinence in these patients usually is secondary to the cognitive deficit, most writers have not separated the treatment of fecal and urinary incontinence in this population. However, in view of the large differences in frequency of occurrence of urination and defecation, it might be more efficient to treat these problems separately.

3. Anal sphincter dysfunction

In 1965, Schuster et al. described a method for evaluating the physiological characteristics of the anal sphincters. Their procedure is illustrated in Figure 1. A brief distension of the rectum will produce a reflex relaxation of the internal and sphincter and a voluntary contraction of the external anal sphincter. Subsequently, Tobon, Reid, Talbert, et al. (1968) showed that this procedure could be used to facilitate the diagnosis of Hirschsprung's disease, a congenital defect characterized by a failure of the internal sphincter and rectum to receive any innervation. In 1974 Engel et al. showed that this technique could be used therapeutically to treat patients with fecal incontinence secondary to dysfunction of the external anal sphincter following rectal surgery. Subsequently, Cerulli, Nikoomanesh, and Schuster (1979) studied a series of 35 post-surgical incontinent patients. Among the 25 patients who had had sphincter surgery, reported in the two studies, 24 became fully continent or at least 90% improved (defined as occasional staining and no soiling). Furthermore, most patients relearned sphincter control within one or two two-hour sessions. In both studies the training procedure comprised three stages: 1) a diagnostic stage in which the threshold of rectal distension necessary to yield reliable sphincteric responses was determined; 2) a treatment phase in which patients were trained to emit effective external sphincter contractions to rectal distensions at near- or subthreshold levels. This method is similar to the behavioral procedure of discrimination training since the main problem these patients must solve is to detect the rectal stimulus and then to respond appropriately; 3) a fading stage in which patients were trained to respond appropriately during periods when visual feedback is withheld.

Although there are no known published reports beyond the two cited above, private communications from other gastroenterologists suggest that this method now is the treatment of choice for these patients.

4. Peripheral nerve dysfunction

Engel et al. (1974) studied three patients with either peripheral neuropathy or spinal cord injury, and Cerulli et al. (1979) reported on 11 patients with spinal cord injury and three patients with peripheral neuropathies. Cerulli et al. (1979) also treated three patients with histories of cerebrovascular accidents and two patients with localized injuries in the ano-rectal region which might have included neuropathy (Crohn's disease, radiation proctitis). In both studies the discrimination training procedure described above was used. All of the post-stroke patients, three of the eight meningomyelocele patients, four of the five post-laminectomy patients and two of the six peripheral neuropathy patients were treated successfully. Elsewhere, Whitehead, Engel, and Schuster (in press) have proposed that the critical variable in determining success among patients with peripheral neuropathy is the capacity of these patients to learn to detect rectal distension. They report that patients who are unable to detect distensions above 15 ml have a significantly poorer prognosis than those who have lower absolute thresholds.

5. Spina bifida

In the previous section it was reported that only three of eight patients with histories of meningomyelocele were able to learn to control their anal sphincters using the discrimination training procedure described above. Furthermore, in the Cerulli et al. (1979) study, one child with a history of imperforate anus also failed to achieve 90% or better continence. Two groups of investigators have reported results which indicate that the discrimination training procedure is inappropriate, and that a response-shaping training paradigm is more effective. Whitehead, Parker, Masek, et al. (in press) have shown that spina bifida children have normal capacities to sense rectal distension. However, in response to adequate rectal stimuli these patients respond with reflex internal anal sphincter relaxation and external sphincter relaxation (see Whitehead et al., in preparation). Apparently, these subjects are initially not able to respond appropriately, to rectal distension. They first must be trained to emit external sphincteric contractions either spontaneously or to sub-threshold or near threshold stimuli; only afterwards can the stimulus intensity be increased. A response shaping procedure also was used by Olness et al. (1980) in training imperforate anus children to become continent. Although no sensory threshold studies were done, it is possible that these patients also have normal sensation, and that they cannot respond appropriately to suprathreshold stimuli.

DISCUSSION AND CONCLUSIONS

The studies reviewed above indicate clearly that a number of patients with chronic, severe fecal incontinence who once were thought to be untreatable can be treated. The treatment varies depending upon the underlying mechanisms, and only a comprehensive, psychophysiological analysis will permit one to choose rationally among the available alternatives (Engel, in press-b). Among the effective behavioral procedures one can include classical conditioning, operant conditioning focusing either on response shaping or discrimination training, or contingency management. If one includes dietary management, i.e. stool softeners, or the selective use of suppositories among these behavioral procedures, then the number of behavioral strategies available to the therapist is very large. Furthermore, as the foregoing analyses have shown, these methods are not equivalent: ultimately, rational therapy depends upon an understanding of mechanism (Engel, in press-a).

REFERENCES

Cerulli, M.A., Nikoomanesh, P., & Schuster, M.M. Progress in biofeedback for fecal incontinence. Gastroenterology, 1979, 76, 742-746.

Engel, B.T. Fecal incontinence and encopresis: A psychophysiological analysis. In R. Holzl & W. E. Whitehead, (Eds.) Psychophysiology of the Gastrointestinal Tract: Experimental and Clinical Aspects. New York: Plenum, (in press-a)

Engel, B.T. Behavioral assessment and treatment of fecal incontinence. In J.H. Sandweiss (Ed.) Biofeedback and Family Practice Medicine. (in press-b)

Engel, B.T. The treatment of fecal incontinence by operant conditioning. Automedica, 1978, 2, 101-108.

Engel, B.T., Nikoomanesh, P. & Schuster, M.M. Operant conditioning of rectosphincteric responses in the treatment of fecal incontinence. New England Journal of Medicine, 1974, 290, 646-649.

Evans, K., Hickman, V. & Carter, C.O. Handicap and social status of adults with spina bifida cystica. British Journal of Preventive Social Medicine, 1974, 28, 85-92.

Foxx, R.M. & Azrin, N.H. Toilet training in the retarded. Champaign, Ill: Research Press, 1973.

Grosicki, J.P. Effect of operant conditioning on modification of

incontinence in neuropsychiatric geriatric patients. Nursing
    Research, 1968 17, 304-311.
Liebman, W.M. Disorders of defecation in children. Post-graduate
    Medicine, 1979, 66, 105-110.
Lorber, J. Results of treatment of meningomyelocele: An analysis
    of 524 unselected cases, with special reference to possible
    selection of treatment. Developmental Medicine & Childhood
    Neurology, 1971, 13, 279-303.
Milne, J.S. Prevalance of incontinence in the elderly age
    groups. In F. L. Willington, (ed.) Incontinence in the
    Elderly. London: Academic Press, 1976.
Olness, K., McParland, F.A. & Piper, J. Biofeedback: A new
    modality in the management of children with fecal soiling.
    Journal Pediatrics, 1980, 96, 505-509.
Rovetto, F. Treatment of chronic constipation by classical
    conditioning techniques. Journal of Behavioral Therapy &
    Experimental Psychiatry, 1979, 10, 43-146.
Schuster, M.M. Constipation and anorectal disorders. Clinics in
    Gastroenterology, 1977, 6, 643-658.
Schuster, M.M. Motor action of rectum and anal sphincters in
    continence and defecation. In C.F. Code, (Ed.) Handbook of
    Physiology, Section 6, Alimentary Canal, Vol. IV, American
    Physiological Society, Washington, D.C., 1968.
Schuster, M.M., Hookman, P. & Hendrix, T.R. Simultaneous
    manometric recording of internal and external anal
    sphincteric reflexes. Bulletin Johns Hospital, 1965, 116,
    79-88.
Tobon, F., Reed, N.C.R.W., Talbert, J.L. & Schuster, M.M.
    Nonsurgical test for the diagnosis of Hirschsprung's disease.
    New England Journal of Medicine, 1968, 278, 188-194.
Welbourn, H. Spina bifida children attending ordinary schools.
    British Medical Journal, 1975 1, 142-145.
Whitehead, W.E., Engel, B.T. & Schuster, M.M. Perception of
    rectal distension is necessary to prevent fecal incontinence.
    Proceedings of the 28th International Congress of
    Physiological Sciences, Budapest, 1980.
Whitehead, W.E., Orr, W.C., Engel, B.T. & Schuster, M.M. (in
    preparation) The phasic external anal sphincter response
    to rectal distension: Learned response or reflex.
Whitehead, W.E., Parker, L.H., Masek, B.J., Cataldo, M.F. &
    Freeman, J.M. Biofeedback treatment of fecal incontinence in
    meningomyelocele. Developmental Medicine & Child Neurology.
    (in press)
Willington, F.L. Introduction. In F. L. Willington, (ed.)
    Incontinence in the Elderly. London: Academic Press, 1976a.
Willington, F.L. The physiological basis of retraining for
    continence. In F. L. Willington (Ed.) Incontinence in the
    Elderly. London: Academic Press, 1976b.
Young, G. C. The treatment of childhood encopresis by conditioned
    gastro-ileal reflex training. Behavioral Research & Therapy,
    1973, 11, 499-503.

POTENTIALITIES OF AUTOMATION AND OF CONTINUOUS RECORDING

AND TRAINING IN LIFE

Neal E. Miller - The Rockefeller University
New York
Barry R. Dworkin - Pennsylvania State College
Hershey, Pennsylvania

Two serious problems preventing biofeedback from achieving its full therapeutic potential are how to devote enough time to training and how to insure transfer from the clinic to life. If imaginatively exploited, rapid advances in electronic technology, microprocessors, and other miniaturized circuits on chips may provide solutions to both of these problems and open up important new possibilities for the prevention or the treatment of a variety of additional disorders.

Enough Training Time

Compared with the amount of time required instrumentally to learn high levels of motor skill, such as juggling, tennis, or playing the violin, the number of hours spent on most forms of biofeedback training is miniscule. Furthermore, the amount of training time involved in most of the all-too-few basic research studies investigating the effects of variables on biofeedback learning is far too short. This may account for their meagre contribution to date to improving the efficiency of learning by biofeedback (Miller, 1982). But the evidence from such skills is that control over the skeletal muscles can continue to be improved by long periods of practice. Sometimes plateaus can make one think that a limit has been reached but, if practice continues, further improvement may be achieved. Similarly, there is evidence that greater control over at least some visceral processes can be achieved by long periods of training. For example, during approximately 250 hours of training, Harris et al. (1973) have taught baboons to produce 30 mmHg increases in systolic blood pressure sustained for 12 hours. Other evidence has been summarized elsewhere (Miller, 1981). Brudny's chapter in this volume illustrates the value of using much longer periods of training with difficult cases of neuromuscular rehabilitation.

But with conventional procedures, long periods of training consume prohibitive amounts of the costly time of skilled therapists. Thus, with such procedures, it often is impracticable to achieve the full potential of therapeutic training. This probably is the main reason why very early results by Shepherd Ivory Franz (1920) showing the potential of training in neuromuscular disorders did not become a part of standard practice.

Automated Therapeutic Training

One of the advantages of biofeedback is that measuring instruments, if properly applied by a therapist with the required knowledge and skills, can take much of the burden off the therapist's sense organs and extend his or her ability to perceive small increments of improvement that gradually can be shaped into larger achievements. Furthermore, displays of the same measurements to the patient -- for example, the movement of a spot on a TV tube indicating a slight contraction of a largely paralyzed muscle or the relaxation of a spastic one -- can substitute for continual observation and instruction by the therapist.

After the skilled therapist has taken the critical steps of correctly diagnosing the patient's problem and assigning a task for him to practice, suitably designed systems for further analysis and display should allow a single therapist to monitor the practice and progress of a group of patients, advancing each one to the next stage as soon as he has shown sufficient progress.

Some imaginative steps toward economically providing additional practice have been taken. In her chapter in this volume, Patel has described a method of using biofeedback combined with relaxation tapes in training groups of patients. For patients with neuromuscular disorders, Brudny (1979) has designed EMG equipment that can be plugged into a TV set, enabling a patient to practice a given assignment at home, economizing on both the patient's and the therapist's time. He also describes this equipment in his chapter in this volume. Kristt and Engel (1975) have experimented with having the patient use a sphygmomanometer to take his or her blood pressure at home. Colgan (1981) has used, for home practice, a device in which the pressure transducer in the occluding cuff of an automatic blood pressure recording device operates through the microchip of a computer game to generate on a TV screen a falling diagonal line that changes temporarily to a horizontal step when the Korotkoff sound is heard by a microphone, indicating the systolic pressure, and then continues falling to become horizontal again when the Korotkoff sound stops, indicating diastolic pressure. The systolic and diastolic pressures from the preceding trial are indicated, respectively, by the upper and lower limits of a lightly shaded horizontal band across the screen. The patient's task is to produce horizontal steps that are below the upper and lower edges of this band. Superior and cheaper devices for automatically measuring blood pressure could materially facilitate the use of home practice to reduce hypertension.

In motivating children recovering from severe injuries or with neuromuscular disorders to exercise certain muscles and joints or to practice skilled movements, use has been made of equipment in which a certain rate of performance of a specified movement or series of movements is necessary in order to keep the sounds of a record player audible or to keep the screen of a TV cartoon show free of snow (Friedlander, 1974). A "Star Wars" TV game has been adapted for similar purposes (Colgan, personal communication). For example, the light pistol used to shoot down space ships was strapped to the head of a spastic child who had difficulty in controlling head movements. The light beam from the pistol was adjusted to be wider than usual. For shaping, it could be narrowed down. This game was found to generate extremely high motivation and hence long periods of intensive practice by such children. Other uses of automation in rehabilitation, such as the Limb Load Monitor, the Step Length Monitor, and a device for monitoring the position of the knee joint, are described by Ince (1976).

The power of many of the foregoing types of applications could be increased still further if part of the burden of shaping the correct response could be shifted from the therapist to a microprocessor or a mini-computer. A step in this direction might be a system in which the therapist periodically monitors graphic displays showing different aspects of progress, such as changes in the central tendency and in the distribution of responses, and, when necessary, types in new constants relevant to the shaping. At present, shaping is largely an art; more basic research is needed to improve that art and to advance it toward a science that could provide a rational basis for fully automated programs of shaping. The extent of our ignorance about what types of changes in a visceral response such as blood pressure should be reinforced to achieve optimal shaping has been pointed out elsewhere (Miller & Dworkin, 1971; Miller, 1982).

Transfer from Clinic to Life?

In some types of biofeedback training, it is fairly obvious whether or not the patient is able to take the crucial step of transferring the control learned in the clinic to the circumstances of his everyday life. Some examples are the patient paralyzed by high spinal lesions who has learned increases in blood pressure that enable them to tolerate a more normal, vertical posture without fainting (Miller & Brucker, 1979); a patient with neuromuscular disorders who is able to walk without braces or to feed himself (Brudny, 1979); or a patient who has learned to control fecal incontinence (Engel et al., 1974). But in other cases, such as headaches, there may be some question about whether a patient is making a verbal report of success in order to please the therapist who has established a personal relationship with that patient. Thus, there is reason to try to collect confirming evidence such as prescriptions for painkillers and whether or not the patient eventually tries some other form of treatment. Verbal reports of relaxation represent another grey area.

The worst problems come with a condition like hypertension, which a patient normally cannot perceive and of which the first event that he senses may be a catastrophic one such as a stroke. For this condition there is a very real possibility that the patient may learn in the clinic to lower his blood pressure either by relaxation or by more direct control and then use this control whenever his blood pressure is being taken, but he may not use it during the many distressing and distracting situations encountered during the major part of his daily life. Thus, there is a very real danger that the physician may be misled into taking the patient off antihypertensive drugs that he really needs (Miller, 1974, 1979).

Patel (1977) has recognized the problem of transfer and taken definite steps to try to improve the transfer to the life situation by a variety of means. She has pasted a red dot on the patient's wristwatch to remind him to practice a mini-relaxation whenever he looks at the time. Similarly, the patient is supposed to practice relaxation in other situations that generally make him tense, such as hearing a telephone ring or encountering a red light. Special techniques are devised to deal with the situations that increase the tensions of the individual patient. Perhaps this approach accounts for her apparent success. Stroebel's (1979) quieting reflex is aimed at a similar problem.

But there is no real way of knowing how successful Patel's patients have been with techniques to transfer their control over blood pressure from the clinic to life. The most convincing way to determine that they have reduced their blood pressure in the life situation would be to use the portable system developed by Bevan et al. (1961) for the continuous recording of blood pressure via a catheter into the artery. While this technique is quite safe, it does involve a slight element of risk which has kept investigators from widely using it. For the present, the method of choice probably is one of the portable devices that automatically take the blood pressure at frequent set intervals such as every 15 minutes during the day and every 30 minutes while the patient is asleep (Harshfield et al., 1981). This system is noninvasive, using a cuff and microphone to take an indirect measure analogous to the traditional clinical one. The patient is alerted by feeling the inflation of the cuff and must remain still for a few moments in order to avoid movement artifacts. Thus, there is a possibility that the patient might specifically relax or use other methods to control his pressure while it is being taken. But if he is able to do that repeatedly under a variety of life situations, this would indicate a considerable degree of control. Such information from the life situation would be much better than none at all.

The Problem of Maintaining Performance

Certain types of skilled performance, such as being able to feed or dress oneself by a patient with a neuromuscular disorder that previously has prevented this, are easily perceived by the patient and ordinarily provide their own rewards. Thus, once achieved, there should be no problem of maintaining the performance unless there is some conflicting strong reward for maintaining the disability, in which case some other means must be found to deal with that conflict. For patients whose activities have been severely limited by orthostatic hypotension, avoiding fainting and achieving a wider range of activities is a reward for the performance of the learned ability to increase blood pressure (Brucker, 1977; Miller & Brucker, 1979). For activities of the foregoing kind, the problem is to achieve a degree of skill that is sufficient to be rewarding; after this, the learned performance should be self-maintained.

We already have pointed out that for the patient with hypertension, the effects of lowering blood pressure are not immediately perspicuous. So, unless such a patient can learn to perceive changes in blood pressure and is motivated to be rewarded by this perception, we will expect problems of maintaining the performance.

If the reinforcement is delayed, as is often the case with responses such as relaxation and hand-warming that are useful in preventing the occurrence of the symptom, we may expect problems of maintaining the activity that produces the therapeutic effect. And, indeed, the evidence suggests that this is the case; to secure the benefits of relaxation, it has to be practiced daily, and there appears to be a problem with maintaining prolonged compliance with such daily practice. Similarly, it is likely that the effect of practicing EEG rhythms ranging from 9 to 20 Hz in reducing epilepsy is an indirect one; the evidence suggests that the patient has to continue such training in order to maintain its benefits (Miller, 1978; Ray, Racynski, Rogers & Kimball, 1979; White & Tursky, 1982).

For the child who is a deep sleeper and treated with Mowrer and Mowrer's (1938) device, in which the first drop of moisture causes a buzzer or a bell to sound, once the use of the device is discontinued the reward of a dry bed and/or punishment of a wet one is considerably delayed and there is some problem of relapse (Lovibond, 1964). However, Finley et al. (1973) have secured evidence that a schedule of partial reinforcement has the expected effect of increasing resistance of the correct response to extinction; in other words, of counteracting the tendency to relapse. Perhaps randomly introducing nights during which the sound of the waking signal is progressively delayed after the first drop of moisture also would help to counteract the problem of relapse.

Carrying Training into Life

One way of solving the two problems of securing sufficient training time and transfer from the clinic to life may be illustrated by an inconspicuous and lightweight posture-training device which the authors have developed for the treatment of idiopathic scoliosis, an S-shaped lateral curvature of the spine which develops approximately five times as frequently in pre-adolescent girls as in boys. At present, the least drastic, while yet effective, form of treatment is wearing a physically restricting and cosmetically disfiguring brace like that illustrated in Figure 1.

Dr. Saran Jonas, one of our collaborators on another project, had suggested that scoliosis might be produced by asymmetrical contractions of the muscles on the two sides of the spine and that it might be possible to use the EMG to measure these contractions and then supply the patient with feedback that would enable her to learn how to correct the condtion. Some brief tests on a patient generously supplied by Dr. Gordon Engler showed that the relevant muscles were deep ones that could not readily be recorded by surface electrodes. So this idea did not seem to be practicable.

Figure 1.   Brace used to treat scoliosis, an S-shaped curvature
            of the spine. (Picture drawn by F. Netter, M.D. and
            reproduced with permission from Scoliosis.   CIBA
            Clinical Symposia, Vol. 24, No. 1, 1972)

     But the authors continued to think about the problem,
shifting to the different tack of trying to conceive of some way
of measuring posture accurately enough to provide feedback for
successful training. After discarding dozens of approaches, we
finally hit upon a harness consisting of a loop through the
crotch around the long axis of the body and attached to the
front   and   back   of   a   little   halter   around   the   neck.
Straightening   the   spine   lengthened   the   loop   by   an   amount
measured by a linear transducer. But when this device was
tested, it was found that the easiest way for the patient to
lengthen this loop was by expanding the chest so that we were
training her to take deep breaths. This was corrected by
subtracting from the loop around the major axis a suitable
component of the signal from a loop around the chest, yielding a
pure measure of lengthening the spine. Additional developments
produced the device illustrated in Figure 2, that is neither
physically restricting nor cosmetically disfiguring. A nylon

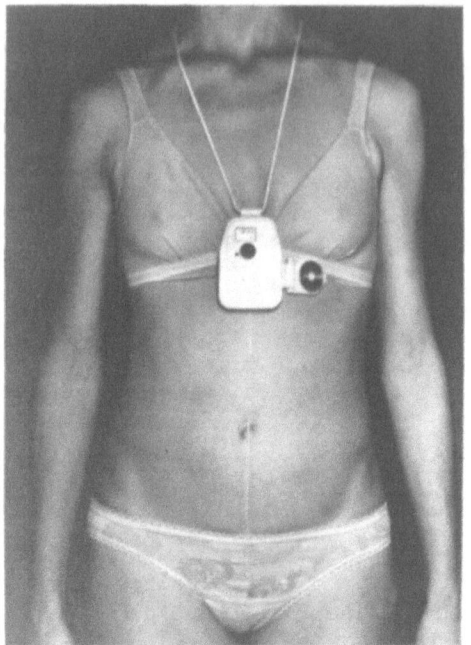

Figure 2. A posture-trainer as a behavioral device for
           scoliosis. Nylon harness around the two dimen-
           sions of the body slide in Teflon tubes. From
           the lengthening of the vertical harness, the
           device substracts a suitable fraction of the
           lengthening of the horizontal one to yield a
           pure measure of lengthening (and hence
           straightening) of the spine. Twenty seconds
           after the patient adopts a poor posture, a
           barely audible warning tone sounds. If she has
           not straightened within another 20 seconds, a
           louder one sounds. An improved posture turns
           off either tone. A digital watch measures the
           time out of good posture. Other refinements
           are described in the text.

fishline slips inside a Teflon tube so there is no friction against the skin.

With further testing, we soon found that having the device sound a tone as soon as the patient adopted a poor posture emphasized the punishment aspect of the tone and made the patient nervous and the device aversive. Introducing a 20-second delay before the tone sounded shifted the emphasis of the task from punishment by the tone for slumping into a poor posture to reward by turning off the tone for straightening up into a better one. This change made the device much more acceptable to the patients. But in this situation there was a tendency to learn the least effortful response, which was momentarily to straighten up every 20 seconds or slightly less. We solved this situation by requiring the patient who had just turned off the tone to earn each second of delay by remaining in a good posture for one second up to a limit of 20 seconds.

Another difficulty was the fact that if the tone was too loud, the girl's friends would hear it which tended to embarrass her, but if it was too weak she would tend to neglect it. We solved this problem by modifying the device so that a tone barely audible to the patient would be sounded at the end of 20 seconds and, if this tone were not turned off by adopting a correct posture within 20 additional seconds, a somewhat louder tone would sound. Under these circumstances, the girls learned to respond to the weak tone so that the louder one virtually never occurred.

One of the problems in the use of such a device is to determine the optimal rate of shaping the patient into the best possible posture. If the criterion for success is made too difficult, the patient will fail and become discouraged; if it is made too easy, the training will be very inefficient. As a check on the patient's performance to be used to guide the process of shaping, we incorporated a digital timer into the device which measures the amount of time that the patient is below the criterion posture. In the future, it should be possible to use a microprocessor chip automatically to adjust the criterion in the light of the patient's performance. Such a device should be designed so that the therapist can set it to move ahead faster for a patient who can work hard and more slowly for one who is easily discouraged. Elsewhere, Dworkin (1981) has discussed in more detail the behavioral principles designed into our posture-training device.

The patient starts by wearing the device two hours a day; as she becomes more skilled in adopting the correct posture and her muscles become strengthened, the period of time is lengthened until she is wearing it all day and all night, with a slight adjustment for the fact that the spine tends to straighten somewhat when the patient is lying down. Wearing the device virtually continuously reduces the reward for leaving it off and also avoids the patient's problem of deciding when to take it off and when to put it on.

The economy achieved by the use of such a device can be realized if one considers the difficulty of having a therapist continually scrutinizing a patient's posture, saying "Straighten up" whenever she adopts a poor posture and "Good" whenever she achieves a better one. But, since the device must be worn in order to be effective and since the criterion must be adjusted properly, we believe that automation will not eliminate the need for the therapist's skills. Instead, it will help the therapist economically to handle more patients.

To date, we have preliminary evidence showing that this device does indeed teach the patient to adopt a better posture, as shown by its measure of lengthening of the spine. Furthermore, the effect on scoliosis as measured by the orthopedist's x-rays of the spine also is encouraging. A report on the results with the first group of 12 patients is being prepared, and we are in the process of initiating the larger studies that will be required to secure completely definitive results.

Other Similar Devices

After we had begun testing our scoliosis device, we were encouraged to learn that Azrin et al. (1968) had used a somewhat similar device successfully to correct a round-shouldered posture. Either two cloth shoulder straps, or, for a more exact measure, two pieces of adhesive tape[1] were attached via an elastic cord to a switch which is closed whenever the cord is stretched by a slouching posture. Closing the switch causes a loud tone to sound. With the successful results reported by Azrin et al., it is remarkable that apparently no one else has followed up his report with a more extensive use of his ingenious device.

---

[1]Actually, we had considered a similar system to measure lengthening of the spine as an index of improved posture, but had discarded attachment via tape because variable movements of the skin would introduce too much error to measure accurately the less extensive movements caused by straightening of the spine.

Other devices have been designed for use either in the life situation or in extensive sessions of automated training. One such device sounds a tone whenever a patient with bruxism shows more than a criterion level of EMG in the masseter muscle (Solberg & Rugh, 1972). Colgan (personal communication) has used a pulse-reading wristwatch to train joggers to be able to estimate their heart rate more accurately. With sufficient ingenuity, yet other devices can and will be devised -- for example, one to measure peripheral circulation by a pair of thermistors, one of which will record skin and the other air temperature, with suitable circuitry to interpret the difference and, if necessary, compensate for the effects of postural changes.

Need for Artifact-Free Recording

The use of a portable device for training in the life situation involves an especially strong requirement for artifact-free recording. But other uses may be somewhat less demanding.

The use of a device for training will motivate the patient to learn to produce artifacts. Ordinarily, he will not distinguish between a signal that is produced by a correct response and one that is produced by an artifact. In many cases, it will be easier to produce the artifact than the correct response. Whenever this is the case, there will be a strong tendency for the patient to learn to produce the artifact unless the response that produces it is so perspicuous that what he is doing can be made obvious to the patient. At best, the artifact is likely to be confusing to the patient; at worst, he is likely to learn to produce it.

For this reason, there is an especial need for progress in developing artifact-free electrodes and transducers as well as superior systems for filtering out or for detecting and discarding artifacts. For these purposes, advances in miniaturized electronic circuitry should be particularly useful. One logical approach is to record the same change in two or more ways that are different enough so that a given movement is unlikely to produce simultaneous artifacts. Then a change in either the correct or the incorrect direction can be signaled only when there is agreement among the different methods of recording. For example, heart rate could be measured by the ECG, by a microphone over the heart, and by a transducer to pick up the pulse with a suitable correction for delay. Then interbeat intervals for which the three measures did not agree could be discarded. Similarly, responses indicating either "stress" or conversely "relaxation" could be detected by appropriate changes in skin temperature, skin resistance, and muscle tension recorded simultaneously from different locations.

For other uses of recording in the life situation there will be no special motivation to learn to produce artifacts. Thus, occasional artifacts are more likely to be random and less likely to be damaging. One such use that has already been discussed is checking up on the transfer from the clinic to life. To avoid learning to produce artifacts during such use, no immediate feedback should be given to the patient.

Another use is to warn the patient that he is responding improperly and needs to take some kind of steps to cope with the situation. Yet another use might be to allow a patient periodically to monitor some relevant variable. Gordon Ball (personal communication) has suggested the utility of a device to allow or to prompt patients with lower back problems to monitor postures that are likely to cause difficulties for them.

Discovering Relationships

Finally, a use for portable recording devices in the life situation is to discover relationships between particular situations and undesirable physiological or skeletal responses. One way of using the device can be to have it give a signal, such as a weak tone or a vibratory stimulus to the skin, whenever the undesirable response occurs. As soon as practicable after receiving the signal, the patient should dictate into a channel of its tape recorder what was happening at that time. As a control, the device should give similar signals to the patient at random times when the undesired response is not occurring. Evidence for the relationship can be used to help the therapist and the patient to discover the kind of situations that cause the problem. If this by itself does not solve the problem, it can motivate the patient to follow suitable therapy; the therapist can try to teach him how to cope better with such situations -- in the extreme case, to avoid or to escape from them.

Evidence from such recordings could be used to detect undesirable effects of certain administrative procedures or of other aspects of the social or physical environment, and to suggest changes that will reduce the undesirable effects. After such changes have been made, additional recording could determine whether or not the changes indeed have been beneficial.

REFERENCES

Azrin, N., Rubin, H., O'Brien, F., Ayllon, T. and Roll, D.

Behavioral engineering: postural control by a portable
operant apparatus. Journal of Applied Behavior Analysis,
1968, 1, 90-108.

Bevan, A.T., Honour, A.J. and Stott, F.E. Direct arterial
pressure recording in unrestricted man. Clinical Science,
1961, 36, 329-344.

Brucker, B.S. Learned voluntary control of systolic blood
pressure by spinal cord injury patients. Ph.D. Thesis, New
York University, 1977.

Brudny, J. Helping hemipiaretics to help themselves. Journal of
American Medical Association, 1979, 241, 814-818.

Colgan, M. Medical uses of biofeedback: principles and case
studies. New Zealand Medical Journal, 1981, 93, 49-51.

Dworkin, B.R. Incorporation of learning principles into
treatment systems. In: Biofeedback: Basic Problems,
Clinical Aspects. E. Richter-Heinrich and N. E. Miller,
(eds.) East Berlin: VEB Deutsche Verlag der
Wissenschaften, 1981, in press.

Engel, B.T., Nikoomanesh, P., and Schuster, M.M. Operant
conditioning of rectosphincteric responses in the treatment
of fecal incontinence. New England Journal of Medicine,
1974, 290, 646-649.

Finley, W.W., Besserman, R.L., Bennett, L.F., Clapp, R.K. and
Finley, P.M. The effect of continous, intermittent, and
"placebo" reinforcement on the effectiveness of the
conditioning treatment for enuresis nocturna. Behavior
Research Therapy, 1973, 11, 289-297.

Franz, S.I. Cerebral-mental relations. Psychology Review, 1920,
28, 81-95.

Friedlander, B.Z. Applications of research in learning and
conditioning. Rehabilitation Psychology, 1974, 21,
142-149.

Harris, A.H., Gilliam, W.J., Findley, J.D. and Brady, J.V.
Instrumental conditioning of large-magnitude, daily,
12-hour blood pressure elevations in the baboon. Science,
1973, 182, 175-177.

Harshfield, G.A., Pickering, T.G. and Laragh, J.H. White coat
hypertension: Elevated blood pressures in patients with
borderline hypertension related to the anxiety of an office
visit. Journal of Behavioral Medicine, 1982, in press.

Ince, L.P. Behavioral Modification in Rehabilitation Medicine.
Springfield, Ill.: Charles C. Thomas, 1976.

Kristt, D.A. and Engel, B.T. Learned control of blood pressure
in patients with high blood pressure. Circulation, 1975,
51, 370-378.

Lovibond, S.H. Conditioning and Enuresis. Oxford: Pergamon
Press, 1964.

Miller, N.E. Introduction: Current issues and key problems.
In: Biofeedback and Self-Control 1973, N.E. Miller, T.X.
Barber, L.V. DiCara, J. Kamiya, D. Shapiro, and J. Stoyva,

(eds.) Chicago: Aldine, 1974, pp. xi-xx.

Miller, N.E. Biofeedback and visceral learning. Annual Review Psychology, 1978, 29, 373-404.

Miller, N.E. General discussion and a review of recent results with paralyzed patients. In: Clinical Applications of Biofeeedback: Appraisal and Status, R.J. Gatchel and K. P. Price, (eds.), New York: Pergamon Press, 1979, pp. 215-225.

Miller, N.E. Some directions for clinical and experimental research on biofeedback. In: Clinical Biofeedback: Efficacy and Mechanisms. L. White and B. Tursky, (eds.) New York: Guilford Press, 1982, in press.

Miller, N.E. and Brucker, B.S. Learned large increases in blood pressure apparently independent of skeletal responses in patients paralyzed by spinal lesions. In: Biofeedback and Self-Regulation. N. Birbaumer and H.D. Kimmel, (eds.) Hillsdale, N.J.: Lawrence Erlbaum Associates, 1979, pp. 287-304.

Miller, N.E. and Dworkin, B.R. Critical issues in therapeutic applications of biofeedback. In: Biofeedback: Theory and Research, G.E. Schwartz and J. Beatty, (eds.) New York: Academic Press, 1977, pp. 129-162.

Mowrer, O.H. and Mowrer, W.M. Enuresis -- a method for its study and treatment. American Journal of Orthopsychiatry, 1938, 8, 436-459.

Patel, C.H. Biofeedback-aided relaxation and meditation in the management of hypertension. Biofeedback and Self-Regulation, 1977, 2, 1-41.

Ray, W.J., Racynski, J.M., Rogers, T. and Kimball, W.H. Evaluation of Clinical Biofeedback. New York: Plenum Press, 1979.

Solberg, W.K. and Rugh, J.D. The use of biofeedback devices in the treatment of bruxism. Journal of Southern California State Dental Association, 1972, 40, 852-853.

Stroebel, C. The Queting Reflex. New York: Guilford Press, 1979.

White, L. and Tursky, B. (eds.) Clinical Biofeedback: Efficacy and Mechanisms. New York: Guilford Press, 1982, in press.

ACKNOWLEDGMENTS

The exploratory work on the posture-training device for scoliosis was supported by Biomedical Research Support Grant 5S07 RR07065. Additional work was supported by research grant MH 28145 from the National Institute of Mental Health, by The National Foundation, and by the Harold and Beatrice Renfield Foundation, Inc.

PATHOPHYSIOLOGY OF HEADACHE FOR BEHAVIORAL THERAPISTS

John R. Graham

The Faulkner Hospital
The Headache Research Foundation
Boston, Massachusetts

My assignment in this chapter is to bring to your attention some features of the pathophysiology of functional headaches which may be useful in planning their management by behavioral techniques. Head pain may occur not only when the head is traumatized or inflamed or distorted by tumor but also when its complicated machinery strains to meet overwhelming mental and emotional distress. It is this "cranial angina" which behavioral therapy may be able to modify, even as help with changes in lifestyle and physical exertion may aid the Type A man with "angina of the heart." The brain really is, amongst its many other functions, the health care system or behavioral therapist for the body. But, as you all know, even behavioral therapists can get tired or have difficult, maybe even painful days.

The various grossly similar areas of the brain contain a wide variety of differing enzymes, vasoactive amines, receptor sites, neurotransmitters, neuropeptides, agonists and antagonists of serotonergic, adrenergic, dopaminergic systems that we are just beginning to know enough about to realize that we are in our infancy of acquaintance with them as yet. Recent information about cranial circulation makes it clear that the blood flow in the brain is greater than had been imagined, that its volume is governed largely by CO_2 and lactate with significant fine tuning by the autonomic nervous system (which previously had been thought to be unimportant) ; a fine tuning reactive to many vasoactive amines such as dopamine, catecholamines and serotonin, which assist in shunting blood about from one region to another, depending on local metabolic activity, and thus creating the possibilities for areas of

relative ischemia and "steal" mechanisms which may result in
malfunction and pain.

We are learning that the limbic system and hypothalamus are
the source of transmitters that govern many of the body's
systemic physiologic activities in relation to stress and mood
and that there are nerve tracts from the voluntary cortex
connected to this system (Bruyn, 1980). The list of releasing
and regulating factors secreted in these areas to effect the
production of all sorts of hormones governing basic homeostasis
and reactions of the body grows almost monthly. The last five
years have seen an incredible growth of knowledge about the
central nervous system pain receptors and the behavior of the
transmitters that affect many bodily functions, doubling also as
important cogs in our central and peripheral pain mechanisms --
the endorphins and enkephlins, substance P, angiotensin and
cholecystokinin, and somatostatin and neurotensin and others
(Sweet, 1980). Having considered for a long time that the pain
of many headaches originated in peripheral structures, such as
the superficial cranial arteries and veins and muscles, we now
are in a quandary as to how much of this process is really based
peripherally and how much is under central control, or both
(Graham & Wolff, 1938).

It is also well to remember that the cranial circulation is
divided into three main categories -- the branches of the
external carotid and those of the internal carotid, and those of
the vertebro-basilar system. Although the branches of the
internal carotid take care of most of the circulation of the
brain, one branch, the ophthalmic artery, leaves the cranium to
go to the eye and to the forehead. On the other hand, the
middle meningeal branch of the external carotid enters the skull
to supply the meninges. The vertebro-basilar arteries take
origin outside the skull and then penetrate the cranium to join
up with the internal carotid system in the Circle of Willis.
The extracranial arteries are different from their intracerebral
counterparts in several ways. They are structured differently
and respond differently to vasoactive substances in a manner
more like the major systemic vessels than the branches of the
internal carotid, which, you will recall, lose their pain
sensitivity as they become the pial vessels. Ergotamine
constricts the external carotid branches markedly but does not
seem to cut down cerebral blood flow.

And there, in the crotch of the bifurcation of the common
carotid in the neck, sits the traffic cop -- the carotid body --
integrating and governing relative flow in the two differently
oriented but integrated systems of extra- and intra-cranial circulation. If he falls asleep or even gets tired or sick, I
could well imagine a traffic jam at this bad corner where the

all-important circulation of the all-important organ of man --
his brain - must be given preferential treatment over all other
areas of the body. This perfect control must obtain whether man
is standing upright with his head on top, or bending over with
his head upside down looking for a collar button under a bureau.
Is it difficult to imagine that this most delicate and sensitive
mechanism which must function perfectly in all positions under
all circumstances, every second, for 70 or more years, may from
time to time have periods in which it temporarily falters with
resulting malfunction and pain?

In addition to the complicated machinery which evolution
has developed for controlling the circulation of the cranium,
another set of reflexes has developed over the ages to protect
the head by tightening its muscular supports to hold it firmly
on its neck against adverse influences of many sorts. These
range from confrontation with a saber-toothed tiger, the pelting
of rain in the rain forest or the snow of a blizzard, the
prickles of the porcupine, the tension of a nightmare, or the
acid accusations of a spouse. We have all felt the tightness of
the muscles in our neck, head and jaws in the dentist's chair,
or while driving against traffic in a snowstorm. This age-old
reflex becomes involved at least as part of the total response
in muscle-contraction or tension headache.

These observations tell us that man's headaches represent a
reaction   to   change   or   stress   involving   several
pathophysiological mechanisms which may operate singly or in
combination.   Sometimes we see head pain appearing in pure
culture as migraine or muscle contraction headaches. More often
we see one of these entities complicated at some stage by
elements of another. In "Combined Headache" we see all these
mechanisms operating simultaneously as important elements of the
clinical experience.

MUSCLE CONTRACTION (TENSION) HEADACHE

The detailed pathophysiology of Tension Headache remains
veiled in considerable mystery. In 1962 the Ad Hoc Committee on
Classification of Headache of the NINDB changed the name of
Tension Headache in an effort to focus diagnostic attention on
the mechanism of the headache which was considered most
important, at that time, in its pathophysiology. The name
Muscle Contraction Headache, therefore, was substituted for the
popular "Tension Headache" monicker. This, however, has never
been widely accepted. Although muscle contraction does appear
to be a fairly constant accompaniment to muscle contraction
headache, it is well recognized that it is probably by no means
the only etiological mechanism. Ischemic changes have also been
noted by some observers to be present during muscle-contraction

headache but other evidence suggests increased blood flow to be
present as measured by tissue radioactive sodium uptake (Tunis &
Wolff, 1952, 1954). More recently it has been pointed out that
some headaches which clinically fit the picture of muscle
contraction headache may or may not be accompanied by
muscle-contraction at all (Anderson & Franks, 1981). It seems
possible that some of these may represent conversion symptoms in
which constant vague head discomfort is reported without a basis
in muscle contraction or whatever other mechanisms may play a
role in muscle-contraction headache.

In general, it seems fair to state that from clinical
observation and physiological experiments contraction in the
head and neck muscles is an important feature in the mechanism
of many of these patients, and one which often bears a relation
to the intensity of the discomfort of the patient. In this
sense, although other factors, as yet unrecognized, may play a
role, it seems reasonable and practical to use muscle
contraction as a measure for response to and guide for
treatment. It is of interest that ergotamine and methysergide,
which are so successful in the management of migraine headaches,
presumably on a vascular basis, have little effect on
muscle-contraction headache, thus suggesting a different
etiology for the two types. On the other hand, it has recently
become obvious that some patients with muscle-contraction and
combined headache respond well to tricyclic antidepressants in
high doses, probably due to their central effects rather than
any direct effect on peripheral muscle contraction. It is also
interesting to note that, clinically, muscle-contraction
headache is prone to occur and increase while the "pressure" on
the individual is mounting -- while the deadlines are drawing
closer -- and to cease when the pressure is relieved, or when
the patient "blows his stack" and settles the issue.

The role played by substances which promote local painful
tissue inflammation, such as prostaglandins, bradykinen, etc.,
in muscle-contraction headache remains, so far as I know,
undetermined. That this may be important is suggested strongly
by the effectiveness of common analgesics in relief of such pain
since their action in many instances seems to depend on their
capacity to neutralize some of these inflammatory substances.

MIGRAINE AND VASCULAR HEADACHE

Let us now examine the pathophysiology of the vascular
headaches of the migraine type. In the first place, migraine
seems to be a familial disease which can plague its victims from
the cradle to the grave. From clinical observation as well as
from physiological tests the migraineur seems to have a
neuro-vascular system which is different in the degree and

probably quality of its response to stress from that of other people. The migraineur often has a very unstable GI tract manifesting itself in colic as a baby, car sickness and cyclic vomiting as a small child, and changing into migraine headache attacks in early schooldays and at puberty and/or at various other milestones of activity and life change as the years go by. A heightened sensitivity to light, noise, smells and vertigo may be demonstrated in these individuals, accentuated at times of attacks (Raffaelli & Menon, 1975; Lehtonen, 1974). Migraineurs are notable for their cold hands and feet, their car sickness, their abnormal vascular responses in relation to body temperature changes (Appenzeller, Davison, & Marshall, 1963) and, later in life, to the development of hypertension, vascular disorders and disorders of collagen (Leviton, Malvea, & Graham, 1974; Miller, Waters, Warnica, Schlachcic, Kreeft, & Theroux, 1981; Coffman & Cohen, 1981).

Although various personality types have been described for migraine the results of such studies do not indicate that all sufferers fall into one group; but they do suggest that an obsessive-compulsive disposition increases the severity and frequency of attacks. By these notations I mean to suggest the migraine- sufferer comes into the world blessed or cursed with the stress-response system of a racehorse -- an Arabian steed -- rather than a truck horse -- a Clydesdale. He or she is an animal equipped with an alertness of mentality and physiological irritability of vascular responsiveness of a heightened sort which responds with abnormal acuteness to stimuli from its external, a physiological and emotional environment. It is interesting that the list of stimuli which provoke migraine in susceptible race horses or, rather, people, is very similar in countries from almost all parts of the world. Most all of the common stimuli call for adjustments in the neuro-vasculature and functions regulated by the limbic and hypothalamic centers, which are proving to be such important managers of the body's response to stress and pain.

In contrast to the muscle-contraction headache patient, whose response to stress seems for the most part to take place while the stress is going on -- in its ascending limb -- and to dissipate when the stress is over, the migraineur has different timing. His reaction seems more than adequate to meet the crisis as it occurs but ends up with a painful readjustment in the descending limb of stress. It is as if all the systems designed to meet the challenge to homeostasis presented by the stressful event or stimulus go into action in exemplary or super-exemplary fashion but leave behind damage or disruption or by-products of the crisis which results in malfunction and pain. The migrainous mother characteristically handles for days or weeks the crisis of the sick child -- but has a headache when

the child recovers. The muscle-contraction-headache secretary
gets a mounting headache as the boss becomes more and more
exasperating - until she blows up and straightens him or her
out. The migrainous secretary handles the problems in the
office all week against many odds, keeping her "cool", but
awakens with a migraine attack after a late sleep on Saturday
morning when the rigors of the work week are all over.

At first, as the experimental method was applied to the
study of migraine, many of its peripheral features occupied the
time and laboratories of the investigators. Although it was
well recognized that the origin of migraine undoubtedly lay in
abnormal function of some areas of the brain, as shown
especially in the prodrome of classic migraine as the centrally
located hemianopsia, aphasia, hemianesthesia and even
hemiplegia, the headache phase was, nevertheless, associated
with visible and testable findings in the periphery - in blood
vessel misbehavior, nasal blocking, rhinorrhea, meiosis, color
changes, and visible vascular dilatation in the head and icy
cold hands and feet in the systemic periphery. It is not
remarkable that some of the peripheral changes were the first to
be checked out for their relation to the attacks of migraine.
Margaret O'Sullivan was photographing the capillary loops of the
migrainous fingers in the 1930s. Wolff and his coworkers were
measuring the amplitude of pulsation of superficial cranial
blood vessels during the headache, and extracting interesting
vasoactive pain-producing kinins from the skin in the location
of the headache. Central mechanisms were also tested but they
were hard to get at for study by early techniques. Although the
evidence was there that central mechanisms were at the root of
migraine, the demonstration that during the headache phase
peripheral structures were intimately involved in and maybe
responsible for the pain overshadowed the migraine scene.

In the past twenty years, as our techniques for studying
intracranial blood flow and cerebral metabolism and physiology
and neurochemistry have improved, the emphasis is shifting back
to where it, I believe, properly belongs, to the central areas
controlling major shifts and adjustments in both brain and
bodily physiology and biochemistry. Nevertheless, we must not
forget that crucial experiments done by Wolff and his group in
the Thirties and Forties, which cannot be done on human beings
in this day and age, did show that the pain of the migraine
headache is mostly related to the misbehavior of the branches of
the external carotid - misbehavior, to be sure, which may
represent a peripheral response mediated by central mechanisms
which may have lowered thresholds to pain as part of a total
response, both inside and outside the head, to deprivation of
blood in critical areas of the brain itself. In this case both
central and peripheral mechanisms may be jointly responsible for

the "ache", and the peripheral abnormalities which are available
for monitoring in biofeedback and other techniques represent
valid talismen of the migrainous disturbance.

I once used to think that the day would come when we should
find one enzyme system that was genetically deficient that would
explain this familial disorder. The more I see of it now, the
more I believe that the trouble will be arising from misbehavior
or, possibly, just excess, of a whole pattern of response in the
neurovasculature, a pattern shaped by many subsidiary side
events even as the now old Cannon concept of the response to
fear, hunger, pain and rage is made up of multiple minor systems
reacting together in a mass formation against a major challenge.

During the past ten years the techniques for studying
cranial blood flow have been perfected so that, without invasive
interference, they show us clearly that in the prodrome of
migraine there is evidence of a general diminution of blood flow
in the brain. In certain areas of the brain, this deprivation
is more marked than in others. In some places, it is reduced
during these periods to 40% or 50% of normal and a malfunction
results -- producing deficits like visual scotomata, aphasia,
hemianesthesia and the like. During the headache phase which
comes on as the deprivation is overcome, the cerebral vessels
and other cranial vessels are hyper-suffused with blood. The
pial vessels -- so dilated -- do not hurt since they are
insensitive to pain. The other cranial vessels, especially the
branches of the external carotid, do become the site of pain
related to their excessive stretching and the local accumulation
of kinins.

It is common to think of this prodromal change in cerebral
blood flow as being due to vasospasm. Some arteriograms have
supported such a view but others have not or have been of
doubtful significance. Cerebral vessels certainly are capable
of severe spasm, especially in response to bleeding around them
in subarachnoid hemorrhage or cerebral hemorrhage. Such spasm
may be a response to serotonin or thromboxane released from
platelets involved in the hemorrhage. Other sources of spasm
may be related to calcium metabolism in the smooth muscle vessel
walls -- a condition possibly common to vessels elsewhere in the
body (as in Prinzmetal's Angina) (Miller et al., 1981; Coffen &
Cohen, 1981).

More attention has been paid recently to the nature of the
blood going through those vessels, its viscosity, its platelet
and red cell aggregates, coagulability, and drop or rise in
certain elements, such as serotonin, histamine, basophilic
components, serotonin-releasing factors, and mono-amine content
of platelets, and fatty acids. Changes in these substances and

globulin components may lead to rouleaux formation or other
sources of increased viscosity and sludging of local areas of
cerebral circulation, causing anoxia. The studies of Mathew,
Largen, Dobbins, Meyer, Sakai, & Claghorn (1980), and Yamamoto &
Meyer. (1980) have shown that the cerebral blood vessels of
migraineurs react differently from normals to CO_2 and O_2, and
their experiments show that these abnormal reactions may be
modified in certain situations by biofeedback training. These
developments certainly begin to tie up with behavioral therapy
(Yamamoto & Meyer, 1980; Mathew et al., 1980).

Other theories about the prodromal phase of migraine relate
to some evidence which has been accumulated by Heyck and
Spierings (1969) and Saxena (1980) that abnormal opening of
arterio-venous shunts in the cranial circulation may account for
the deprivation of blood in the tissues of the brain and,
through ischemia, the subsequent dilatory headache. There is
evidence to suggest that ergotamine, so effective in stopping
the migraine pain, may do so by closing such shunts.

Another theory that may have a bearing on the prodrome of
migraine involves a phenomenon known in the cat brain as the
"spreading depression" of Leao (Milner, 1958). In the cat,
stimulation of the visual cortex creates a brief spreading wave
of excitation followed by a prolonged wave of inhibition of EEG
activity. The spreading scintillating lights of the migraine
scotoma in man, followed by the inhibition of sight, follows the
same timing as this cat phenomenon. Whether this actually takes
place in man, and if so, what triggers it, is unknown.

The platelet is appearing as an interesting component of
the migraine attack. This minute particle proves to be a
storehouse for powerful chemicals which affect blood flow,
viscosity, clotting and blood vessel constriction and
dilatation. During the migraine attack we know now that
substances are present in serum that cause the platelet to
discharge its serotonin, ATP, ADT, Thromboxane A_2, a heparin
neutralizing agent, and a prostacyclin inhibitor (Gawel,
Burkitt, & Rose, 1979). In the process of supplying these
agents affecting vessel size and blood flow, the platelet
changes its shape into that of a starfish and itself becomes an
aggregate in the beaver-dam that stems hemorrhage. We know that
this excessive aggregation happens during migraine. What we do
not know for sure is whether this is peculiar to the migraine
attack or whether it also happens in normal controls if they are
subjected to the same stress as must have just happened to the
migraine patient to bring on an attack.

There are other experiments to show that the migraine
patient, whether in an attack or not, has unusual sensitivity

to, and supplies of, catecholamines and serotonin and their precursors (Sicuteri, 1966). Most recently, evidence is accumulating to suggest that the endorphins in the spinal fluid are unusually low during the migraine attack and other investigations show abnormal behavior of other vasoactive agents during menstrual migraine attacks (Anselmi, Baldi, Cassaci & Salmon, 1980) and unusual aberrations of the renin-angiotensin system at the time of the headache (Nattero, Bisbocci, & Ceresa, 1979). Italian investigators are demonstrating abnormal sensitivity of human hand veins to injections of minute amounts of vasoactive amines -- sensitivity which changes as a migraine attack happens (Bianco, Anselmi, & Sicuteri, 1980).

While these biochemical changes are being studied in an effort to determine whether they are the cause or effect of migraine and whether they are specific for migraine or just part of a general reaction to stress, another great territory related to migraine is being explored -- namely, sleep. We have already mentioned a very important feature of migraine -- its occurrence in relation to stress, especially in the descending limb of the stressful experience. It is not surprising then, that migraine tends to arise frequently during sleep, during or a short time after REM sleep. Already we know of a headache occurring in those who have a problem with obstructive sleep apnea.

Of special interest is the particular variety of headache called Cluster Headache. Clinically, this form of vascular headache, more common in men, occurs with major changes of pace from intense to relaxed activity. Individual attacks of this very severe headache, related to migraine in our concept, occur on the way home from work, during even a short nap, following the relaxation of an alcoholic drink, even during cessation of alert thinking, and especially during sleep -- REM sleep. It is as if these headaches occur when the alerting system is turned off, or perhaps when one half of the adrenergic system -- the beta-adrenergic half - is turned off. Keeping such a patient awake all night may prevent his attacks. In addition to the timing of the individual attacks, the timing of the whole cluster of attacks occurring every day or night, over a period of several weeks or months, is often related to a vacation starting after a period of great Type A activity, the onset of sickness which causes a legalized enforced rest, or after the solution of a prolonged perplexing life problem (Yasue, Touyana, Shimamoto, Kato, Tanaka, & Akiyama, 1974).

During such attacks and also the prolonged cluster period, there is evidence of a change in vasomotor or autonomic tone - in the form of pupillary meiosis, excessive tearing, nasal stuffiness, excessive salivation, gastric hypersecretion and bradycardia -- all suggestive of parasympathetic dominance over sympathetic activity. During the cluster period a single drink or an injection of histamine or a pill of nitroglycerine will result in a severe attack. But when the patient has gone back to his "normal" status these triggers no longer produce their effect. The similarities between this type of headache and Prinzmetal's Angina -- so common in men -- occurring somewhat in clusters and coming on at rest or during sleep rather than during exertion -- is striking. In Prinzmetal's Angina an alpha-adrenergic stimulus at a time when the patient is under beta-adrenergic blockade will produce an attack -- during which coronary artery spasm may be demonstrated. In one case reported by Ekbom, Jr., an attack of cluster headache was associated with a spasm in the carotid artery in the region of the siphon (Ekbom & Greitz, 1970). These are reasons why we need to study the autonomic tone of patients in cluster and in migraine headache in addition to those having Prinzmetal's Angina. And these are states of human physiologic behavior that should be of interest to the behavioral therapist -- possibly subject to alteration by his techniques.

A study of these human physiologic states during which the terrain apparently changes for a while, longer or shorter, in the direction of making one more or less subject to the stimuli that produce vascular headache seems very worthwhile. This is important from the point of view of headache in itself -- but it also holds importance for the management or possibly prevention of other serious disorders with which migraine headache and autonomic over-activity are key factors. Our studies and those of others suggest a close relation between migraine and cardiovascular disease, hypertension and heart attack, and disorders of collagen.

Since pregnancy is the one normal physiological state during which common migraine temporarily goes away, much study needs to be devoted to it in an effort to learn the body's own secret for this relief. Toxemia of pregnancy, more common in migraineurs, looks like an area for research (including the physiology of the placenta) which may shed light on abnormal behavior of the neurovasculature, of relevance not only to the etiology of migraine but also to factors of growing importance in the regulation of circulation in general (Rotton, 1959; Ferris, 1977).

Migraine is recognizable early in life, appearing often in relation to milestones of increased responsibility - when the youngster first hits school, puberty, senior high school, college, the first job or marriage. Efforts to affect favorably over-activity of the autonomic system in the young may lead to prevention not only of many days of discomfort from migraine and muscle-contraction headache but also more serious disturbances in later life that end in organic tragedy. Our modern pharmaceutical approaches to altering human physiology and pathophysiology have proved very helpful. Their effectiveness depends on their capability of really changing physiological responses. By this very token, however, unplanned and undesirable side effects frequently develop since these drugs often create changes not just in one regulatory system but also in several others not related to the problem at hand. Furthermore, sitting at our desks in the office, we cannot expect to be as clever in manipulating these delicate mechanisms appropriately as the health care system we have built into our brain which is constantly receiving such accurate and timely information about what is going on in the body. If the behavioral therapist can teach the brain some useful tricks in managing certain physiological responses to the stresses around us, or correct some of the inappropriate reactions we have already developed, or even teach us how best to use the energies and equipment we have, even though we may be imperfectly endowed, then the details of management can be left to the organ systems which are in a position to deal in a coordinated fashion with the constantly occurring changes and readjustments which we all must make, all the time.

Headaches is a wonderful area for the work of the behavior therapist. It involves not only training a given organ how to behave under various circumstances. In a way it is training the therapist in our central nervous system how to train behavior throughout the body. It is like teaching teachers to teach.

REFERENCES

Anderson, C. D. & Franks, R. D. Migraine and tension headache: Is there a physiological differences? Headache, 1981, 21,63-71.
Anselmi, E., Baldi, F., Cassaci & Salmon, S. Endogenous opiods in cerebrospinal fluid and blood in idiopathic headache sufferers. Headache, 1980, 20, 294-299.
Appenzeller, K. D., & Marshall, J. Reflex vasomotor abnormalities in the hands of migrainous subjects. Journal of Neurosurgical Psychiatry, 1963, 26, 447-450.
Bianco, B., Anselmi, B., & Sicurteri, F. Supersensitivity of the vein smooth muscle during morphine abstinence, migraine and central panalgesia. (Abstract, full text in press, International Congress). Headache, Florence, March, 1980.
Bruyn, G. W. The biochemistry of migraine. Headache, 1980, 20, 235-246.
Coffman, J. D., & Cohen, R. A. Editorial: Vasospasm - ubiquitous? New England Journal of Medicine, 1981, 304, 780-781.
Ekbom, K., & Gritz, T. Carotid angiography in cluster headache.

Acta Radiologia Diagnosia, 1970, 10, 177–186.

Ferris, T. P. Toxemia of pregnancy: A model of human hypertension. Cardiovascular Medicine, 1977, 2, 877–895.

Gawel, M., Burkitt, M., & Rose, F. C. The platelet release reaction during migraine attack. Headache, 1979, 19, 323–327.

Graham, J. R., & Wolff, H. G. Mechanism of migraine headache and action of ergotamine tartare. Archives of Neurology & Psychiatry, 1938, 39, 737–763.

Heyck, H. Pathogenesis of migraine. Research and Clinical Studies in Headache, 1969, 2, 1–28.

Lehtonen, J. B. Visual evoked cortical potentials for single flashes and flickering light in migraine. Headache, 1974, 14, 1–12.

Leviton, A., Malvea, B., & Graham, J. R. Vascular diseases, mortality and migraine in the parents of migraine patients. Neurology, 1974, 24, 669–672.

Mathew, R. J., Largen, J. W., Dobbins, M. A., Meyer, J. S., Sakai, F., & Claghorn, J. L. Biofeedback control of skin temperature and cerebral blood flow in migraine. Headache, 1980, 20, 19–28.

Miller, D., Waters, D. D., Warnica, W., Schlachic, J., Kreeft, J., & Theroux, P. Is variant angina the coronary manifestation of a generalized vasospastic disorder? New England Journal of Medicine, 1981, 304, 763–766.

Milner, P. M. Note on the possible correspondence between the scotomas of migraine and spreading depression of Leao. EEG Clinical Neurophysiology, 1958, 10, 705.

Nattero, G., Bisbocci, D., & Ceresa, F. Sex hormones, prolactin levels, osmolarity, and electrolyte patterns in menstrual migraine - relation with fluid retention. Headache, 1979, 19, 25–30.

Raffaelli, E., & Menon, A. D. Migraine and the limbic system. Headache, 1975, 15, 69–78.

Rotton, W. N. Migraine and eclampsia. Obstetrics and Gynecology, 1959, 14, 322–330.

Sicuteri, F. Vasoneuro-active substance in migraine. Headache, 1966, 6, 109–126.

Spierings, L. H., & Saxena, P. R. The action of ergotamine on the distribution of carotid blood flow - the migraine shunt theory revisited. Headache, 1980, 20, 143–145.

Sweet, W. H. Neuropeptides and momoaminergic neurotransmitters: Their relation to pain. Journal of Royal Society of Medicine, 1980, 73, 482–491.

Tunis, M. M., & Wolff, H. G. Analysis of cranial artery pulse waves in patients with vascular headache of the migraine type. American Journal of Medical Science, 1952, 224, 565–568.

Tunis, M. M., & Wolff, H. G. Mechanism of migraine headache and action of ergotamine tartare. Archives of Neurology & Psychiatry, 1938, 737-763.

Yamamoto, M., & Meyer, J. S. Hemicranial disorder of vasomotor adrenoceptors in migraine and cluster headache. Headache, 1980, 29, 321-335.

Yasu, H., Touyana, M., Shimamoto, M., Kato, H., Tanaka, K. S., & Akiyama, F. Role of autonomic nervous system in pathogenesis of Prinzmental's variant form of angina. Circulation, 1974, 50, 534-539.

BIOFEEDBACK IN THE TREATMENT OF MIGRAINE

Jackson Beatty

Department of Psychology and Brain Research Institute
University of California
Los Angeles, California

ABSTRACT

This paper considers the use of biofeedback methods for the treatment of migraine headache. The incidence of headache as a clinical problem is first examined. The pathophysiological basis of migraine and the preferred methods of treatment are presented. Next, basic principles of controlled treatment evaluation are discussed. Two biofeedback procedures for the treatment of migraine are then considered: hand warming and cephalic pulse amplitude reduction. The therapeutic rationale for each treatment is presented and data from all relevant controlled experiments are summarized. It is concluded that there is no evidence for specific effects of learned hand warming, but that a specific component might be present for learned cephalic vasoconstriction.

INTRODUCTION

Headache is a symptom rather than a disorder and may result from a variety of physiological conditions. This review focuses upon migraine, one of the most common types of primary headache. The incidence of migraine is relatively high, at least 5% and most probably about 20% (e.g. Childes and Sweetnam, 1961; Lennox, 1941; Lyght, 1966). Thus migraine headache constitutes a significant health problem, resulting in both personal discomfort and the social cost of work absence. Further, there is no completely satisfactory pharmacological treatment for migraine (Dalessio, 1980). Both the high incidence of migraine and the absence of satisfactory pharmacological treatment have encouraged the development of behavioral approaches to migraine management.

In this review, I shall first briefly describe migraine and the behavioral therapies that have been proposed for its treatment. Next several problems in treatment evaluation are discussed, which are relevant to the behavioral treatment of headache. The relevant literature is then summarized. These data lead to a more realistic view of the present status of behavioral headache therapy.

MIGRAINE: A DESCRIPTION OF THE DISORDER

The migraine headache is only part of a generalized functional disturbance that characterizes a migraine attack. The headache itself is usually unilateral and is generated in the extracranial vasculature of the head. The temporal, frontal or supra-orbital arteries on the affected side are prominently enlarged. Often accompanying the head pain are gastrointestinal symptoms such as nausea, vomiting, constipation or diarrhea. Very often the extremities are cold and cynosed, the face is pallid and sweating and tremors are present. The migraine attack is characterized by widespread autonomic disturbance.

In classical migraine, the attack is preceded by sensory auras or visual field effects. These focal neurological signs are the direct result of intense vasoconstriction of the intracranial arteries supplying the cerebral cortex. The extent of vasoconstriction is sufficient to induce a transient ischemia, usually in the occipital lobe. In common headache, no focal symptoms precede the headache.

In both classical and common migraine, the initial headache is a dull, pounding pain of increasing intensity. After an hour or so, the quality of the pain changes, becoming steady and intense. These subjective reports correspond closely to the pathophysiology of migraine. Migraine is a vascular headache characterized by a unilateral loss of tone in the extracranial vasculature. The arteries are then unable to withstand the pressure wave of blood generated with each heartbeat, so that they become severely stretched as blood is passed through them. This distension activates pain-sensitive fibers around the affected vasculature, giving rise to the initial pulsatile quality of migraine pain. As time passes, the repeated insults to the vasculature create a sterile edema, rendering the arteries stiffened and tubelike. The edema marks the beginning of the steady pain of the later migraine attack (Lance, 1978; Dalessio, 1980). The standard treatment for migraine is to restore vascular tone by administering a vasoconstrictive agent before the sterile edema is established. By inducing vasoconstriction, the abnormal distensions of the arteries are prevented and the source of the migraine pain is eliminated. Ergotamine tartrate is the most commonly employed vasoconstrictive drug used in migraine, although many migraineurs refuse the treatment because of unpleasant side effects of the ergotamine (Dalessio, 1980).

BEHAVIORAL TREATMENT OF MIGRAINE HEADACHE

A variety of behavioral methods have been proposed in the treatment and management of migraine headache. A number of these are clearly non-specific, in that they are not uniquely tailored to the pathophysiology of headache. Stress management techniques constitute one large and popular class of non-specific interventions. Autogenic training (Luthe and Schultz, 1969), progressive relaxation procedures (Jacobson, 1938) and the relaxation response (Benson, Beary & Carol, 1974) are major examples of contemporary stress management methods. Inasmuch as these techniques are utilized without adaptation or modification in dealing with a wide range of pathophysiological processes, they may not be considered to represent specific behavioral interventions for the treatment of migraine. Two procedures, however, have been offered as specific behavioral treatments for migraine headache. The first, proposed by Sargent and his coworkers (1973) is the use of operant methods to induce hand warming in migraine patients. The rationale for this procedure is that the hands and feet of migraineurs are often cynosed during a headache, with normal peripheral circulation returning after the conclusion of the attack. Hand warming was seen as an attempt to behaviorally induce a cardiovascular response characteristic of headache termination. The second intervention is to utilize operant procedures to mimic the pharmacological effects of ergotamine tartrate in an attempt to restore normal tone to the extracranial vasculature during the initial phase of the migraine headache. This procedure involves training vasoconstriction of the temporal artery of the head. Evidence for and against the efficacy of these procedures as behavioral medicine in the treatment of migraine will be considered below.

EVALUATION OF THERAPEUTIC EFFECTIVENESS

In evaluating the therapeutic effectiveness of biofeedback procedures in the treatment of migraine, it is not sufficient to simply apply the proposed treatment to a number of patients and observe clinical improvement; effects produced by other factors may be improperly attributed to the treatment. Particular attention must be paid to the problems of spontaneous remission and placebo effects, two phenomena that plague clinical evaluations of all types.

The problem of spontaneous remission in evaluating a treatment results from the fact that few disorders are permanent. Most have a course of deterioration followed by improvement. Moreover, many disorders are sporadic, coming and going rather unexpectedly. This means that the condition of a patient undergoing a treatment may not only depend upon the therapy, but also upon other factors that vary with time. These problems are even more pronounced in a clinical setting. People tend to seek help when they are feeling worst; the very time that patients present themselves for treatment is likely to be when the disorder is most disabling. One implication of this non-random selection of patients for treatment is that, given time, the disorder is more likely to diminish rather than progress. Any therapy, even one with no actual value, if administered at this time would appear to be efficacious. Thus, the facts that the body tends to heal itself and that disorders tend to be transitory mean that spontaneous remission effects must be carefully controlled in evaluating the therapeutic efficacy of any new treatment.

One useful method for separating spontaneous remission from treatment effects is the use of a waiting list control group. By randomly assigning some patients to the treatment group and others to a waiting list for later treatment, any improvement shown by the controls while waiting may be taken to represent the beneficial effects of time, not treatment, whereas the patients in the treatment group(s) reflect both factors.

Placebo effects are more difficult to evaluate, but very much as important as spontaneous remission effects. The placebo effect is the magic effect of being treated, as distinct from the specific therapeutic benefits produced by a therapy. Placebo effects tend to be short-lived, therefore offering little help in the treatment of chronic disorders. A number of things are known about placebo effects. First, placebo effects, being products of social interaction, grow with the authority of the practitioner and the impressiveness of the treatment procedure. For this reason, biofeedback procedures might be expected to elicit a substantial placebo effect; sophisticated-looking biofeedback devices may appear to represent to the patient the very promise of modern technology to cure age-old problems of disease and ill health. Second, the magnitude of the placebo effect is known to increase with the enthusiasm of the practitioner. Thus, common medical cynicism holds that every new treatment works at least for a few years, while those who believe in the method are testing it on their patients. Biofeedback therapies are not without their ardent champions. Finally, the size of the placebo effect often increases with the cost of the treatment. For each of these reasons, placebo effects may be expected to play a major contributing role to any positive treatment effects obtained when biofeedback procedures are employed in the treatment of migraine.

The problem of controlling for placebo effects in assessing a new therapeutic procedure is difficult indeed. What is needed is to compare the results of the treatment in question with those obtained with another procedure that is known to lack any specific effect. Here, problems arise from the dependence of the placebo effect on social factors. If the control treatment is in itself unimpressive, or if offered unenthusiastically, differences in the outcomes of the two treatments may reflect only differing strengths of the placebo effect between groups. Thus, experimental and control procedures must be very carefully equated. One seemingly elegant approach to this problem is the use of double-blind experimental designs, in which neither the practitioner nor the patient knows which treatment is being administered. Double-blind evaluations are possible in biofeedback research, but are, in fact, seldom employed. Yet even with a double-blind design there can be problems. If the treatment and control procedures produce detectably different effects on the patient, the best laid plans for a double-blind study may be foiled. It is not uncommon, particularly when a study is conducted in a hospital ward, for all the patients to know who is receiving the active treatment, while all the staff remain blind. Thus, any study attempting to separate placebo from treatment effects must be intelligently designed and carefully executed.

There is, by now, a rather large literature reporting the results of biofeedback therapies in the treatment of migraine, but very few of these reports are of use in evaluating the effectiveness of these behavioral methods. Most are case reports or single group outcome studies, which offer no help in separating treatment from placebo and remission effects. It is only the controlled experimental investigations that can provide evidence as to the efficacy of biofeedback in migraine.

BIOFEEDBACK IN THE TREATMENT OF MIGRAINE: HANDWARMING

The most common form of biofeedback training in the treatment of migraine is hand warming, following the early suggestion of Sargent, Green and Walters (1973) of the Menninger Clinic. The popularity of this method is partially attributable to its simplicity; temperature sensing devices are inexpensive to construct and manufacture. But what evidence is there that the procedure is therapeutically effective? There are only five controlled studies testing the specific efficacy of learned hand warming as treatment for migraine. Their results provide little evidence of specific therapeutic effectiveness.

The earliest report of a controlled experiment was that of Andreychuk and Skriver (1975), who compared the effects of hand-warming feedback and autogenic relaxation, self-hypnosis and relaxation training, and relaxation training and EEG alpha-feedback on a combined index of headache hours and intensity in three groups of 11 migraineurs. All subjects received ten weekly training sessions. A comparison of pre-training and post-training headache scores revealed a significant reduction in headache hours/intensity in each of the three groups. However, there was no significant difference between groups. These data suggest that the hand-warming feedback training exerts no specific effect in the treatment of migraine headache. Also indicative of a non-specific effect was the fact that highly suggestible subjects tended to show more positive responses in each of the three treatment groups.

Similar results were obtained by Mullinix, Norton, Hack and Fisher (1978), who compared true and false feedback of hand temperature for 12 patients divided into two groups. Each subject was given nine training sessions. During training, subjects receiving accurate temperature feedback showed significantly higher hand temperature. Although both groups showed a reduction in headache symptomology after treatment, there were no significant differences in treatment efficacy between groups. Moreover, subjects who were best at hand warming were not the most improved. The simplest interpretation of these data is that there is no specific effect of learned hand warming on migraine headache.

Similar conclusions were reached by Blanchard, Theobald, Williamson, Silver and Brown (1978). Thirty migraineurs were assigned to one of three procedures: hand temperature feedback with autogenic training, progressive relaxation, or waiting list control. Subjects in both experimental groups showed improvement on a number of headache variables, differing significantly from subjects in the waiting list control. The control subjects were then added to one of the two experimental conditions, increasing the sample size of the treatment groups. With this addition, subjects in the relaxation group showed a significantly greater improvement in headache than did subjects in the biofeedback group at the end of training. However, at a followup evaluation three months after training, there were no significant differences between groups; all patients tended to revert to pre-training headache levels. Data such as these are indicative of placebo effects.

The fourth controlled study was conducted by Cohen, McArthur and Rickles (1980), who compared the effects of four experimental biofeedback procedures in 42 migraine patients. Each patient, after eight weeks of headache charting, was assigned to one experimental group for 24 sessions of training. Training was either to increase hand temperature, decrease frontalis EMG activity, decrease pulse amplitude in the temporal artery or increase occipital-parietal alpha frequency activity. The results were not encouraging for the use of biofeedback in the treatment of migraine. Subjects in all groups except the alpha feedback group showed statistically significant changes in the controlled variables over the course of training. However, the magnitude of the changes was exceedingly small. With respect to headache symptomology, only the number of headaches per week showed a significant reduction following training. Subjects in all groups reported about 20% fewer headaches after training; there were no significant differences between groups. None of the other headache parameters showed significant modification. Finally, there was no significant relationship between the degree of learned control and the reduction in headaches. This pattern of data suggests that only a placebo effect is operating in all the treatments tested.

The most recent controlled study evaluating the effects of learned hand-warming on migraine was reported by Elmore and Tursky (1981), who compared the effects of nine sessions of hand warming training with cephalic vasoconstriction training in a group of twenty-three migraineurs. Hand-warming feedback produced significant increases in hand temperature that could also be elicited after training without feedback. Concomitant recording of superficial temporal artery and digital pulse volume showed that hand-warming training tended to produce modest temporal artery vasoconstriction and digital vasodilation. However, for subjects receiving hand- warming training, there were no significant decreases in any of the headache measures in the month following training.

Other evidence supports the idea that trained hand warming may exert only a placebo effect on migraine headache. Price and Tursky (1976), for example, tested 40 migraineurs and 40 normal subjects in one of four experimental treatments: true feedback for hand temperature, false hand temperature feedback, taped relaxation instructions, and a neutral tape control (instructions on avocado growing). Both cephalic and digital pulse and blood volumes were recorded. Although there were marked differences in the response of the migraineurs and control subjects, the expected effects of treatment were not present. There was no difference in peripheral dilation among any of the three relaxation groups. Thus, no specific effect of feedback training could be discerned. Perhaps most damning for a specificity hypothesis was the finding that in both migraineous and normal subjects, digital and temporal pulse volume were positively correlated. All known specific treatments for migraine act to reduce, rather than augment, pulse volume in the cranial arteries of the headache patient.

BIOFEEDBACK IN THE TREATMENT OF MIGRAINE: CEPHALIC VASOCONSTRICTION

The second biofeedback approach to the treatment of migraine is the attempt to operantly mimic the pharmacological effects of the primary drug utilized in treating migraine, ergotamine tartrate (Delassio, 1980). Ergotamine tartrate acts to restore vasomotor tone in the affected cranial arteries by inducing vasoconstriction. This terminates the painful distention of the cranial arteries, which is the proximal cause of migraine pain. The rationale behind the cephalic vasoconstriction approach is to train migraineurs to restore vasomotor tone during the period of pulsatile migraine pain by previous operant training to reduce the amplitude of pulsation in the temporal arteries of the head.

Friar and Beatty (1976) provided the first test of this hypothesis. A sample of 19 young, otherwise healthy migraine patients, all of whom responded positively to ergotamine tartrate, were selected for participation. The ergotamine tartrate criterion was particularly important, in that it seemed unreasonable to expect operant procedures to elicit a clinically meaningful vasoconstriction if a potent pharmacological agent was not effective. Subjects were assigned to one of two training groups that differed only in the site of the response to be modified: the experimental group was trained to reduce the amplitude of pulsation in the extracranial branch of the temporal artery whereas the control group was trained to reduce pulse amplitude in the arterial bed of the index finger. All subjects maintained a headache log for at least 30 days before and after training. Training consisted of eight one-hour instruction sessions and a final ninth testing session, in which the subject's ability to induce vasoconstriction without feedback was measured. Both temporal and digital pulse volume and skin temperature were measured in all subjects.

Using standard operant procedures and a computer-based pattern recognition algorithm to eliminate movement artifacts from the pulse volume recordings, subjects were able to learn to control the arterial bed of interest. Subjects in the control group were able to produce a 33% reduction in finger pulse amplitude when tested without feedback; no reduction in temporal pulse amplitude occurred in these subjects. In contrast, subjects in the experimental group showed a 20% decrease in temporal pulse amplitude, which was accompanied by a 30% reduction in finger pulse amplitude. These differences were all statistically significant, as was the difference in temporal pulse amplitude between experimental and control subjects. Thus some degree of learned control of the migraine-relevant vasculature was achieved by experimental, but not control, subjects.

Evidence from the headache logs provides some support for the presence of a specific effect of the experimental treatment on migraine. Subjects in the control group showed modest (14%) but statistically non-significant decreases in both the total number of headaches and the number of major migraine attacks (headaches of more than three hours). For the experimental group, the reduction was larger in magnitude and marginally significant statistically. These subjects reported a 36% decrease in number of headaches ($p < .10$, one-tailed) and a 44% decrease in major headaches ($p < .05$, one-tailed). There was no reduction in mean headache intensity for either group. These data are more in accord with a specificity than a placebo hypothesis. The effect of the experimental treatment should be to abort headaches, like the vasoconstrictor ergotamine tartrate, and the reduction in major headaches is in accord with this prediction. Further, the treatment is not intended as an analgesic. The failure to find a decrease in rated headache intensity is supportive of a specific, non-placebo interpretation of the effect. One might expect a placebo treatment of affect rated headache intensity as a part of a general positive response to relaxing treatment.

Bild and Adams (1980) have obtained similar data in another controlled experimental evaluation of the cephalic pulse volume procedure. They assigned a group of 21 diagnosed migraine patients to a cephalic pulse volume training (N = 7), frontalis EMG training (N = 6) or a waiting list control (N = 6) group. Patients in the two training groups kept a headache log for six weeks before and after training. Each patient in treatment received ten one-hour sessions of training.

An analysis of the training data indicated that subjects in the pulse volume group achieved voluntary control of the targeted response by the end of training, but that pulse volume amplitude was unchanged in subjects receiving EMG training. Conversely, subjects receiving EMG training showed reduced frontalis tension, whereas EMG amplitude did not change in subjects receiving pulse volume training.

In comparing pre-training and post-training headache logs, all groups showed a decline in headache frequency. However, a simple effects analysis indicated that this decrease was significant only for the subjects receiving pulse volume feedback. The authors describe these data as follows: "If improvement is defined as one half the frequency of headaches per week and one half the hours of headache activity per week, six of the seven subjects in the (pulse volume) group, three of the six subjects in the EMG group, and one of the six subjects in the waiting list group showed improvement (p. 55)." Data such as these are suggestive of a specific effect.

A third controlled evaluation of the cephalic vasoconstriction hypothesis was provided by Elmore and Tursky (1981) in their comparison of hand warming and cephalic vasoconstriction training. Subjects trained to reduce pulsation amplitude in the superficial temporal artery showed significant reductions in temporal pulse amplitude during training and when tested without feedback. These responses were similar to those obtained by subjects given hand-warming training. However, unlike the hand-warming subjects, these patients exhibited a significant decrease in hand temperature when controlling cephalic pulse amplitude. This intervention, unlike hand warming, produced significant changes in both the frequency of headache and the use of headache medications.

In summary, Friar and Beatty (1976), Bild and Adams (1980), and Elmore and Tursky (1981) demonstrate that subjects who were able to voluntarily regulate a parameter reflecting vasomotor tone of the cephalic vasculature showed a modest and apparently specific reduction in headache duration. The only other attempt to test the vasomotor hypothesis was that of Cohen et al. (1980), who failed to obtain evidence that their subjects had learned a significant degree of vasomotor control. Thus, their failure to find a specific therapeutic effect of the cephalic vasomotor training is to be expected.

SOME CONCLUSIONS

The experimental data reviewed above make several points. The first is that there is no evidence for a specific effect of hand-warming procedures on migraine. All available evidence points to a placebo interpretation of the positive effects obtained. The second conclusion is that hand warming is no more effective than other, simpler relaxation procedures. Both parsimony and honesty indicate that the simpler relaxation procedures should replace hand warming biofeedback if a placebo type treatment of the disorder is desired. The third conclusion is more tenuous: learned control of cranial vasomotor tone might produce a specific beneficial effect in the treatment of migraine. But even here, caution is indicated. The training procedures employed to date are difficult at best and the magnitude of the obtained therapeutic effect is not large. At present, the case for biofeedback in the treatment of migraine is not strong.

REFERENCES

Andreychuk, T. & Skriver, C. Hypnosis and
    biofeedback in the treatment of migraine.
    International   Journal   of   Clinical   and
    Experimental Hypnosis, 1975, 23, 172-183.
Benson, H., Beary, J.F., & Carol, M.P.
    The relaxation response. Psychiatry, 1974,
    37, 37-46.
Bild, R. & Adams, H.E. Modification of migraine
    headaches by cephalic blood volume pulse and EMG
    feedback. Journal of Consulting and Clinical
    Psychology, 1980, 48, 51-57.
Blanchard, E.B., Theobald, D.E., Williamson, D.A.,
    Silver, B.V., & Brown, D.A. Temperature
    biofeedback in the treatment of migraine
    headaches. Archives of General Psychiatry, 1978,
    35, 581-588.
Childes, A. & Sweetnam, M. Study of 104 cases
    of   migraine.   British   Journal   of   Internal
    Medicine, 1961, 18, 243-
Cohen, M.J., McArthur, D.L. & Rickles, W.H.
    Comparison of four biofeedback treatments for
    migraine headache: Physiological and headache
    variables. Psychosomatic Medicine, 1980, 42,
    463-480.
Dalessio, D. Wolff's headache and other
    head pain (4th edition). New York: Oxford
    University Press, 1980.
Elmore, A.M. & Tursky, B. A comparison of two
    psychophysiological approaches to the treatment
    of migraine. Headache, 1981, 21, 93-101.
Friar, L.R. & Beatty, J. Migraine: Management
    by   trained   control   of   vasoconstriction.   Journal
    of Consulting and Clinical Psychology, 1976, 44,
    46-53.
Jacobson, E. Progressive relaxation (2nd edition).
    Chicago, University of Chicago Press, 1938.
Lance, J.W. Mechanism and management of headache.
    London: Butterworths, 1978,
Lennox, W. Science and seizures. New York: Harper,
    1941.
Luthe,   W.   &   Schultz,   J.H.   Autogenic   therapy:
    medical applications. (Vol. 2). New York: Grune
    & Stratton, 1969.
Lyght, C. The Merck Manual. Rahway, New Jersey:
    Merck, Sharpe & Dohme, 1966.
Mullinix, J.M., Norton, B.J., Hack, S. & Fishman,
    M.A. Skin temperature feedback and migraine.
    Headache, 1978, 17, 242-244.

Price, K.P., & Tursky, B.   Vascular reactivity
    of migraineurs and non-migraineurs: A comparison
    of responses to self-control procedures.
    Headache, 1976, 16, 210-217.
Sargent, J.D., Green, E.E. & Walters, E.D.
    Preliminary report on the use of autogenic
    feedback training in the treatment of migraine
    and tension headaches. Psychosomatic Medicine,
    1973, 35, 129-135.

BEHAVIORAL TREATMENT OF MIGRAINE

Niels Birbaumer and G. Haag

Department of Clinical & Physiological Psychology
University of Tubingen

Behavioral concepts regarding migraine headaches ignore psychodynamic and other speculative theories about the psychological origin of the disorder. There is good reason for this omission. Virtually nothing that can be used in the systematic analysis or non-pharmacological treatment of this illness has resulted from psychoanalytic thinking. Yet psychoanalytic thinking emphasized early "parental" restrictions of emotional behavior as one major cause of many psychosomatic disorders, including migraine, long before behavioral medicine took over the dominant position in psychotherapeutic research (Alexander, 1950). The position adopted in this paper tries to rehabilitate some of these early ideas on the basis of recent empirical data, which suggest that a more comprehensive view of the physiological, emotional-psychological, and social-behavioral bases of migraine headaches would be valuable. It is argued in particular that biofeedback studies have over-emphasized the causal role of self-regulation of a specific physiological system (i.e. digital vasodilation or cranial vasoconstriction). Most of the studies reviewed here, including our own research, reveal only minor therapeutic effects of biofeedback itself, aside from the influence of other behavioral factors (such as causal attribution, patient-therapist relationship, etc.). These factors have conventionally been subsumed under the category of "placebo-effects" in controlled studies on behavioral treatment of migraine. However, it has become increasingly clear that many such placebo effects can be analyzed with behavioral methodology, and that much of the success of behavioral interventions can be attributed to them. The reason for this is that the factors maintaining migraine headache are mainly to be found in the environmental contingencies operating on the patient, rather than in physiological malfunctioning.

SYMPTOMS AND MEDICAL CLASSIFICATIONS

Although the evidence is incomplete, it seems highly probable that the various diagnostic categories of migraine respond differently to behavioral treatment.

Four main types of migraine headache are usually distinguished: classic migraine (with prodromata), common migraine (without prodromata), cluster headache (brief, appearing in groups, interval between groups may be long) and complicated migraine, also known as hemiplegic or ophthalmoplegic migraine (with neurological signs) (see Ad Hoc Committee, 1962, for a thorough description of symptoms). The differentiation between these four categories, and their distinction from tension headaches, is difficult to diagnose. Most studies show overlapping categories, using the common criteria for classic migraine, such as unilateral onset, presence of a prodrome, nausea and/or vomiting, and a family history of migraine. There is evidence that cluster headaches differ in origin from the other migraine syndromes (Ekbom, 1970). The mechanism assumed to underlie these four migraine syndromes corresponds reasonably well with the symptom descriptions: most of the psychophysiological and neurological studies support Wolff's (1963) theory of vasolability in migraine. Nevertheless, the mechanisms responsible for the vasolability remain unclear.

The caliber of the extracranial head arteries, in particular the temporal artery, is substantially decreased during the prodromal phase. Cold hands and feet are typical subjective correlates of the reduced extracranial vasodilation. A similar pattern has been found for intracranial blood flow (Sakai and Meyer, 1978). However, Sakai and Meyer (1978) did not observe abnormal intracranial blood flow in patients during remission, although most studies on extracranial blood supply report a tendency towards vasoconstriction even in the headache free periods. Yet a study done in our laboratory on ninety-six migraine patients with Doppler ultrasound measurement of extracranial blood flow, demonstrated a significant increase of blood flow velocity in the external carotid artery even in pain-free intervals (Schroth et al. 1980). Some of the patients had a migraine attack during the Doppler ultrasound measurement and showed a further increase of blood flow velocity.

There is general agreement that extra- as well as intracranial blood flow increases during the pain phase, together with dilatation in the main arteries. Again, however, some extracranial arterioles and the supratrochlearis artery exhibit decreased blood flow amplitude (Heyck, 1975; Haag, Gerber, Birbaumer, Mayer, Lutzenberger & Schroth, 1981). Heyck has put forward a theory of arteriovenous shunts to account for the simultaneous dilation and flow increase in the main extracranial arteries, and constriction of other arterial connections, particularly those affecting the blood supply to the brain. Physiological analyses of the regional changes in blood flow before, during, and after migraine attacks are extremely important for the development of adequate biofeedback procedures (see below).

Operant self-regulation of the arterial tone and/or blood flow may be another useful diagnostic tool for predicting the outcome of biofeedback therapy. We were able to demonstrate that the migraine patients exhibit relatively poor performance in pulse plethysmographic self-regulation of the temporal artery (Birbaumer, Reuben, Ehlers, Schrode & Haag, 1981). A comparison of results of two biofeedback experiments involving patients with psychsomatic disorders illustrates this differential ability of patients to control autonomic and central nervous activity. In the first experiment eighteen patients suffering from diverse psychosomatic disorders and twenty healthy controls were given visual feedback of slow cortical potentials (SCP), recorded unipolar precentrally from both hemispheres and frontally using a linked earlobe reference. Subjects were instructed to steer a small "rocket", whose flight path on a television monitor was determined by the linear integral of the cortical SCP DC shift, into one of two goals on the screen, corresponding to either positive or negative potential variation. Feedback trials alternated with test trials without visual feedback; in the latter, subjects heard the signal tones and were instructed to imagine themselves steering the rocket into the appropriate goal. After six seconds a white noise stimulus was presented which the subjects were required to interrupt by pressing a button as quickly as possible. In the second experiment, a group of nine migraine sufferers, comparable with a migraine subgroup in the first experiment, and nine controls were given visual and acoustic feedback of arterial pulse amplitude, by means of an infrared photoplethysmograph placed directly over the temporalis superficialis, next to the ear on the side of the head most prone to migraine attacks. The size of a square displayed on a T.V. monitor gave direct analog feedback of the momentary pulse amplitude. Subjects were told on contraction trials to reduce the size of the square. After reaching a preset success level, a white noise signal was automatically extinguished (binary feedback).

In the first experiment the patient group showed a significantly greater average difference in potential shift

between trials requiring positive and negative variation of left
hemispheric precentral activity (p < 0.01) and a similar tendency
for frontal activity (p < 0.01) in comparison to normal controls.
Very pronounced differences occurred in the test trials without
feedback (p < 0.01). In contrast to these results, patients in
the second study, were given peripheral biofeedback pertinent to
their particular disorder, showed a significantly inferior
ability to master arterial contraction as compared with control
subjects (p < 0.05). These results support the hypothesis that
psychosomatic patients show a particular lack of visceral
control in the affected system. The marked ability of cortical
self-control may impair corticovisceral regulation mechanisms.

Guglielmi, Roberts, Tellegen and Zimmermann (1981)
investigated digital vasodilation in healthy volunteers with
labile or stabile vasomotor responses. The group with labile
vasomotor responses demonstrated better performance; this
suggests that patients with a rigid system without variance will
be inferior in self control. These data are confirmed and
extended by a controlled study of Harris, Ray and Stern (1978)
with seven gastric and six migraine patients: gastric patients
were inferior in a biofeedback task influencing gastric
motility, while the migraine patients could not control hand
temperature readily. In their classic study on vascular
reactivity in migraineurs and nonmigraineurs, Price and Tursky
(1976) found the same difficulty for extracranial arteries. To
conclude, these data indicate a response specificity in the
migraine patient, with a rigid vascular tone on cephalic and/or
digital locations that prevents self-regulation of vascular
changes at the same speed as normals or patients with other
psychosomatic syndromes (Cohen, Richles & McArthur, 1978).

On the other hand, it cannot be inferred from these studies
that patients who show impaired self-regulation capacity during
early treatment sessions, will not profit from biofeedback
therapy. All authors agree that there are many patients in the
hand-warming training or the temporal artery biofeedback therapy
who do not show any physiological change during self-regulation
therapy, but nevertheless profit from the biofeedback procedure
(Knapp and Florin, 1981).

Perhaps the most important empirical task of the next
decade will be the search for reliable predictors of outcome for
each therapy that take into account the characteristics of the
individual patient.

BEHAVIORAL TREATMENTS OF MIGRAINE

In a review of the comparative effectiveness of migraine
treatments Baust (1979) concluded that psychotherapy is the most

effective form of therapy independent of the specific therapeutic procedure. Since lower side effects are reported for psychotherapy than for medication, its effectiveness is even more impressive. Moreover, as chronic medication, particularly with ergotamines and phenacetins, can produce headaches (together with renal disturbance), psychotherapy must be the method of choice in the treatment of migraine. The fact that most patients today are still chronically treated with toxic medication highlights the ineffectiveness and ignorance of our public and private health systems.

The relative effectiveness of the different psychotherapies is not clear, despite the increasing number of controlled studies that have been reported, particularly in the field of behavioral medicine (see chapters by Kroner and Beatty). The large number of studies on HW have been reviewed by Diamond et al. (1978) and Birbaumer (1980). Only one double-blind study by Kewman and Roberts (1980) showed no difference between hand warming, cooling and the monitoring of headache frequency. The therapeutic effects reported by Kewman and Roberts were small, in that no change in the frequency of attacks was demonstrated. However, seven subjects did not learn the task; therefore, it is difficult to draw a definite conclusion from this study. The relatively small therapeutic benefits point to the effects of unspecific factors.

It is extremely difficult in a double-blind setting for the subjects to develop a consistent causal attribution of their self-regulatory efforts; it is impossible for most of the patients to attribute the feedback of a self-regulatory attempt (i.e. warming or cooling) correctly. The interaction of the specific feedback effect (hand warming) with the unspecific effects (attribution, expectancy, sympathy) may be the secret of the success of clinical studies. On the other hand, the comparable effects of relaxation and biofeedback techniques underline the uselessness of hand warming biofeedback machine for migraine patients.

Elmore and Tursky (1980) have compared temporal pulse amplitude reduction biofeedback and hand-warming biofeedback, and this has led them even to doubt the concept of sympathetic overarousal of migraine sufferers. For the successful pulse amplitude group showed increasing levels of sympathetic activity in other physiological parameters. In contrast, an aggravation of symptoms was found in the HW group. An aggravation of symptoms has also been observed in some of the HW patients treated in our laboratory compared with those trained in temporal artery control. Gauthier and Bois (1980) compared four groups: temporal artery cooling (TAC), temporal artery warming (TAW), hand warming (HW), and hand cooling (HC). The only

difference between the groups was due to better effects in the
TAC group.

With the exception of the study of Cohen et al. (1980), all
controlled studies to include temporal artery pulse amplitude
reduction as the biofeedback signal have demonstrated superior
effects to various comparison conditions, such as HW (Elmore and
Tursky, 1980), false feedback and placed respiration (Quintanar
et al., 1979), vasoconstriction of the hand (Friar and Beatty,
1974), or vasodilation of the temporal artery (Christie and
Kotses, 1973).

Thus   the   majority   of   studies   investigating
biofeedback-induced reduction of the temporal artery pulse
amplitude (TA) confirm the effectiveness of this procedure as
compared to various control groups. However, all the positive
reports have involved only a few sessions (rarely more than ten
sessions), and none has scheduled follow-up periods exceeding
more than four months. The only exceptions are the study by
Cohen et al. (1980) and our own studies (see below). Neither of
these groups have found a difference between the TA and control
groups after therapy or in follow-up sessions. Similarly, both
groups have failed to observe any lasting ability to control the
physiological variables. It is tempting therefore to speculate
that TA produces fast "placebo" responses which are more
impressive than those generated by other methods, perhaps
because the procedures have an authenticity and logical
rationale that are very convincing to the patient. The
"placebo" mainly consists of a change from external to internal
attribution of self-regulatory abilities. The internal
attribution of control, together with a generally increased
feeling of control, may contribute to the release of anti-pain
substances (endorphines). However, the physiological and
psychological stimuli that cause the disease are unaffected, and
eventually override these first short-lived effects. Belief in
the efficacy of therapy, and the confidence in the competence of
the therapist, is higher at the beginning of the therapy, and
facilitates the change from external (medical) to internal
(psychological) attribution of the cause of the disease.

These conclusions are supported by our data (see below),
and by the fact that other psychotherapies that aim at changing
the controlling psychological stimuli result in equal or even
better long-term effects. Mitchell and Mitchell (1971)
published two controlled studies which investigated systematic
desensitization combined with assertion training. Relaxation
and medication served as control conditions. Behavior therapy
consciously reinforces beliefs of internal control, and in
addition teaches the patient to change those aversive social or
cognitive stimuli that contribute to the disorder on a

behavioral level. It is amazing that these pioneering studies have not been replicated and extended during the last decade. Other psychological treatments of migraine include hypnosis and self-help groups; both methods are reported to provide good effects but no studies employing satisfactory control procedures have been described.

Another "non-specific" factor, the pain-producing effect of the medication, has been neglected in most psychological studies. If the credibility of treatment and trust in the therapist is high, subjects are more willing to reduce medication, particularly at the beginning of therapy. In some cases this reduces headache frequency and intensity dramatically. A thorough documentation of the medication change or a complete withdrawal of medication before therapy is, therefore, mandatory if therapeutic effects are to be evaluated correctly.

STEPS TOWARD A PRESCRIPTIVE BEHAVIORAL TREATMENT

The two studies done in our laboratory have not been based on the problem of comparative effectiveness. Differential effectiveness of psychotherapies for different kinds of patients is, of course, implied in the fact that we use different techniques. But is it really necessary to have a variety of psychological treatments, especially in behavioral medicine? Or is it merely sufficient, as some large-scale comparisons of different psychotherapies (psychoanalytic, behavioral, verbal, etc.) suggest, to do something regardless of what it is? (Luborsky, Singer & Luborsky, 1975). As long as a warm and friendly therapist talks with his patient for an extended period of time, a positive effect will occur, and this effect may exceed the benefits of a placebo pill. The short overview of psychological treatments given above suggests this may be the case for migraine headache. All the behavioral techniques (relaxation, desensitization, assertion training, hypnosis, self-help groups, biofeedback of different physiological systems) produce substantial improvements for 50 to 70% for a short period, but long-term effects are not clearly documented; there is no reason in the available literature to hope that such impressive improvements will be maintained for more than one year.

STUDY 1

Rationale and Methods

Study 1 did not concentrate on migraine but on psychosomatic disorders in general, and challenges the suggestion of non-specificity in treatment made above (Haag et al., 1981). Thirty-six patients with different psychosomatic disorders (ten of them with migraine headaches, the rest with essential hypertension, insomnia, asthma, tachycardia, hyperhidrosis) were assigned to two out of three selected behavioral treatments. Twelve sessions of each treatment were scheduled, the treatment sequence being varied randomly within patients. Each individual therefore received twenty-four sessions in toto. Follow-ups were scheduled after six months and eighteen months. Only the data from the first follow-up are available at present.

The three treatment modalities were social skills training, with its main emphasis on the expression of positive and negative emotions in vivo, cognitive therapy (Meichenbaum, 1974), and biofeedback (heart rate or blood pressure or frontalis EMG biofeedback; the particular biofeedback procedure was chosen individually for the different symptoms). Jacobson's relaxation training combined with some autogenic and meditative formulae was administered to all patients. All treatment modalities were applied by highly experienced clinical psychologists with a behavioral and/or psychoanalytic background. An extensive diagnostic battery was given before therapy, after the first twelve sessions, and after the follow-ups. The diagnostic data were categorized by four independent raters according to their predictive value for the three treatment modes: each patient therefore received a weighted score indicating which of the three treatments was the most appropriate for this individual. Concordance between the four raters exceeded 0.7. After the rating procedure each patient received two of the three treatment methods in a random order; one of the two treatments always constituted the best technique according to the raters. Patients had to complete a diary with daily ratings of symptoms, depression, motivation for therapy and medication. The diary started forty days before therapy and was extended on to the follow-ups. The data of the diaries have been evaluated by ARIMA models (auto-regressive-integrated-moving average, Box and Jenkins, 1970). ARIMA models eliminate serial dependency from time series and allow the parametric analysis of means and variance. The z-values of the differences between the forty day baseline, therapy 1, and therapy 2 constitute the basis for the group statistics.

If there is any specificity in behavioral treatment, the administration of the ideal treatment must result in greater success than the treatment that was second best in the prognostic rating of the diagnostic battery.

Table 1
Results of Study 1

		statistical measures		
		reduction of symptoms (1)	reductions of medication (2)	mean of (1) and (2)
first treatment	n =	31	21	31
	sign.	6	6	8
	z-value	-5.40	-5.71	-6.86
second treatment	n =	25	17	25
	sign.	11	13	15
	z-value	-10.77	-17.43	-15.54
best available treatment	n =	28	19	28
	sign.	9	12	14
	z-value	-9.85	-14.13	-13.33
treatment ranking second according to prognostic rating	n =	28	19	28
	sign.	9	12	14
	z-value	-6.01	-7.30	-8.58

n =   number of patients in the particular cell

sign =   number of patients with significant effects (z larger
           1,96) in the desirable direction (reduction of symptoms)

z-value   is significant at 1.96, negative sign. indicates reductions
           of symptoms or/and medication.

Results.   Results are reported for thirty-one patients.
The responses of the migraine patients were no different from
those with other disorders. (The N is different for the groups
because   not   all   thirty-six   patients   have   completed   both
therapies at the time of this report.)   The values in Table 1
represent the ARIMA-filtered values of the symptoms and weighted
medication from the diaries, computed as differences from the
forty days baseline.   After the second treatment, effects are
more pronounced in all three parameters, reflecting a main
effect for the duration of therapy (number of sessions).   In
addition, the preferred treatments produced significantly larger
effects than the treatments rated second.   Nevertheless, these
effects were smaller than the effects of treatment duration.
All   the   values   reported   here   are   independent   of   mood

(depression) and expectancy of success; mood was lifted over time, while expectancy first increased but then decreased again. There is no significant difference between the three treatment methods (social skills, cognitive therapy and biofeedback).

STUDY 2

Rationale and Methods

Only migraine patients participated in Study 2. Two different treatments were compared in the cross-over design: biofeedback of the temporal artery amplitude, and a combined social skill and cognitive treatment package called "concordance therapy" (CT): concordance of the psychological, behavioral, and physiological responses of the patients is the aim of this therapy. A relaxation group served as a third group. In addition to the measures described for Study 1, a detailed exploration of attributional processes was made using questionnaires.

As in Study 1, follow-ups have not been completed, but ARIMA-statistics from the diaries do not reveal any differences between the three groups after the second therapy. A marginally significant effect on reduction of headache activity was found for the group starting with biofeedback compared with the two other therapies. However, intensity and frequency of attacks were significantly reduced for all three treatment groups. Table 4 presents the number of subjects with improvements according to mean changes (first line) compared to significant improvements according to ARIMA. Effectiveness is overestimated if the evaluation is based only on mean changes (see Table 2).

As in Study 1, there is a significant series effect for all measures (p < 0.005), demonstrating that the second ten sessions result in larger improvements than the first ten sessions, regardless of the therapeutic technique.

Most interestingly, causal attribution exhibited dramatic changes over therapy: in all three groups pain was largely explained as a medical problem before therapy and as a psychological one afterwards. Attribution of control changed significantly from medical (pills) to psychological self control. The reduction of medical attribution correlated with decreasing headache intensity (r = .37, p <.03). As in Study 1, expectancy had no effect; however, patients whose partners were willing to join them in therapy did show better responses.

Table 2

Proportion of patients improving, according
to different criteria

	Attacks	Medication	Pain Rating
"improvement" (mean change)	71%	76%	87%
sign. improvement (ARIMA)	32%	52%	48%

SUMMARY

To summarize the results of these two studies: Duration of treatment (number of sessions) is the main contributor to success, irrespective of the therapeutic technique involved. On the other hand, patients respond differently to different behavioral treatments, demonstrating the need for prescriptive psychotherapy even within a single group of disorders such as migraine. Biofeedback seems to have slightly superior effects within the first few sessions. The results demonstrate that comparisons of mean changes, and the use of classical inferential statistics (ANOVA), overestimate the therapeutic success by up to 200 %. Attributional changes (from medical to psychological, from external to internal, from no control to control) probably constitute one of the major effects of all the psychological treatments involved here. Objective prediction of which patient will benefit from what form of therapy, will be the principal target of research over the next decade in our laboratory, as it is in other fields of behavioral medicine.

Note

Work reported from our laboratory was supported by the German Research Society (DFG). The described studies were done

together with Professor Dr. K. Mayer and Dr. D. Gerber from the
Neurological Clinic of the University of Tubingen. The
statistical analysis and the development of the biofeedback
procedure was done by Dr. W. Lutzenberger.

REFERENCES

Ad Hoc Committee on Classification of Headaches. Journal of
    American Medical Association, 1962, 179, 717-718.
Alexander, F. Psychosomatic Medicine. Norton, New York, 1950.
    Baust, W. Der vasomotorische Kpofschmerz. Med. Welt, 1979,
    30, 761- .
Birbaumer, N. Psychosomatische Storungen. In W. Wittling,
    (ed.) Handbuch der Klinischen Psychologie, Band 5. Hamburg:
    Hoffmann und Campe, 1980, 139-176.
Birbaumer, N., Reuben, C., Ehlers, A., Schrode, M., & Haag, G.
    Self-regulation of central and peripheral processes in
    patients with psychosomatic disorders. Proceedings of the
    Biofeedback Society of America. Twelfth Annual Meeting,
    Louisville, Kentucky, 1981.
Box, G.E.P. & Jenkins, G.M. Time Series Analysis.
    Forecasting and Control. San Francisco: Holdenday, 1970.
Christie, D.J.& Kotses, H. Bidirectional operant conditioning
    of the cephalic vasomotor response. Journal of
    Psychosomatic Research, 1973, 17, 167-170.
Cohen, M.J., Richles, W.H., & McArthur, D.C. Evidence for
    physiological sterotype in migraine headaches.
    Psychosomatic Medicine, 1978, 40, 344-354.
Cohen, M.J., McArthur, D.C., & Richles, W.H. Comparison of four
    biofeedback treatments for migraine headache:
    Physiological and headache variables. Psychosomatic
    Medicine, 1980, 42, 463-480.
Diamond, S., Diamond-Falk, J., & Deveno, T. Biofeedback in the
    treatment of vascular headache. Biofeedback and
    Self-Regulation, 1978, 3(4), 385-404.
Ekbom, K. A clinical comparison of cluster headache and
    migraine. Acta Neurologica Scandinavia Suppl., 1970, 41,
    5-48.
Elmore, A.M. & Tursky, B. A comparison of the
    psychophysiological and clinical response to biofeedback
    for temporal pulse amplitude reduction and biofeedback for
    increases in hand temperature in the treatment of migraine.
    Headache, 1980, 20, 162.
Friar, L.R., & Beatty, J. Migraine: Management by trained
    control of vasoconstriction. Journal of Consulting &
    Clinical Psychology, 1976, 44(1), 46-53.
Gauthier, J., & Bois, R. Comparison between peripheral and
    cephalic temperature biofeedback in the treatment of
    migraine headache. Proceedings of the Biofeedback Society
    of America, Colorado Springs,1980, 46-49.

Gerber, W.D., Haag, G., Birbaumer, N., & Mayer, K. Studie zur differentiellen Indikation fur Psychotherapie bei Migrane-patienten. Unveroffentlichter Forschungsbericht, Tubingen, 1981.

Guglielmi, R.S., Roberts, A.H., Tellegen, A., & Zimmermann, R.L. Vasomotor lability and voluntary control of peripheral skin temperature. Psychophysiology, 1981, 18, 178.

Haag, G., Gerber, W.D., Birbaumer, N., Mayer, K., Lutzenberger, W., & Schroth, G. Differentielle Indikation zur Psychotherapie der Migrane. In H. Huber, (ed.) Migrane. Urban and Schwarzenberg, Munchen, in press.

Harris, M.E., Ray, W.J., & Stern, R.M. Symptom related differences in biofeedback performances of gastric and migraine patients. Proceedings of the Biofeedback Society of American, Denver, 1978, 32-35.

Heyck, H. Der Kopfschmerz. Differentialdiagnostik, Pathogenese und Therapie fur die Praxis. Stuttgart: Thieme, 1975.

Kewman, D.G., & Roberts, A.H. Skin temperature biofeedback and migraine headaches. A double-blind study. Biofeedback and Self-Regulation, 1979, 4, 257.

Knapp, T.W., & Florin, I. The treatment of migraine headache by training in vasoconstriction of the temporal artery and a cognitive stress-coping training. Behavior Analysis & Modification, 1981, 4, 267-274.

Luborsky, L., Singer, B., & Luborsky, L. Comparative studies of psychotherapies: Is it true, that "everybody" has won and all must have prizes? Archives in General Psychiatry, 1975, 32, 995-1008.

Meichenbaum, D. Cognitive Behavior Modification. Morristown, New Jersey: General Learning Press, 1974.

Mitchell, K.R. & Mitchell, D.M. Migraine: An exploratory treatment application of programmed behavior therapy techniques. Journal of Psychosomatic Research, 1971, 15, 137-157.

Price, K.P. & Tursky, B. Vascular reactivity of migraineurs and non-migraineurs: A comparison of responses to self-control procedures. Headache, 1976, 16, 210-217.

Quintanar, L.R., Caccioppo, J.T., Monyak, N., Alvarez, L., & Snyder, C.W. Effects of cranial vasoconstriction and paced respiration on migraine headache. Paper presented at the Annual Meeting of the Society for Psychophysiological Research, Cincinnati, Ohio, October, 1979.

Sakai, F. & Meyer, J.S. Regional cerebral hemodynamics during migraine and cluster headache measured by the 133Xe inhalation method. Headache, 1978, 13, 122-132.

Schroth, G., Gerber, W.D., & Langohr, H.D. Dopplersonographische Untersuchungen zur Migrane. Paper presented at the 2nd Doppler-Sonographie-Tagung, Tubingen, 1980.

Wolff, H.G. Headache and other headpain. New York: Oxford University Press, 1963.

PSYCHOPHYSIOLOGY OF MIGRAINE AND TENSION HEADACHE

Birgit Kröner

University of Bochum
Physiological Institute
West Germany

PSYCHOPHYSIOLOGY OF MIGRAINE AND TENSION HEADACHE

According to the Ad Hoc Committee Report (1962) migraine and tension headaches are the most common functional headaches. It is estimated that 6-12% of the general population suffers from chronic headaches impairing their mental and physical health to a considerable degree. Quite paradoxically not only the pain itself but also its treatment -- mostly the prescription of analgesics or drugs containing ergotamine tartrate -- has adverse effects on the patient's well-being. Long-lasting consumption of these drugs produces severe somatic side effects and can easily lead to drug abuse and addiction. These risks give a special motivation to the investigation of alternative intervention techniques, hopefully replacing pharmacological treatment.

Before reviewing the state of the art in the field of behavioral treatment techniques, I want briefly to characterize the body of knowledge bearing on the etiology especially of tension headache, and the rationale underlying treatment (for migraine see chapters by Graham and Beatty). Replacement of the ambiguous term "tension headache" by "muscle contraction headache" suggests a specific factor in the causation of pain, namely an abnormally high degree of muscular contraction. It is generally believed that sustained muscular contraction leads to the production of certain pain-inducing metabolic substances (catabolites) which are not carried away by the blood because of ischemia in the contracted muscles (Dalessio, 1972). Though at the time there was only very little evidence on which muscles (head, neck, shoulders) were involved in the pain-producing process, the frontalis muscle was chosen as the target of biofeedback therapy.

Migraine, on the other hand, is thought to be a vascular dysfunction. This theory, assuming two different etiological processes for migraine and tension headache, is widely accepted, though evidence supporting these hypotheses is rare, especially in respect to tension headache (see Philips, 1978). Recently a number of studies have begun to systematically investigate muscular contraction level in different muscles as the supposedly crucial factor in tension headache. Without going into too many details, the results can be summarized, as follows: In most but not all cases an elevated muscular tension level is found in patients, particularly in the frontalis and neck muscles, in some cases in the temporalis (Philips, 1977; Vaughn, Pall, & Haynes, 1977; Boxtel & Roozeveld, 1978; Pozniak-Patewicz, 1976). Sometimes a high tension level is observed under rest conditions, sometimes only during or after stress. But the most surprising outcome is that migraine patients show even higher levels of tension in the tested muscles than the so-called tension headache patients (Philips, 1977; Bakal & Kaganov, 1977). These facts cast doubt on the assumption that "tension" headache is specifically caused by extreme contraction of muscles in the head, neck or shoulder.

In a study on hyperemia in migraine (Sakai & Meyer, 1978) some tension headache patients were included. They showed no sign of intracranial hyperemia like the migraineurs but the small number of participants does not allow a final conclusion on the specificity of pathophysiological vascular processes in migraine. Some authors suggest that migraineurs are characterized by a general vasomotor lability not confined to the cranial area. Very few studies on vasomotor activity have included tension headache patients, so we have no firm ground for deciding whether migraine and tension headache patients behave differently in this functional area (see Cohen, 1978). As a matter of fact, two studies showed a sustained digital vasoconstriction in reaction to orienting stimuli in both groups of patients, whereas headache-free subjects reacted by vasodilation (Tunis & Wolff, 1953, 1954; Bakal & Kaganov, 1977).

We may summarize as follows: In a considerable number of cases migraine and tension headache patients possess an elevated level of muscular tension in different head and neck muscles. This symptom seems to be more pronounced during or after stress. Whether it is a cause or consequence of pain, or both, is still undecided. Both patient groups seem to react to orienting stimuli differently in their vascular system than headache-free persons, i.e. their behavioral pattern is determined by sympathetic activity (vasoconstriction). It seems likely that abnormal <u>cerebral</u> vasomotor processes play an important role in migraine, reflecting a disturbed mechanism of feedback control, and are less important in tension headache.

There is a traditional consensus in the literature that headache is a psychosomatic disease. Without going into the details of different theories, it is generally considered that specific personality traits, like suppressed hostility, ambitiousness, rigidity, compulsiveness, etc. are predisposing factors in the development of headaches, especially migraine (Wolff, 1937). Furthermore, it is assumed that acute emotional problems and conflicts, partly deriving from the described habitual maladaptive reaction styles, are at least contributory factors in triggering an attack. In the framework of a learning theory model, headache may be viewed as a learned avoidance response or as a behavior reinforced by positive consequences.

In my opinion it is very doubtful whether any of the specific hypotheses about the "headache personality" have found confirmation in studies using objective tests instead of clinical interviews (see Harrison, 1975). Nevertheless, evidence shows that headache patients have a distinct personality pattern which differs from that of the headache-free population. It is marked by neurotic and psychosomatic disorders (Kudrow & Sutkus, 1979). Reliable differences in this pattern between the two types of patients have not yet been found (Henryk-Gutt & Reese, 1973; Philips, 1976; Kudrow & Stukus, 1979). We conclude that headache is indeed a psychosomatic syndrome including physiological and psychological dysfunctions.

Before we finally come to our main topic, the treatment of headache, I would like to say some words about the problems of diagnosis or patient classification. Since there are no objectively defined characteristics for the two different types of headache, we have to rely upon the subjective reports of the patients describing their symptoms. The classification system offered by the Ad Hoc Committee is based on clinical experience and is not yet substantiated by statistical analysis of symptom covariance (Ziegler, Hassanein & Hassanein, 1972). Every researcher trying to classify a particular patient finds himself in conflict because the patient's symptoms never match the model. The common solution to this problem is to rely on the judgment of a specialist -- whose criteria are mostly not made explicit -- or to name a few more or less arbitrary symptoms as criteria for classification. Therefore, differences between results can easily be attributed to differences in patient selection. So I think it is very necessary to spend more research effort on the issue of syndrome analysis, as long as there are no distinct and easily measurable objective variables defining different types of headache.

CLINICAL STUDIES ON TENSION HEADACHE: A REVIEW

Though at the time there was very little evidence to support the "muscle tension" hypothesis, Budzynski et al. (1973) chose the frontalis muscle as therapeutic target, and gave the impulse for intensive research on frontalis EMG biofeedback as a therapy for tension headache. Within 16 biofeedback sessions their patients learned to reduce their muscle tension and experienced considerable relief from pain. Headache activity tended to decrease further in the follow-up three months later. In the yoked control (false feedback) no improvement could be observed. It was concluded that feedback-induced lowering of frontalis EMG was responsible for the improvement. Whether this assumption was justified is still doubtful since all patients were instructed to regularly practice relaxation at home. Therefore, we have a confounding of therapeutic factors.

Wickramasekera (1972, 1973) too reported successful application of frontalis feedback to tension headache patients. Headache activity decreased when false feedback was replaced by contingent feedback. In the other case headache improvement and reduction of tension level accelerated when feedback was given instead of training in general muscular relaxation.

Philips (1977) used frontalis and temporalis feedback depending on base level of tension before therapy. She observed a considerable decrease of muscular contraction, headache activity and medication intake, which remained low on follow-up. Using a set of three symptoms to define the type of headache, Philips differentiated between pure tension headache and patients with combined migraine and tension symptoms. The last group profited less from the training even though they had a high level of muscular tension.

Kondo and Canter (1977) confirmed the efficacy of contingent vs. "false" feedback of the frontalis muscle. However, negative results were reported by Epstein and Abel (1977). No reduction of muscular tension could be demonstrated. Furthermore, there was no correlation between decrease of tension and decrease of headache intensity during training. Nevertheless some patients experienced a certain amount of pain relief outside the laboratory.

In order to evaluate the results of biofeedback relative to other techniques, we will have a look at studies where other therapeutic strategies were used and in some cases directly compared with biofeedback. Fichtler and Zimmerman (1973) and

Tasto and Hinkle (1973) both demonstrated successful application of a rather short form of relaxation training (three to four weeks). When testing alpha feedback, Ehrisman interestingly found no decrease of muscular tension in the improved patients. However, most comparative studies indicate no superiority of one particular technique, but do show the efficacy of all treatments compared to control groups (Haynes et al., 1975; Cox et al., 1975).

In some cases, feedback and relaxation training were combined and compared with the effects of each technique alone. Here results are more or less contradictory. In Chesney and Shelton's study (1976) the combined treatment was superior to the others, relaxation training being more effective than feedback when compared to a control. Hutchings and Reinking (1976) report better results with feedback alone or in combination with relaxation than with relaxation training alone.

A very different intervention strategy was attempted by Seregny (1979). A so-called "re-education group" was told that headache is a psychophysiological reaction to stress and was taught how to approach stress-provoking situations in a new manner. Only the patients given this re-education treatment demonstrated a significant decrease of headache activity, whereas relaxation training had no effect either alone or when added to the cognitive treatment. A similar cognitive treatment strategy was pursued by Holroyd et al. (1977) and Holroyd and Andrasic (1978). Their treatment aimed at a better discrimination of stressors, sensitization to thoughts preceding or during stress, and development of cognitive mechanisms for coping with stress. It was compared to EMG feedback combined with relaxation training. Controls were a free discussion group and a self-monitoring group. Only the last group showed no improvement; the others were almost equally successful. The cognitive treatment was slightly more efficacious than the others, but even the control group where the patients freely discussed their problems without any therapeutic guidance showed some success.

CHRONIC HEADACHE AND ITS TREATMENT BY BIOFEEDBACK AND RELAXATION TECHNIQUES: RESULTS OF A RECENT STUDY

At the time our study was planned there was still much optimism about the specific psychophysiological effects of biofeedback therapy. Our main interest therefore was to compare different forms of biofeedback and relaxation in the treatment of different headache syndromes. Additionally, we were interested in the problem of diagnosis or classification of headache types. Our aim was to collect a comprehensive set of data about our headache patients. Before coming to some of our main results I shall briefly describe the design of the project.

Patients were collected through a local newspaper announcement. Three hundred and two persons came to our laboratory, all suffering for at least two years from chronic headache. All had to fill out a comprehensive questionnaire about the symptoms of headache, possible etiological factors and precipitating events. One hundred fifty-two patients went through all stages of our research program. Only those patients were admitted whose history showed no sign of physical causation of headache (head injury or disease) or neuralgia, who had at least several attacks per month, and were willing and able to comply with the experimental regimen. Only about ten patients resigned from the study during the therapeutic stage.

Before therapy began all patients were given a number of personality inventories, including measures of anxiety, neuroticism, stress susceptibility, psychosomatic disorders and some others. Above that, all subjects completed three baseline sessions where different physiological functions (frontalis and neck EMG, SRL, SRR, HR, breathing frequency) were registered under stress and rest conditions. Additionally, a test of habituation to non-signal tone stimuli was carried out. For at least four weeks patients had to monitor their baseline headache activity, i.e. note for every waking hour the intensity of headache, and kind and quantity of medication taken. Prior to that, the rating procedure was practiced for two weeks so that we could discuss any problems arising from the task.

After finishing the baseline, subjects were randomly assigned to the different therapeutic programs, i.e. biofeedback and relaxation therapy and the particular treatment groups. Our endeavor to define homogenous patient groups by means of cluster analysis of the symptom-questionnaire before treatment and assign them in a systematic way to the treatments failed. The observed solutions of cluster analysis were not reliable. We tried a different approach to the problem but we were obliged to begin with therapy before the results of this analysis were available. Four relaxation procedures were applied: autogenic training, as originally developed by J.H. Schultz and described by W. Luthe; progressive relaxation, in an abbreviated version as generally used in behavior therapy; and a combination of both. All of these procedures included specific instructions devised to enhance mechanisms of active self-control, whereas a fourth group (combined technique) was told that regular performance of relaxation exercises automatically leads to a reduction of headache. All relaxation groups (n=10-12) met twice per week (one hour) for six weeks. Headache rating was carried out throughout this time until four weeks after termination of therapy. It was resumed three to four months later, again for four weeks. The biofeedback groups (n=14-18) were run on the same time schedule. One group was trained to control muscular tension in the frontalis, and was instructed from the start to regularly practice the learned skill at home and to use it as a coping reaction against pain. The second group was trained in the same manner without explicit suggestion from the therapist for home practice. The third group was treated by feedback from the neck muscles. The fourth and fifth groups were trained to control an autonomic function, i.e. skin resistance level. One of these groups used portable instruments for home training and came to the laboratory only every fortnight for an additional feedback session.

After treatment all patients once again filled out the personality questionnaires and completed one more session where physiological variables were measured under conditions of rest and relaxation. All treatment groups were to be compared with two headache control groups, a waiting control and a false feedback group (placebo control). We had one more control group consisting of headache-free subjects who participated in all three sessions of physiological measurement and carried out the personality tests.

The following research issues will be dealt with in this outline of the results:

A. Empirical analysis of headache symptoms
B. Headache and personality
C. Outcome of therapy in terms of headache activity and medication intake.

The analysis of phyisological data is still in progress, so that results cannot yet be communicated.

A. Empirical Analysis of Headache Symptoms

All researchers engaged in the field of headache research find themselves confronted with the problem of diagnosis or categorization of headache. In one of the rare studies on symptom analysis, Ziegler et al. (1972) could not corroborate the "notorious" headache types using factor analytic methods. We wanted to find out whether reliable syndromes can be isolated by factor analyzing our questionnaire, and whether these syndromes hold any resemblance to the described clinical types. Out of the 82 questionnaire items identified in the literature as being relevant to the classification of headache, 52 were chosen for factor analysis. Thirty items were rejected because they were answered by 90 or more percent of the 302 subjects in the same direction. Two factors were judged to be reliable (5%

criterion) and were amenable to interpretation. Each factor explained 15% of the total variance. They fulfilled the simple structure criterion and did not correlate with each other (confirmed by results of an oblique rotation). The first factor was interpreted as "migraine" syndrome. It was defined by gastrointestinal disturbances and sensoric oversensitivity during pain, symptoms emphasizing the considerable distress experienced during the attack. Unilaterality and attack-like pain onset complement the picture of this syndrome. The second factor is characterized by muscular phenomena and a more constant type of pain.

We may conclude that our data support the existence of two independent headache types, clinically known as migraine and tension headache. We did not find the frequently differentiated forms of classical and common migraine. Furthermore it is noteworthy that our factors did not include high-loading items referring to visual disturbances in the beginning of headache; these phenomena are generally considered to support the theory of biphasic vasomotoric processes in migraine. Our satisfaction with these results confirming to a certain degree the clinical classification system must not make us blind to the fact that most of the inter-individual variance is not explained by the two factors. There is a considerable specificity in the individual headache profile.

We used the syndrome scales for a post-hoc analysis of therapy outcome to investigate the syndrome-specific efficacy of different treatment procedures. We found no difference in headache improvement between the four types of headache patients -- those showing a clear-cut migraine or tension-type headache, those having symptoms of both types and those showing low scores on both scales. This result supports the assumption of a relative unspecific process of therapy.

B.   Headache and Personality

Despite the belief that headache patients are characterized by a specific premorbid personality pattern, there are very few studies where patients with different types of headache have been compared with each other and normal controls. We wanted to investigate whether our four types of headache patients, as defined by their scores on the two factor scales, could be differentiated from each other and a normal control in terms of personality traits. As a first step we analyzed differences between the headache sample and controls. Differences were found on various scales, mainly showing greater neuroticism, depression, excitability, anxiety and more psychosomatic complaints in the headache patients. In a questionnaire of stress susceptibility, headache patients proved to be more vulnerable to stressors than headache-free controls.

However, these results do not resolve the question of which variables discriminate the groups and with what degree of accuracy. Discriminance analysis was therefore applied to our data and only four variables emerged with discriminant value. The proportion of correct classifications is rather low (44%), with headache-free subjects being classified with the smallest error rate (23.5%). Taking the first two canonical factors as coordinates, it is evident that the headache groups lie rather close together; the groups with high scores on both scales and with low scores on both are the most distinct.

In summary, we may conclude that headache patients are characterized by personality variables reflecting a high psychophysiological arousal level. The reduced ability to cope with environmental and emotional problems coincides with a great number of complaints about somatic disturbances. This picture is frequently found in patients whose physical disorders are supposed to be psychologically based, but is not specific to headache patients. The data support the argument that any treatment supposed to reduce the habitual arousal level and to strengthen the ability to react with less psychophysiological excitement to stressors, can at least be valuable in reducing the headache activity.

A closer look at the results of the discriminance analysis shows that the scales containing headache symptoms discriminate between headache patients and the normal control. They should not be interpreted in terms of structural personality traits. Therefore we find no support in our data for the theory of a particular personality structure predisposing the individual to the development of headach disorders. Furthermore, there is no evidence for a particular structure of personality in the four headache groups, suggesting that there is no specific psychological etiology for migraine and tension headache.

C. Outcome of Therapy

There are some major differences between the two basic therapeutic strategies used in this study. The relaxation training was carried out in groups and supervised by one therapist. This permitted a close and intensive patient-patient and patient-therapist interaction. Patients could freely discuss their experiences and attitudes towards therapy and their symptoms with each other and the therapist. These conditions may facilitate cognitive restructuring and modeling effects. On the other hand, biofeedback patients performed their training alone in the laboratory so occasions for communication with other patients were rare. The feedback sessions were run by different experimenters, instructions were partly given in written form, and interaction was confined to technical rather than therapeutical aspects. This was done in order to distinguish biofeedback-specific processes from unspecific effects. The relaxation groups probably had a clearer picture of <u>what</u> they were supposed to learn, i.e. relaxation, and <u>how</u> they could accomplish this. The biofeedback groups were only told that they were going to learn to control a physiological function responsible for their headache by listening to the feedback signal. The term relaxation was never used, neither was the feedback function explained.

These factors may account for the observed differences in therapeutic outcome. There were four measures of therapeutic outcome: mean intensity and duration of headache per week, and number of migraine drugs and analgesics taken. Differences between pre-treatment, post-treatment and follow-up period were used in the analysis. Three of the four relaxation groups showed a significant reduction of headache intensity and duration. The improvement increased from the post-therapeutic to the follow-up period. Only patients trained with autogenic training failed to improve. The symptom reductions were significant at the 10% level in comparison with the placebo and waiting controls. There was a small but non-significant reduction in the consumption of analgesics. The migraine drug intake changed very little. Apart from the failure of the autogenic training group to improve there were no differences between treatment groups.

Four of the five biofeedback groups showed a significant decrease of headache intensity which was most pronounced in the follow-up period. The change was significantly larger than that of the placebo group, or the non-successful feedback group (trained to increase skin resistance level in the laboratory). But only the patients trained in frontalis muscle control with instructions to regularly practice at home showed a decrease in duration of headache. With one exception there were no significant changes in drug intake. An increase in migraine drug consumption could be observed in some groups.

Overall, relaxation training was more successful than biofeedback treatment. For example, 44% of the relaxation training participants reduced their headache intensity to less than 50% of baseline level, whereas only 29% of biofeedback subjects did so.

The level of therapeutic success was not overwhelmingly impressive in this study, though in a considerable number of cases improvement was substantial. With some exceptions, differences between treatments could not be observed. These results confirm that the effective therapeutic elements are not specific, either in respect to the physiological functions modified or in the type of headache treated. Since the physiological results are not yet available, we are not able to confirm our assumption that the common element of all therapies is the learning of a psychophysiological relaxation response. But it seems relatively unimportant which relaxation technique is learned. Equally important is the attitude towards one's symptoms. Patients trained with relaxation or biofeedback techniques learn to discriminate bodily sensations and become more sensitive to situations inducing headache. They feel less helpless and are able to counteract their headache, rather than being condemned to endure an unchangeable destiny. The superiority of the relaxation training is possibly based on these kinds of cognitive change, which were facilitated by the setting in which treatment was carried out.

SUMMARY AND EVALUATION

Most of the studies reviewed meet the criteria for well controlled clinical experiments, including controls, multiple measurements at different stages of the therapeutic process, and follow-up measures. Therefore, the conclusions offered by the data stand on relatively firm ground. The first conclusion we might draw is that frontalis EMG-feedback and verbally directed relaxation training are effective techniques in the treatment of tension headache, and lead to a considerable degree of symptom decrease. Blanchard et al. (1980) support this view on the basis of statistical evaluation of the outcome of 19 different studies. They found no differences in the efficacy of frontal EMG-feedback, relaxation alone and combined with feedback, all of which were superior to the medication placebo and headache-monitoring groups used as controls. These control conditions do not, of course, represent a very hard test; at least they give no reason for arrogance on the part of supporters of the behavioral medicine approach. Yet, on the other hand, it seems likely that the patients treated in these studies had tried numerous treatments before without success.

Taking the efficacy of biofeedback and relaxation therapy as proven, the most important question is how does this improvement come about. What therapeutic processes are initiated by these techniques? In our opinion, there are many reasons to believe that the reduction of frontalis muscle tension is neither necessary nor sufficient for pain relief. Since high muscle tension in the frontalis is probably not the cause of so-called tension headache, a decrease in contraction cannot be the _cause_ of symptom relief. Several studies have demonstrated a lack of correlation between symptom intensity and degree of muscular tension. Likewise, considerable improvement in reported headache activity has been found in the absence of changes in muscular tension. Therefore, we conclude that EMG-feedback does not work specifically via learned decrease of frontalis tension.

Moreover, the assumption that therapy operates through the modification of physiological functions is probably too limited. The headache syndrome is accompanied by psychological instability. There are some indications that pain relief is accompanied by improved coping reactions to stress, reduction of anxiety and other psychosomatic disorders, etc. (Cox et al., 1975; Holroyd & Andrasic, 1978). This seems to be at least indirect evidence for the importance of psychological processes in therapy, even if the treatment claims to be a physiological modification technique.

Cognitive reappraisal of stress and headache-inducing events, and the reinforcement of coping strategies seem to play important roles in therapy. It is possible that the physiologically oriented treatments also instigate, albeit indirectly and unsystematically, modifications of the patient's cognitive and emotional attitude to himself and his symptom. In our opinion, it is not yet clear what happens in so-called cognitive therapy, and whether it operates purely by cognitive mechanisms. Improvement could also be brought about by physiological reactions similar to the ones trained in relaxation techniques. The blocking of negative emotions in stress situations which Holroyd and Andrasic (1978) required of their patients could in fact have been accomplished by relaxing in these situations. Holroyd et al. (1980) view the relaxation behavior as one effective response for coping with events inducing headache or exacerbating them.

Returning to the evaluation of biofeedback and relaxation therapy, we suggest that a systematic modification of cognitive variables can be achieved more easily in group therapy, since this permits more intensive patient-patient and patient-therapist interaction than the laboratory setting. Additionally, relaxation therapy is more economical, as it is applied in groups and requires no costly instrumentation.

I believe there is no need for more comparative studies on different techniques at present; rather there is a need for multimodal treatment and multimodal measurement, so as to get a closer view of the processes responsible for therapeutic success. The answer may lie in a combination of physiological, cognitive, behavioral and emotional modifications in the headache patient.

REFERENCES

Ad Hoc Committee on classification of headaches. Journal of American Medical Association, 1962, 179, 717-718.

Bakal, D.A. and Kaganov, J.A. Muscle contracton and migraine headache: Psychophysiologic comparison. Headache, 1977, 17, 208-214.

Blanchard, E., Andrasik, F., Ahles, T. and Teders, S. Migraine and tension headache: A meta-analytic review. Behavior Therapy, 1980, 11, 613-631.

Boxtel, van, A. and Roozeveld van der Ven. Differential EMG activity in subjects with muscle contraction headaches related to mental effort. Headache, 1978, 17, 233-237.

Budzynski, T.H., Stoyva, J.H., Adler, G. and Mullaney, D.S. EMG-biofeedback and tension headache: A controlled outcome study. Psychosomatic Medicine, 1973, 35, 484-496.

Chesney, M.A. and Shelton, J.L. A comparison of muscle relaxation and electromyogram biofeedback treatments for muscle contraction headache. Journal of Behavior Therapy & Experimental Psychiatry, 1976, 7, 221-225.

Cohen, M. Psychophysiological studies of headache: Is there similarity between migraine and muscle contraction headaches? Headache, 1978, 18, 189-196.

Cox, D.J., Freundlich, A. and Meyer, R.G. Differential effectiveness of electromyographic feedback and verbal relaxation instructions and medication placebo. Journal of Consulting & Clinical Psychology, 1975, 43, 892-898.

Dalessio, D.J. (Rev.) Wolff's headache and other head pain. Oxford University Press, New York, 1972.

Epstein, L.H. and Abel, G.G. An analysis of biofeedback training effects for tension headache patients. Behavior Therapy, 1977, 8, 37-47.

Fichtler, H. and Zimmermann, R.R. Changes in reported pain from tension headaches. Perceptual Motor Skills, 1973, 36, 712- .

Friedman, A.P., Pool, N. and Storch von, T.J. Tension headache. Journal of American Medical Association, 1953, 151, 174-177.

Harrison, R.H. Psychological testing in headache: A review. Headache, 1975, 14, 177-185.

Haynes, S.N., Griffin, P. and Mooney, D. Electromyographic biofeedback and relaxation instructions in the treatment of muscle contraction headaches. Behavior Therapy, 1975, 6, 672-676.

Henryk-Gutt, R. and Reese, W. Psychological aspects of migraine. Journal of Psychosomatic Research, 1973, 17, 141-153.

Holroyd, K.A., Andrasik, F.J. and Westbrook, I. Cognitive control of tension headache. Cognitive Therapy Research, 1977, 1, 121-133.

Holroyd, K.A. and Andraski, F. Coping and the self-control of
    chronic tension headache. Journal of Consulting & Clinical
    Psychology, 1978, 46, 1036-1045.
Holroyd, K.A., Andrasik, F. and Noble, J. A comparison of EMG
    biofeedback and a credible pseudotherapy in treating
    tension headache. Journal of Behavior Medicine, 1980, 3,
    29-39.
Hutchings, D.F. and Reinking, R.H. Tension headaches: What form
    of therapy is most effective? Biofeedback & Self
    Regulation, 1976, 1, 183-190.
Johnson, W.G. and Turin, A. Biofeedback treatment of migraine
    headache: A systematic case study. Behavior Therapy, 1975,
    6, 394-397.
Kondo, C. and Canter, A. Time and false electromyographic
    feedback: Effect on tension headche. Journal of Abnormal
    Psychology, 1977, 86, 93-95.
Kudrow, L. and Sutkus, B.J. MMPI pattern specificity in primary
    headache disorders. Headache, 1979, 19, 18-24.
Philips, C. Headache and personality. Journal of Psychosomatic
    Research, 1976, 20, 535-542.
Philips, C. The modification of tension headache pain using EMG
    biofeedback. Behavior Research & Therapy, 1977, 15,
    119-129.
Philips, C. Tension headache: theoretical problems. Behavior
    Research & Therapy, 1978, 16, 249-261.
Pozniak-Patewicz, E. Cephalic spasmus of head and neck
    muscles. Headache, 1976, 15, 261-266.
Sakai, F. and Meyer, I.S. Regional cerebral hemodynamics during
    migraine and cluster headaches measured by the 133Xe
    inhalation method. Headache, 1978, 18, 122-132.
Sargent, J.D., Green, E.E. and Walters, E.D. Preliminary report
    on the use of autogenic feedback training in the treatment
    of migraine and tension headaches. Psychosomatic Medicine,
    1973, 35, 129-135.
Seregny, J.F. A comparison of relaxation training and
    re-educative training in the treatment of chronic
    muscle-contraction headache. Dissertation & Abstract
    Intern., 1979, 39,(11-B), 5586.
Tasto, D.L. and Hinkle, J.E. Muscle relaxation treatment for
    tension headaches. Behavior Research & Therapy, ,1973, 11,
    347-349.
Tunis, M.M. and Wolff, H.G. Studies on headache: Long term
    observation of the reactivity of the cranial arteries in
    subjects with vascular headache of the migraine type.
    Archives of Neurology & Psychiatry., 1953, 70, 551-557.
Tunis, M.M. and Wolff, H.G. Studies on headache: Cranial
    artery vasoconstriction and muscle contraction headache.
    Archives in Neurology & Psychiatry, 1954, 71, 425-431.

Vaughn, R., Pall, M.L. and Haynes, S.N. Frontalis EMG response to stress in subjects with frequent muscle contraction headache. Headache, 1977, 16, 313–317.

Wickramasekera, J. Electromyographic feedback training and tension headache: Preliminary observations. American Journal of Clinical Hypnosis, 1972, 15, 83–85.

Wickramasekera, J. The application of verbal instructions and EMG feedback training to the management of tension headache. Preliminary observations. Headache, 1973, 13, 74–76.

Wolff, H.G. Personality features and reactions of subjects with migraine. Archives in Neurology & Psychiatry, 1937, 37, 895–921.

Ziegler, D.K., Hassanein, R. and Hassanein, K. Headache syndromes suggests by factor analysis of symptom variables in a headache prone population. Journal of Chronic Disease, 1972, 25, 353–363.

PAIN: ITS PATHOPHYSIOLOGY, A BRIEF REVIEW

Blaine S. Nashold and J. Ovelmen Levitt
Department of Surgery
Duke University Medical Center

Durham, North Carolina

> "Physical pain is not a simple fact of
> nervous impluses traveling over a nerve
> at a predetermined gait. It is the result
> of conflict between the stimulation and
> the individual."
>
> Sir Charles Sherrington 1947

Introduction

Much of human existence is devoted to the avoidance of pain or seeking relief from persistent pain. Pain is an important part of the human protective mechanism, but when pain becomes chronic, it may occupy all of a human's existence. Scientifically, pain can be defined as an unpleasant sensory and emotional experience associated with actual or potential tissue damage or it can be described in terms of such tissue damage. Nociception denotes the experimental approach to the study of neuronal representation of noxious stimuli in humans and animals. The nature of pain has permeated human thought since ancient times; even the book of Genesis reminds us that pain came to man through the fall. Ancient myths record the painful suffering of gods and certain theological dogma has centered on the role of pain and suffering in human existence.

Historical Perspective

The beginning of modern conceptual thought on pain mechanisms was first expressed by Rene Descartes (1644). His simple diagram of a kneeling man beside a fire (stimulus), with a tube connecting the toe to the brain, evokes the first conceptual model of a painful stimulus evoking a response in a receptor located in the foot and transmitting it through the nerves to the brain where it is analyzed and appropriate action takes place. Descartes did not have the anatomical or physiological data to prove his concept, but later anatomists and physiologists, using the Cartesian model, developed our modern conceptual models for pain mechanisms. In 1826, Mueller put forth the idea of the specificity theory of sensory receptors. This followed von Frey's (1894) observations demonstrating specific anatomical structures mediating pain, cold, warmth and touch. It is interesting that since free nerve endings were the most common, they were considered the pain receptors. Later, Goldscheider (1885) de-emphasized the specific receptor idea and proposed the idea of stimulus intensity as the controlling factor in the sensation of pain. In 1920, Sinclair, Weddell and Zander (1952) and later Wollard (1935) proposed the pattern theory of pain, suggesting that activation of groups of receptors or nerves produces a pattern within the central nervous system of the perceived sensation. Later, based on clinical observation, Head (1920) attempted to divide sensation into two specific types: protopathic sensibility -- in which pain was strong, nonlocalized and radiating, and produced by sensations from stimulations of hairs and extreme temperature -- and sensation designated as epicritic.

In 1965, Melzack and Wall (Melzack & Wall, 1965) proposed the "gate control" theory in which they proposed that there are specific fibers and receptors for pain and there is some dependence on the number and types of fibers activated and the temporal dispersion within the spinal cord. They emphasized the importance of the dorsal root entry zone and the substantia gelatinosa as an important synaptic region in the dorsal cord for sensory integration, and proposed a presynaptic pain mechanism in which the activity of large afferent fibers maintain a dorsal root negativity, and the action of small unmyelinated fibers result in a dorsal root positivity. A continuous barrage of large primary afferent (A) fibers maintains negativity of the dorsal roots that decreases or limits action of small pain fibers. This is followed by a reciprocal action on the "T" cells in the substantia gelatinosa. The sum of activity of these cells is related directly to the afferent input and the perception of pain. Therefore, a gate controlling mechanism localized in the dorsal root entry zone was of key importance. In the clinical setting, if one stubs a toe, the sharp pain is due to activation of delta fibers and the dull aching pain is due to C-fibers. Vigorous rubbing of the toe, therefore, stimulates A fibers and the pain is blocked or lessened. The exact neurophysiological details of the gate control theory have not been fully realized. However, it did reactivate neurophysiological studies of pain mechanisms and was responsible for the application of electrical stimulation for treatment of chronic pain conditions in man.

Peripheral Receptors and Nerve Fibers

There are three types of unmyelinated fibers which convey pain. The free-ending fibers are responsive to high intensity mechanical deformation of the skin and conduct at about 5 to 10 m/sec. The other two types of fibers are not only activated by mechanical stimulation but also by heat and chemicals and are considered polymodal and may also respond to both noxious and non-noxious stimuli. The physiological effect of chemicals that produce noxious sensation are complex and are often associated with varying degrees of tissue injury and may activate mechanical or thermal receptors in local tissue injury. Mechanoreceptors and thermoreceptors may discharge under the influence of certain chemicals such as prostagalandin E, histamine, serotonin, and bradykinin. Intra-arterial or intraperitoneal bradykinin results in pain in both man and animals. Experimental observation suggests increased peripheral C-fiber discharge causing heightened activity of the neurons in the dorsal root entry zone.

Peripheral Nerve

It is well known that a direct relationship exists between the conduction velocity and fiber diameter of afferent nerves, and this has been used by the physiologists to classify the various types of nerve fibers. A mixed peripheral nerve with motor and sensory fibers shows a range of fiber size from 1 to 20 μm. Pain conducting fibers, the C- and A-delta fibers are 2 μm and conduct slowly at 0.5 to 4 m/sec. Some B fibers (8 to 14 μm) conduct faster and they also transmit noxious sensations. The largest fibers are proprioceptive (1A) and are not involved in nociception.

Dorsal Roots and Ganglion

Bell(1811) and Magendie (1822) proved that the spinal root was composed of a motor ventral division and a sensory dorsal division. Recent observations, however, show that afferent fibers may pass through the ventral root and explain the paradoxical results following dorsal rhizotomy in man to relieve pain.

Dorsal Root Entry Zone

Anatomically the dorsal root entry zone (DREZ) can be divided into six lamina as proposed by the neuroanatomist Rexed

(1952). These anatomical divisions also have functional characteristics. Rexed designated six lamina with lamina I made up of dense dendritic fibers running horizontally along the longitudinal aspect of the cord. The substantia gelatinosa is probably represented by lamina II and III which receive very substantial afferent and efferent input. The neurons in this region develop prolonged discharge after stimulation which suggests that this area could be the origin of the intractable pain associated with brachial plexus avulsion and post-herpetic pain. Both have been relieved after localized lesions of the dorsal root entry zone. Pharmacological studies reveal a number of chemical agents in this region. They are synaptic excitatory transmitters such as glutamic acid, substance P, as well as enkephalin, neurotension, cholecystokinin, neurophysin, oxytocin, and glucagon, but the role of these agents in the genesis of pain is unknown.

Afferent Spinal Cord Tracts

The spinothalamic tract can be considered the primary pain pathway in man where it reaches its highest phylogenetic development. It is made up of small myelinated fibers which arise from second order neurons in the dorsal root entry zone (layers I & V). Ascending pathways for pain include the Lissauer's tract which is an ipsilateral pathway. The paleothalamic pathways are older phylogenetically and represent diffuse ipsilateral contralateral roots for pain transmission. The duality and diffuseness of these pain tracts probably explain the difficulties encountered when surgical section, such as a cordotomy, fails to relieve the pain. The diffuse spinal reticular pathways have extensive interconnections with the brainstem where the protective reflexes are organized. These pathways also make connection with regions of the mesencephalon in the periaqueductal gray and the interlaminar nuclei of the thalamus. The lateral spinothalamic tract also makes collateral connections on its rostral passage to the thalamus where it ends in the ventrolateral thalamic nucleus.

Mesencephalon

The mesencephalon, although the smallest portion of the brain, measuring 15 to 20 mm. in length, contains one of the most complex neuroanatomical and physiologic regions to be found within the central nervous system. The midbrain is a primordial organ for the correlation of sensory impulses and these impulses are received by the midbrain from the eyes, ears, skin, muscles, joints and bones and are there correlated in the interest of producing the most effectual motor responses.

The organization of certain instinctual reflexes for fear, flight, food and sexual response is mediated through circuits in the mesencephalon, and the state of the sentient consciousness is unique, dependent upon its integrity. The pain pathways are located in the more dorsal aspect of the midbrain in the region of the tectum and the dorsal tegmentum. The full length of the mesencephalon is traversed by the Sylvian aqueduct, connecting the cerebral ventricles with those of the posterior fossa and surrounding it is the gray matter, a core of neural tissue that appears to be related to complex affective and emotional activities of the organism. The ascending pain and temperature paths traverse the dorsal part of the mesencephalon with the lateral spinothalamic and trigeminal thalamic pathways representing the primary direct paths. The lateral spinothalamic tract conveys pain and temperature sensations from the entire contralateral surface of the body, excluding the face, and it lies in the dorsolateral angle of the dorsal tegmentum. Within it is represented a specific somatotopic schema of the body, the lower portion of which lies lateral to the more cephalad portions. The cross sectional area of the lateral spinothalamic tract at the superior colliculus is extremely small, 0.65 sq. mm. in area, and approximately 1500 fibers make up the tract at this level, although the cervical segment contains approximately 15,000 fibers, indicating that between the cervical and upper midbrain region a large number of fibers leave the main tract, making connections below the level of the rostral midbrain. The trigeminal thalamic pathways convey pain and temperature sensations from the facio-oral region and lie medial to the lateral spinothalamic tract at the lateral edge of the central gray adjacent to the aqueduct. The exact organization of this pathway is still not completely known. The mandibular division appears to be situated more laterally than the ophthalamic division, and there is intermingling of these facial fibers at this level with those of the 7th, 9th, and 10th cranial nerves. In other words, two body schemes related to pain perception are represented at this level of the midbrain. The surface and appendicular parts of the body are represented in the lateral spinothalamic tract, while the facial, oral and internal body cavities are represented more medially. This has been well demonstrated by direct stimulation of the mesencephalon in awake man. In addition to these direct pathways, there are diffuse multisynaptic spinoreticular pathways, and these include the spinotectal and the spinocollicular pathways which also convey noxious sensation and are older phylogenetically. The ascending reticular fibers nearby the periaqueductal region are responsible for the origin of affective reaction to painful stimuli. There is also additional anatomical evidence of pathways conveying painful sensations from the facial areas which may be ipsilateral as well as contralateral, and there appears to be a large

concentration of ipsilateral, trigeminal thalamic fibers in the
region of the central tegmental tract.

Thalamus

The thalamus is an important region for somatosensory
integration. The spinothalamic pathways synapse in the ventro-
posterior and ventro-lateral thalamic region while the
spinoreticular fibers gain access to the midline nonspecific
thalamic nuclei, parafascicularis, and centralis lateralis and
posterior complex of the medial geniculate. There are no
specific thalamic neurons responsive to nociceptive stimulation.
However, pain is experienced in man with stimulation of the
central gray of the midbrain as well as the nonspecific thalamic
nuclei in the posterior lateral thalamic regions. The thalamus
appears to be diffusely organized and neurons in the
ventrolateral nuclei may develop spontaneous epileptiform
discharges following deafferentation of the dorsal root input in
animals. Stereotactic lesions of these various thalamic regions
only temporarily reduce pain in man.

Cerebral Cortex

Only an occasional cortical cell will respond to
nociceptive input. The primary sensory thalamic nucleus (VPL
and VPM) project to cortical sensory areas SI and SII. Pricking
pain is relayed to the sensory cortex. Burning pain has less
direct cortical input. Pain projections via the hypothalamus
and frontal cortex are associated with production of pain
associated with affective and motivational responses.

Neuropharmacology of Pain

Changes in cerebral monamines are affected by nociception.
Certain of these effects, especially in the midbrain, can be
reversed by serotonin or L-dopa. Pain in persons with cancer
has been reported to have been reduced after the administration
of L-dopa. Stimulus-produced analgesia in animals may be
significantly increased by the depletion of noradrenalin.
Morphine analgesia can be modulated by levels of monamine and
reserpine antagonizes morphine analgesia in experimental
animals.

Substance P and glutamate may be important in nociception.
The dorsal roots, the dorsal root entry zone, hypothalamus and
substantia nigra contain high amounts of substance P. Section
of a dorsal root results in loss of substance P in the
substantia gelatinosa and an increase occurs with dorsal root
stimulation. Excess amounts of substance P in the dorsal root
entry zone may result in hyperactive discharge of certain

neurons. Substance P cannot be considered a neurotransmitter at this point, but its importance to nociception is clear. The role of glutamate is unclear, although it is found in high concentrations in the dorsal roots and ganglia.

Endogenous Morphine-Like Substances

The periaqueductal injection of morphine results in an analgesic effect five hundred times greater than if administered systemically. Other brain areas are nonresponsive to similar morphine injections. Spinal subarachnoid or epidural morphine injections also result in profound analgesic effects. An endogenous ligand of morphine has been isolated from the brain. It is a pentapeptide with amino acid sequences and has been called enkephalin. The enkephalins are thought to be the breakdown products of endorphins which are the active agents in the central nervous system. The B-endorphin is 18 to 33 times as potent an analgesic as morphine when tested in rats. These substances are in high concentration in nervous regions of the brain which are involved in nociception. There is strong evidence that analgesia produced by stimulation of the periaqueductal gray, morphine and enkephalin analgesia all have a common physiologic substrate.

Electrical Stimulation-Produced Analgesia

Based on the conceptual model of the gate control theory, electrical stimulation was applied to injured peripheral nerves of man with reduction of pain. Following this, larger areas of the body involved in pain were treated by direct electrical stimulation on the dorsal surface of the spinal cord, the so-called dorsal column stimulation. The mechanism of pain relief is still unknown, but it is agreed that it is not via a "gating mechanism".

Central stimulation of the periaqueductal dorsal gray and thalamus can result in painful sensations as already noted. Of great interest were the observations in rats that analgesia also resulted when the raphe nuclei were stimulated in the midbrain. These central areas are closely contiguous to the mesencephalic central gray. The analgesic effect appears to be mediated via a serotonin pathway projecting caudally as far as the spinal cord and dorsal root entry zone. This type of central stimulation has been used in a limited manner in humans suffering from intractable pain by implanting chronic depth electrodes in or near these central structures of the thalamo-mesencephalic junction.

Deafferentation Syndrome: A Chronic Animal Pain Model

A condition of chronic pain or dysaesthesia has been

produced in different species of animals with peripheral nerve, dorsal root, spinal cord anterolateral column lesions and with certain supraspinal lesions. This deafferentation (a time lag) syndrome typically takes days to weeks to develop and occurs in only certain percentages of subjects (it may be genetically determined) and takes the form of excessive grooming and licking, scratching and biting to the point of autotomy and persists for lifetime. The preparation can be studied behaviorally, electrophysiologically, pharmacologically and histologically, and subjected to additional surgical procedures to alter the time course or the quantitative aspects.

The dorsal roots of cervical plexus in the rat have been sectioned either pre- or post-ganglionically to produce the deafferentation syndrome. The appearance of the syndrome is not reliable when the lumbosacral roots are cut. In order to produce the syndrome most or all of the roots to the brachial plexus must be sectioned or avulsed. Basbaum (1974) describes a "deliberate, persistent and delicate" process, beginning after one day, the animals first removing the nails of the foot ipsilateral to the rhizotomies and gradually removing the surface skin and eventually the digits may be amputated.

There are three postulated neurophysiological bases for the chronic pain of deafferentation. The first has been related to peripheral nerve injuries, especially with neuroma formation. The damaged ends of afferent C-fibers are thought to generate impulses giving rise to sensations of pain. There is a release of norepinephrine associated with the impulse formation. Two other bases, central in origin, have been proposed -- either the deafferentation could produce disinhibition in certain pathways, with unmasking of a primitive multisynaptic pain pathway, or the deafferented structures in pain pathways may become supersensitive and discharge in an epileptogenic manner. Both of these hypotheses have been tested in animals.

Both Duckrow and Taub (1977) and Albe-Fessard, Nashold, Lombard, Yamoaguchi and Boureau (1978) performed multiple rhizotomies or root avulsions in rats with results similar to those of Basbaum, Duckrow and Taub (1977) showed that the incidence of self-mutilation was reduced from 89% to 40% when the rats were given doses of Diphenylhydantoin greater than 10 mg/kg/day, while Albe-Fessard, et al. (1978) found that the latency of the appearance of the deafferentation syndrome could be delayed by repetitive electrical stimulation of the side contralateral to the rhizotomies. Basbaum (1974) found that in bilateral cervical rhizotomized rats (C5-T2), a unilateral spinal cord hemisection or lesion of the lateral funiculus at C4 prevented chewing of the limb ipsilateral to the lesion. A related observation was made by Lombard, Nashold, Albe-Fessard, Salman and Sakr (1979) who found that the appearance was later in those rats which had avulsion of the dorsal roots than in those whose roots were sectioned. Avulsion was found to produce a localized lesion of the spinal cord and enable more shrinkage of the dorsal horn and gliosis.

Heybach, Levitt and Brodish (1979) have shown that the deafferentation syndrome after multiple cervical rhizotomies occur in genetically blind mice and also that corticosteroid levels are elevated during periods of autonomy with the weight of the adrenals elevated by threefold.

A deafferentation syndrome can be produced by peripheral nerve section in which regeneration to the foot is prevented. The behavioral manifestations are similar in that the autotomy begins with the toenails and proceeds proximally. However, Wall (Wall and Gutnick, 1974) suggests that there is a generation of nerve impulses in the cut and sprouting peripheral nerve fibers which causes the autotomy. An antisympathetic drug (Guanethidine) prevented the appearance of autotomy. It was suggested that the local release of norepinephrine by sympathetic fibers in the neuroma may be one of the factors producing sufficient peripheral afferent discharge to cause autotomy. Therefore, these authors are suggesting a peripheral origin in chronic pain.

A related study by Wiesenfeld and Hollin (1980) in which rats with peripheral nerve sections were exposed to cold, showed that under these conditions the animals exhibited signs of extreme discomfort which included licking and biting the operated foot and jumping around the cage. Removal from the cold caused behavior to markedly decrease and eventually vanish. Cold has been shown to enhance peripheral synthesis and release of catecholamines.

Wiesenfeld and Lundblom (1980) compared three types of lesions causing autotomy in the rat -- multiple dorsal rhizotomy, spinal nerve section just distal to the dorsal root ganglia (DRG), and section of the median, ulnar and radial nerves. They found that the mean time to autotomy was similar for peripheral nerve section and dorsal rhizotomy and longer than for postganglionic section. Also, the peripheral nerve section and rhizotomies had the most severe autotomy. They stated that peripheral nerve injury may be due to abnormal activity in nociceptive afferents while pain due to rhizotomies may be caused by disinhibition of interneurons.

Inbal, Devor, Tuchendler and Lieblich (1980) reported that the extent of autotomy varies greatly in genetically different populations of rats and suggested that genetic differences may be the basis for the fact that after seemingly identical nerve injuries some humans develop chronic pain syndromes and others do not.

Levitt and Sedivec (unpublished observations) noted the development of wounds on proximal limb parts of cats after complete lumbosacral dorsal rhizotomy, and Levitt and Nashold (Levitt & Nashold, unpublished observations) observed, also in cats, the presence of wounds several weeks after complete brachial plexus, dorsal and ventral rhizotomy and brachial plexus, dorsal root avulsion.

A deafferentation syndrome can be produced in animals with spinal cord lesions. Ingebitsen (1933) has described forepaw self amputation after cervical hemisection in rats and Basbaum (1974) observed self-mutilation in the contralateral hindpaw after thoracic hemisection in rats.

A deafferentation syndrome has also been produced in the monkey with spinal cord lesions. Christiansen (1966) reported that a thoracic hemisection and 50% of the anterolateral lesions in macaques resulted in biting of the contralateral hindpaw. The pattern was similar to what has been described for the rat with nailbed lesions followed by wounds of toes with progression to autotomy. Levitt and Levitt (1981) observed that certain species of macaques were much more prone to develop dysaesthesias, supporting the concept of a genetic influence, that involvement of the spinal cord ipsilateral to the affected foot can prevent the syndrome, and that the development and persistence of the syndrome is uninfluenced by morphine administration.

Abnormal motor activity has been found and abnormal electrical activity can be recorded from deafferented neurons. Teasdall and Stavraky (1952) have described hypersensitivity of cat spinal cord neurons after posterior rhizotomy. The deafferented neurons were more sensitive to descending corticospinal impulses than on the intact side as evidenced by lower threshold, bilateral involvement, and an after discharge of flexor contractions of the hindlimb. They observed these findings in a sensitization of the deafferented spinal neuron to chemical stimulating agent.

Loeser and Ward (1967) recorded spontaneous hyperactivity in the dorsal horn of cats with cordotomies or rhizotomies or combinations of these lesions.

Anderson, Black, Abraham and Ward (1967) reported that retrogasserian rhizotomy was followed after 10 to 20 days by generalized spontaneous hyperactivity of single cells in the spinal trigeminal complex which varied, depending on chronicity from rhythmic high frequency firing to continuous high frequency firing. Kjerulf, O'Neal, Calvin, Loeser and Westrum (1973) also described bursting spontaneous discharges in the lateral cuneate nucleus after deafferentation as did Millar, Basbaum and Wall (1976) in the study of the partially deafferented Gracile nucleus. Lombard, Nashold and Pelessier (1979) recorded in the thalamus of rats with complete cervical rhizotomies which were behaviorally showing a deafferentation syndrome and recorded spontaneous bursting activity after 195 days.

Abnormal discharges have been recorded in the human midbrain after deafferentation (Nashold & Wilson, 1976) and dysesthesias have been described in humans after brachial plexus avulsion (Zorub, Nashold & Cook, 1974), dorsal rhizotomy (Tasker, Organ & Hawrylyshyn, 1980), anterolateral cordotomy (Nordenbos, 1959) and brainstem lesions (Dejerine & Roussy, 1906).

During the past ten years a renewed interest in pain physiology has occurred and this has been accompanied by increased clinical interest in the evaluation and treatment of persons in chronic pain. The establishment of Pain Clinics throughout the world has brought together neurologists, neurosurgeons, anesthesiologists, psychologists, and psychiatrists to form a multidisciplinary approach to this serious medical problem.

> "Not everyone has a soul of fire, and, in actual
> human life, even in the case of the great mystics,
> the struggle against pain exacts a high price."
>   Leriche

REFERENCES

Albe-Fessard, D., Nashold, B. S., & Lombard, M.C. Yamaguchi, Y. & Boureau, T. Rats after dorsal rhizotomy. A possible animal model for chronic pain. Pain Abstracts, 1978, 1, 268.

Anderson, L. S., Black, R. C., Abraham, J., & Ward, A. A. Neuronal hyperactivity in experimental trigeminal deafferentation. Journal of Neurosurgery, 1967, 35, 444-452.

Basbaum, A. I. Effects of central lesions on disorders produced by multiple dorsal rhizotomy in rats. Experimental Neurology, 1974, 42, 490-501.

Bell, C. Idea of a New Anatomy of the Brain. Submitted for the Observations of His Friends, Strahan & Preston, 1811.

Christiansen (Levitt), J. Neurological observations in

macaques with spinal cord lesions. Anatomical Record,
1966, 154, 330.

Dejerine, J., & Roussy, G.   Le syndrome thalamique.
Review Neurology (Par), 1906, 14, 521-532.

Descartes, R. L'homme, translated by M. Foster, Lectures on the
History of Physiology during the 16th, 17th, and 18th
Centuries. Cambridge University Press, 1644.

Duckrow, R. B., & Taub, A. The effect of diphenylhydantoin
on self-mutilation in rats produced by unilateral multiple
dorsal rhizotomy. Experimental Neurology, 1977, 54, 33-41.

Goldscheider, A. Neue Thatsachen uber die Haut sunnesverven.
Pfleugers Archives Gesnerus Physiology, Suppl.,1885,
1, 1-110.

Head, H. Studies in Neurology. London: H. Froude, Inc.,
1920.

Heybach, I.P., Levitt, M., & Brodish, A. Pituitary-adrenal
activity associated with central pain following complete
forelimb deafferentations in rats. Social Neuroscience
Abstracts, 1979, 5, 611.

Inbal, R. M. Devor, Tuchendler, O., & Lieblich, I. Autotomy
following nerve injury: Genetic factors in the
development of chronic pain. Pain, 1980, 9, 327-377.

Ingebritsen, O. C. Coordinating mechanisms of the spinal
cord. Genetic Psychology Monographs, 1933, 13, 485-555.

Levitt, M., & Levitt, J. H. The deafferentation syndrome
in monkeys: dysaesthesias of spinal origin. Pain,
(in press).

Levitt, J., & Nashold, B. S. (Unpublished observations).

Levitt, J. & Sedivce (Unpublished observations).

Loeser, J. D., & Ward, A. A. Some effects of deafferentation
on neurons of the cat spinal cord. Archives of Neurology,
1967, 17, 629-636.

Lombard, M. C., Nashold, B. S., Albe-Fessard, D., Salman, N.
& Sakr, C. Deafferentation hypersensitivity in the rat
after dorsal rhizotomy: A possible model of chronic
pain. Pain, 1979, 6, 163-174.

Kjerulf, T. D., O'Neal, J. T., Calvin, W. H. Loeser, J. D.
& Westrum, L. E. Deafferentation effects in lateral
cuneate nucleus of the cat: Correlation of structural
alterations with firing pattern changes. Experimental
Neurology, 1973, 39, 86-102.

Lombard, M. C., Nashold, B. S., & Pelessier, T.   Thalamic
recordings in rats with hyperalgesia. Advances in Pain
Research and Therapy. Vol. 3, J. J. Bonica, et al., New
York: Raven Press, 1979.

Magendie, F. Experiences surles functions de racenes des neurf
nachidiens. Journal of Physiology & Experimental Pathology,
1822, 2, 276-279.

Melzack, R., & Wall, P. D. Pain Mechanisms: A New Theory.
Science, 1965, 150, 971-979.

Millar, J. A., Basbaum, I., & Wall, P. D. Restructuring of
    the somatotopic map and appearance of abnormal neuronal
    activity in the Gracile Nucleus after partial
    denervation. Experimental Neurology, 1976, 50, 658-672.
Mueller, J.   Zur vergleichenden Physiologie des Gesichssennes
    des Menchen und der Thiere nebst einen Versuch uber du
    Bewegungen der Augen und uber den menschlichen Blick.
    Knobolch, 1826.
Nashold, B.S., & Wilson, W. P. Observations in man with
    chronic implanted electrodes in the midbrain tegmentum.
    Confinia Neurology, 1974, 27, 30-44.
Nordenbos, W. Problems pertaining to the transmission of nerve
    impulses which give rise to pain. Preliminary statement.
    Pain. Amsterdam: Elsevier, 1959.
Rexed, B. The cytoarchitecctonic organization of the spinal
    cord in the cat. Journal of Comparative Neurology, 1952,
    96, 415-495.
Sinclair, D.C., Weddell, G., & Zander, E. The relationship of
    cutaneous sensibility to neurohistology in the human.
    Pennsylvania Journal of Anatomy, 1952, 86, 402-411.
Tasker, R. R., Organ, L. W., & Hawrylyshyn, P. Deafferentation
    and causalgia. In J. J. Bonica (ed.), Pain. New York: Raven
    Press, 1980.
Teasdall, R. D., & Stavraky, G. W. Responses of deafferented
    spinal neurons to corticospinal impulses. Journal of
    Neurophysiology, 1952, 16, 367-375.
Von Frey, M. Beitrage zen Physiologie des Schmerzsennis.
    Berlin Sachs., Ges. Wiss. Math. Phys. Kl. 1894, 46,
    185-196, 283, 296.
Wall, P. D., & Gutnick, M. Ongoing activity in peripheral
    nerves: The physiology and pharmacology of impulses
    originating from a neuroma. Experimental Neurology, 1974,
    43, 580-593.
Wiesenfeld, Z., & Hollin, R. G. Stress related pain behavior
    in rats with peripheral nerve injuries. Pain, 1980, 8,
    285-298.
Wiesenfeld, Z., & Lindblom, U. Behavioral and electrophysio-
    logical effects of various types of peripheral nerve
    lesions in the rat: A comparison of possible models for
    chronic pain. Pain, 1980, 8, 285-298.
Wollard, H. H. Observations on the termination of cutaneous
    nerves. Brain, 58, 352-367, 1935.
Zorub, D. S., Nashold, B. S., & Cook, W. A. Avulsion of the
    brachial plexus. A review with implications on the
    therapy of intractable pain. Surgical Neurology, 1974, 2,
    347-353.

BEHAVIORAL ASSESSMENT OF PAIN

Bernard Tursky and Larry D. Jammer

Departments of Political Science and Psychology
State University of New York at Stony Brook
Stony Brook, New York

Pain is a universal human experience; every person, with the exception of the unfortunate few that are born with congenital pain perception deficiencies (Sternbach, 1963), experiences severe pain in his/her lifetime. Bonica (1981) estimates that sixty-five million Americans suffer from chronic pain of such severity as to cause them to seek therapy by physicians and other health professionals. Given the impact that pain imposes on our lives, it is disturbing to find that the dimensions of pain have not been clearly defined and that the assessment of pain has not been standardized. Sternbach (1978) and Wolff (1978, 1980) argue that our current pain measurement techniques are inadequate, while Hendler (1981) suggests that pain is not a measurable experience.

A major problem in pain assessment is that there is no generally accepted definition of pain. This creates the dilemma of trying to quantify an unspecified variable. A major stumbling block towards a definition rests with the fact that the pain experience involves a myriad of factors which combine to provide each individual with his/her own unique characteristic perception and response to pain. A partial composite of the factors which contribute to the experience of pain, and hence affect the measurement process, include cultural factors (Sternbach & Tursky, 1965; Tursky & Sternbach, 1967), social modeling influences (Craig, 1978), personality factors (Mersky, 1978), and learned behaviors (Fordyce, 1976, 1978). The diversity of these contributing factors adds to the complexity of the measurement process.

One impediment towards a working definition (and thus an adequate measurement methodology) stems from the idea that pain can be regarded either as a sensation or as an emotion. Each philosophy may dictate a measurement approach quite different from the other, although not necessarily mutually exclusive. Melzack (1980) notes that since the beginning of the century, most of the research on pain has centered around the idea that pain is primarily a sensory experience. This is demonstrated by the earlier work of classical psychophysicists (i.e. von Frey) who focussed on the measurement of stimulus parameters and emphasized the measurement of thresholds, and by the later works of Hardy, Wolff, and Goodell (1952). Other measurement techniques falling in this category include the method of limits, a superior technique for obtaining threshold values (Wolff, 1978).

More recently, psychophysics has evolved an orientation that postulates that one can make quantitative estimations of one's sensory experiences. These methods include magnitude estimation, cross-modality matching (Stevens, 1975) and, most recently, sensory decision theory (SDT) (Green & Swets, 1966).

At the other end of the scale, pain may be considered as a psychological phenomenon where sensation is influenced by emotional and personality factors. In this instance the assessment procedures usually focus on obtaining and evaluating a psychological profile for each patient.

These measures, which include the widely employed MMPI, the Eysenck Personality Test and symptom check lists add little to our understanding of the degree of pain being experienced by the individual, but may provide explanatory information about the individual's high or low tolerance for pain. A more productive approach toward the definition and assessment of pain may be derived from a hybrid approach, the utilization of proven psychophysical procedures (sensory) in conjunction with the qualitative descriptions reflecting the motivational-affective component of the pain response. This paper will briefly examine the more commonly employed procedures in the behavioral assessment of pain, with an emphasis on the developments leading up to and including the use of psychophysical principles to scale verbal pain descriptors. Suggestions as to the utility of such procedures in the clinical setting will be proposed.

Toward a Definition of Pain

Sternbach (1968) refers to pain as an abstract term which is indicative of a variety of phenomena from which one makes a selection based upon one's orientation; pain is given either a neurological, physiological, behavioral, or subjective description. In the physiological/neurological context, pain is viewed as a signal of disregulation in the body. Psychologically, pain functions as a symbol of either emotional disturbance or behavioral deficits. Degenaar (1979), in discussing the philosophical function of pain states that "... it can be seen as a 'significance' which warns of a disharmony in the moral order which calls for an intentional act of personal involvement" (pp. 294).

Although such descriptions of pain may advance our understanding of the concept of pain, research methodology demands more precise and operationalized conceptualizations. Recently, various dimensions of pain have been identified and manipulated. According to Melzack (1973) "... pain represents a category of experiences, signifying a multitude of different causes, and characterized by different qualities varying along a number of sensory and affective dimensions" (p. 46). Tursky (1976) has identified and scaled three important dimensions of pain: an intensity dimension, a reactive dimension, and a descriptive dimension.

Pain has also been defined as an unpleasant experience which hurts a person and which is primarily associated with tissue damage (Hardy, Wolff, and Goodell, 1952). Mersky (1975) also refers to pain as an unpleasant experience which we primarily associate with tissue damage or describe in terms of tissue damage or both. Fordyce (1978) adds a useful qualifier to the end of Mersky's description: ". . .and the presence of which is signalled by some form of visible or audible behavior." Indeed, clinically, if an individual did not engage in pain behaviors (verbal, facial expressions, medication intake), the pain would remain a private experience and there would be no pain problem. This emphasis on defining pain in terms of the emitted behaviors associated with its presence allows us to describe and measure the pain experience in an overt "objective" manner. This assessment procedure is referred to as the behavioral analysis of pain.

Through these definitions several dimensions of pain emerge (i.e. sensory, affective, and behavioral). The identification and definition of the dimensions of the concept to be measured are the precursors to the development of the necessary measurement tools to accurately assess the human pain experience.

Pain Response Parameters

Three points in the process of measurement have been identified: (1) pain threshold, (2) pain tolerance, and (3) pain sensitivity range.

Historically, the pain threshold has been studied the most extensively, although it has been criticized as being unrelated to clinical pain (Beecher, 1959). Threshold refers to the point at which an individual first perceives stimulation as painful during an ascending series of stimuli, or at which pain disappears during a descending series of nociceptive stimulation.

Pain tolerance is defined as the point at which an individual will withdraw or refuse to accept stimulation of a higher magnitude or continue to endure stimulation at a given level of intensity. Wolff and Horland (1967) point out that the tolerance level should not be considered an absolute threshold, in that every individual can tolerate greater noxious stimulation if encouraged to do so. Wolff (1978, 1979, 1980) has indicated that the tolerance measure is perhaps the most useful parameter, having more clinical relevance than the pain threshold. Indeed, evidence indicates that the tolerance level may be associated with psychological factors such as attitudes and motivation, while pain threshold tends to be loaded with physiological factors. Wolff (1979) has demonstrated that pain tolerance is quite sensitive to the analgesic effects of moderate and potent analgesics such as codeine, meperidine (demerol[r]) and propoxyphene (darvon[r]). It also appears that the pain tolerance level is an insensitive measure of the effects of mild analgesics such as aspirin and acetominophen. Paradoxically, the pain threshold, which is considered a poor measure of the effects of potent analgesics, may be an effective indicator of the effects of mild analgesics such as aspirin.

The pain sensitivity range (PSR) is defined as the difference between the pain tolerance and the pain threshold. Some studies indicate that although threshold and tolerance levels appear to be correlated, the PSR is correlated to tolerance, and yet is unrelated to threshold (Weisenberg, 1976; Wolff, 1980). It appears that PSR is a relatively stable measure in comparison to the tolerance and threshold measures, being not as susceptible to modulation from intrinsic and extrinsic factors. Wolff (1980) suggests that the PSR is an indirect measure of a "pain endurance" factor. This PEF was derived by a factor analysis of a variety of experimental and clinical pain responses by Wolff (1971) and was replicated independently by Timmermans and Sternbach (1974, 1976). The pain endurance factor is relatively stable for a given individual and is regarded as a personality trait (Wolff, 1980). It appears to be a measure of a patient's ability to tolerate pain. In a three-year longitudinal study with chronic arthritics who underwent corrective surgery, the PSR was the sole preoperative factor that significantly predicted post-operative clinical ratings at a six-month follow-up (Wolff, 1971).

It appears then that the directly measured pain parameters
such as pain threshold and tolerance are sensitive to the
influence of internal and external modulators (e.g. drugs) and
thus may be well suited for the evaluation of analgesics and
other pain-alleviation interventions. However, they lack in the
ability to predict an individual's reaction to and tolerance of
endogeneous pain. On the other hand, the PSR and pain endurance
factor may prove to be valuable predictors of an individual's
ability to tolerate pain and hold promise as useful clinical
measures.

Verbal Rating and Visual Analogue Scales

The most frequently used pain evaluation procedure is
self-report. Patients can usually provide detailed information
about the location and strength of their pain during their
examination period. The most commonly employed pain measures
are categorical rating scales that consist of either a set of
pain descriptors (e.g. weak, moderate, severe, intense)
comprising a verbal rating scale (VRS), or a series of numbers
(such as 0,1,2,3,4) indicative of some pain intensity continuum.
One of the more recent developments has been the application of
the visual analogue scale (VAS) to the measurement of pain. The
VAS is a direct scaling technique in which the subject or
patient is asked to indicate the intensity of his/her pain on a
straight line (either vertical or horizontal) that usually
represents a continuum ranging from "no pain" to "the worst pain
possible".

While both VRS and VAS have been found to have validity
(Wolff, 1978), Gracely (1979) warns that both of these
categorical measures of pain "must be used and interpreted in
light of biases and response behaviors associated with these
methods" (p. 821). One specific drawback of these judgment
scales is that responses are usually analyzed with the
assumption that each number, descriptor, or portion of a line
possesses interval or ratio qualities in that each unit is taken
to represent an equal psychological unit of pain. This
assumption is unfounded. In a recent study, Lodge and Tursky
(1979) demonstrated conclusively that category scales at best
provide ordinal information. Gracely (1979) suggests that the
scaling data may be appropriately analyzed by the Thurstonian
method of successive categories. For a more detailed discussion
of verbal rating and visual analogue scales, the interested
reader is referred to Huskisson (1974), Scott and Huskisson
(1976), and Ohnhaus and Adler (1975).

Another major drawback of such categorical scaling
techniques (analogue, numerical, and descriptor) is that they

assess pain as though it were a unidimensional variable, varying only in intensity. There can be no doubt that pain is multidimensional; it can be described as pounding, splitting, nagging, stabbing, or burning and as intolerable, awful, distressing, unbearable, or excruciating. The relationships among these different expressions of pain become especially important when dealing with clinical pain, where physicians often rely only upon patients' self-report to understand their pain.

Melzack and Torgerson (1971) were among the first to develop a multidimensional pain descriptor scale. They had subjects categorize 102 pain descriptor words obtained from the clinical literature into subgroups representing different aspects of the pain experience. They identified three major classes of descriptors: (1) sensory quality descriptors, in terms of temporal, spatial, thermal, pressure and other properties such as pounding, burning and aching; (2) affective quality descriptors, in terms of tension and fear and autonomic properties such as nauseating, awful and nagging; and (3) evaluating terms that describe the overall intensity of the experience, such as miserable, agonizing, and torturing. These classes were further divided into sixteen subclasses that contained from two to six descriptors each. Subsequently, groups of doctors, patients, and students were asked to assign an intensity value ranging from 1 (mild pain) to 5 (excruciating pain) to each term within each subclass. Substantial agreement in classifying the many different terms was obtained, and although the precise intensity values varied somewhat among the groups, they all agreed on the relative positioning of the descriptor terms, despite the fact that the subjects had a variety of different cultural, socioeconomic and educational backgrounds (Melzack, 1975).

These scales were then developed into a clinical assessment instrument, known as the McGill Pain Questionnaire (MPQ) (Melzack, 1975, 1980). In addition to the twenty classes of descriptors which comprise the pain rating index portion (PRI) of the questionnaire, the MPQ contains the additional measures of the spatial distribution and temporal aspects of the pain, as well as a measure of the present pain intensity (PPI).

The pain rating index, employing the twenty classes of pain descriptors, can be scored in two fashions. In one, the sum total of all scale values of all the words chosen in a given category (sensory, affective, and evaluating) is assessed for all categories, based upon the mean scale values obtained for each word by Melzack and Torgerson (1971). An alternative method is based on the rank values of the words; the words in each subclass are assigned consecutive integer values indicating only their relative ranking, with 1 being assigned to the "least painful" word. The sum of the values associated with the words chosen by the patient/subject is then obtained for each category.

The questionnaire can be administered prior to and following some form of intervention and differences can be expressed as percentage change from initial values. Melzack (1975) suggests that t-tests may be employed to determine whether a statistically significant change has taken place.

Although the descriptor terms used in the MPQ were chosen carefully, Melzack and Torgerson utilized category scaling procedures to evaluate these words, thus providing inaccurate information concerning the relative strengths of the descriptors. This detracts from the utility of the device. Despite this deficiency, The McGill Pain Questionnaire has demonstrated value as a diagnostic instrument. Duboisson and Melzack (1976) administered the MPQ to 95 patients suffering from one of eight pain syndromes and, employing a multiple group discrimination analysis, were able to identify unique verbal response patterns for each of the eight syndromes which were significantly different. In addition, the MPQ has been employed successfully in monitoring the effects of a number of interventions, such as the analgesic effects of drugs (Melzack, 1980) and hypnotic training and alpha-feedback training (Melzack, 1975), as well as identifying the patterns and differences in the verbal descriptions of a number of pain syndromes (e.g. dental, menstrual, and back pain). Eventually it should be possible to reliably categorize distinct pain syndromes with the expectation of more appropriate diagnosis and treatment being administered to the pain patient.

Psychophysical Assessment of Pain

To overcome the constraints imposed by category scaling procedures, the major dimensions of pain perception should be evaluated using the psychophysical evaluation techniques developed by psychologists to quantitatively evaluate the strength and constancy of stimuli in a number of physical stimulus modalities such as loudness (Stevens, 1964) and brightness (Marks & Stevens, 1966). Psychophysics allows for one to make quantitative estimations of the magnitude of one's sensory experiences. The two major psychophysical evaluation procedures that have been employed in the assessment of human pain in the laboratory are the use of Signal Detection Theory (SDT) or Signal Decision Theory, and magnitude estimation (ME) or cross-modality matching (CMM).

Signal Detection Theory

Signal detection theory was introduced to the area of pain measurement by Clark (1969, 1976) after having been applied to numerous measurement problems in sensory psychology. When applied to the problem of pain measurement, SDT is designed to distinguish between a person's report of pain and his/her sensory experience induced by noxious stimuli. Signal decision theory stresses the distinction between the sensory aspect of the pain experience and an individual's criterion for reporting pain. Basically, SDT divides the traditional pain threshold into two components. The first component, d', provides a relatively pure measure of sensory discriminability, which is believed to be unaltered when psychological variables such as expectancies, attitudes, and motivation are manipulated. An individual who demonstrates a low d', is failing to discriminate between higher and lower stimulus intensities. This may result if the intensities of the stimuli are close together or when the individual's sensory system is insensitive.

The second component of an individual's performance is the response criterion, Lx, which is indicative of the individual's response bias; that is, his/her willingness to report pain. A high Lx value indicates an unwillingness to report a sensory experience as painful. In sensory decision theory, if two individuals have the same d', they endure a similar amount of pain, but the person with the higher Lx does not report it as such (Clark, 1976).

If the assumptions of SDT are held to be true, it appears that SDT offers a very important advance to the measurement of pain, as it (SDT) claims that it is possible to decompose a complex pain response into sensory and attitudinal dimensions. As a result of this attractive prospect, SDT has been increasingly employed in the measurement of pain.

SDT has been reported to be sensitive to the effects of a variety of pain modulators including acupuncture (Chapman, Chen & Bonica, 1977; Chapman, 1975; Clark & Yang, 1974), drugs (Chapman, Murphy & Butler, 1973; Chapman, Gehrig & Wilson, 1975), age (Harkins & Chapman, 1976; Clark & Mehl, 1971), transcutaneous stimulation (Chapman, Wilson & Gehrig, 1976; Bloedel, McCreery & Erikson, 1976), social modeling (Craig & Coren, 1975) and suggestion and placebo (Clark, 1969; Clark & Goodman, 1974). The general consensus among these studies is that physical modulators of pain, such as drugs, transcutaneous stimulation, and acupuncture resulted in changes in d' and on occasion Lx, while psychological factors such as social modeling, age and suggestion tended to change the criterion (Lx) without an alteration in d'.

A number of other researchers (Rollman, 1977, 1979; Wolff, 1980; Gracely, 1979; Bonica, 1980) question the applicability of SDT to the measurement of human pain. Rollman (1977) indicates that SDT measures discriminability, but not pain. Wolff (1980) states that it is incorrect to speak of d' as a measure of the sensory component and of Lx as indicative of the attitudinal component. Wolff also believes that the d' component is actually a complex phenomena which is comprised of many other factors (i.e. psychological) in addition to the sensory component. Gracely (1979) also purports that discriminability (d') is not just a measure of sensory performance, but is also a measure of cognitive performance. Furthermore, discriminability between pain sensations fails to measure the intensity of the sensations. In support of this point are the results of a study conducted by Rollman (1979), in which determinations of d' were made before and after a reduction of sensory input was achieved (by an attenuation in the intensity of the noxious stimuli, electrical stimulation). Contrary to the basic assumptions of SDT, d' remained constant while the criterion to report pain shifted to a more conservative position. For a detailed discussion of the current controversy surrounding SDT, refer to the exchange between Chapman (1977) and Rollman (1977). It appears that the use of SDT in the pain research arena still requires a great deal of investigation and systematic exploration.

Magnitude Estimation and Cross-Modality Matching

A more productive psychophysical approach may be the use of S.S. Stevens' magnitude estimation (Stevens, 1957) and cross-modality matching (Stevens & Mach, 1959) procedures to develop quantitative, cross-modality validated, bias-free information that can describe the relationship between pain stimulation and subjective judgment of the strength of that stimulation. In this process, subjects/patients are required to assess each dimension of their pain experience by producing an estimate related to the strength of sensation using some carefully calibrated response modality. Numbers, lines, sound level, light intensity, and strength of handgrip have all been used to assess experimentally induced or natural pain (Elmore & Tursky, 1981; Cross, Tursky & Lodge, 1975; Sternbach & Tursky, 1964).

In the cross-modality matching and validation procedures, subjects/patients are asked to adjust the strength of two or more modalities (e.g. brightness and line length) for each pain judgment so that the strength of each response matches the pain intensity. As a test of the validity of the scales on the different modalities, subjects are then asked to adjust the magnitude of one modality (e.g. line length) in response to

another (e.g. brightness); that is, the subject matches the
scales to one another. When the values for the two modalities
(stimulus/response) are plotted against one another, a power
function should be obtained whose exponent is equal to the ratio
of the exponents of the power functions of each modality against
stimulus (i.e. pain) intensity (Stevens, 1975).

These quantitative evaluation procedures may be extremely
useful in the assessment of experimental pain, but in the
clinical setting patients tend to communicate information about
pain suffering by the use of more qualitative language which
varies from patient to patient. There would be great clinical
utility in an assessment device which would allow for the
quantification of these verbal descriptors, in which one given
pain descriptor implies two, ten, or fifty times the intensity
or reactivity for pain than does another given descriptor.

Although verbal descriptors of pain are not physically
measurable, psychophysical scaling procedures can be used to
scale and verify the relative magnitude of the dimensions
implied by these descriptors. Tursky (1976) utilized a
cross-modality matching paradigm to develop and evaluate three
sets of pain descriptors that could be used to describe the
intensity, reactive, and sensory dimensions of the human pain
experience. The intensity scale is defined as a measure of how
much the pain hurts in units of intensity, the reactive scale is
defined as a measure of how unpleasant the pain feels in units
of reaction, and the sensory scale is defined as a measure of
what the pain feels like in units of sensation. The selection
of the descriptors for each scale was based on the sorting of a
large number of descriptors, including many used by Melzack and
Torgerson, into unpleasantness, intensity, and sensory
categories by students, lay people, and social science
professionals. Further classification employed simple, verbal
magnitude estimation procedures to eliminate descriptors that
displayed too much variability in classification or estimation.
The intensity descriptors ranged from "just noticeable" to
"excruciating", the reactive descriptors ranged from "bearable"
to "agonizing", and the sensation descriptors ranged from
"tingling" to "piercing" (see Figure 1).

Fifty-six students were then asked to make judgments of the
magnitude of the intensity, reactive, and sensation descriptors
and of apparent line length using the three psychophysical
response modalities of numerical magnitude estimation, handgrip,
and sound pressure. In each modality, the subject produced a
response proportional to the magnitude of each stimulus (i.e.
pain descriptor) relative to a standard. Number estimates were
given verbally, sound level was adjusted by the subject using a

INTENSITY	UNPLEASANTNESS	FEELING
Moderate	Distressing	Stinging
Just Noticeable	Tolerable	Grinding
Mild	Awful	Squeezing
Excruciating	Unpleasant	Burning
Very Strong	Unbearable	Shooting
Very Intense	Uncomfortable	Numbing
Severe	Intolerable	Throbbing
Intense	Bearable	Stabbing
Very Weak	Agonizing	Itching
Strong	Miserable	Aching
Weak	Distracting	Cramping
Not Noticeable	Not Unpleasant	None
		Pressure

PRESCRIBED MEDICATION	HOW OFTEN	DOSE SIZE	HOW MANY

Date___7-28-77 Thursday___

Type of Discomfort	TIME		DISCOMFORT RATING				MEDICATION		
	Dur.	Begin	Intensity Word	Unpleasant Word	Feeling Word	Numb.	Name	How Many	Dose
headache	5:30 AM	Strong	distracting	throbbing	8	(off check)		
same headache	12:45 PM	Moderate	tolerable	throbbing	4			
same headache	6:45 PM	Strong	distracting	throbbing	8			
RATING FOR THE DAY		strong	distracting	throbbing	8				

AVERAGE DISCOMFORT NUMBER | 10

HOURS SLEPT LAST NIGHT___5 hrs (Intermittent) 12:30 PM to 5:30 AM___

STRESSFUL EVENTS & COMMENTS___Nausea, Diarhea, dizziness (morning)
Went to sleep with headache (strong) Sweated___

Figure 1. ' Sample page of pain diary.

tone potentiometer, and a hand dynamometer was used to produce handgrip responses.

Figure 2 shows a plot of the functions generated for the intensity dimension descriptors. As these words do not have a known metric, responses cannot be plotted directly against these words. Instead, the calculated scale values were used as a surrogate metric. The geometric means of the magnitude estimation, handgrip, and sound pressure should approximately produce their expected theoretical functions when plotted against the derived scale values. The actual exponents of the generated functions generated across all groups were 0.68 for sound pressure, 1.03 for magnitude estimation, and 2.01 for handgrip. The theoretical exponents are 0.67 for sound pressure, 1.0 for magnitude estimation, and 1.7 for handgrip. The product-moment correlations between response modes (i.e. handgrip, magnitude estimation, and sound pressure) for the stimuli presented were extremely high, ranging from 0.98 to 0.99. The responses to the reactive and sensory descriptor words produced similar results.

It may be argued that this psychophysical procedure is of
limited utility, in that the use of handgrip and sound pressure
as scaling procedures is too complex or cumbersome for use in
the pain clinic or physician's office. In an attempt to develop
an assessment device with clinical potential, the descriptor
evaluation study was replicated using line length production and
numerical magnitude estimation. These measures are relatively
simple psychophysical response measures that can easily be
employed in the field or clinician's office to calibrate
individual pain descriptor responses.

The geometric means of line production responses were
computed and plotted against the geometric means of the
magnitude estimation responses for the same stimuli. The
theoretical relationship between these quantities is a power
function with an exponent of one. The obtained exponents were
1.11 for the intensity scale, 1.08 for the reaction scale, and
1.3 for the sensory scale. The product-moment correlations for
these measures ranged from 0.993 to 0.997, thus demonstrating
the validity and reliability of the procedure.

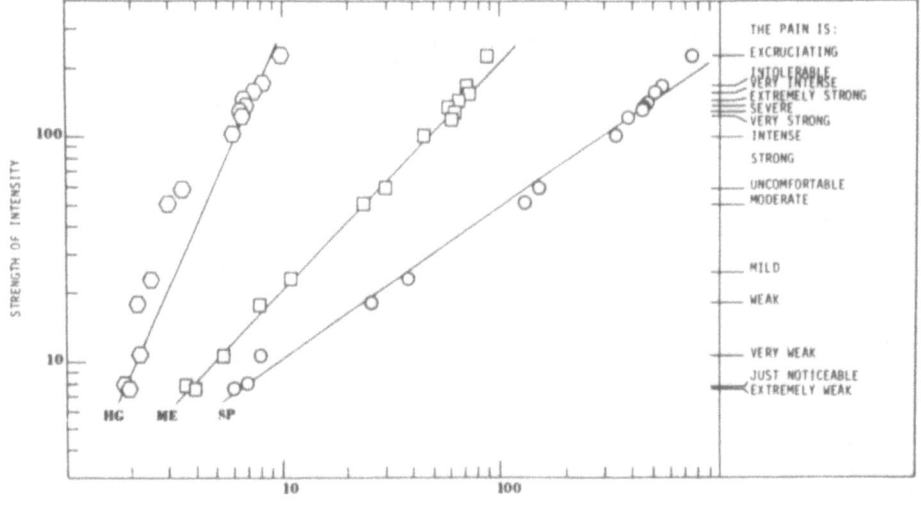

RELATIVE MAGNITUDE OF RESPONSE

Figure 2.   Strength of Intensity v. Relative Magnitude of
             Response. Handgrip (HG), Magnitude Estimation
             (ME), and Sound Pressure (SP) responses (x-axis)
             as a function of the strength of intensity of
             pain (y-axis) based on all three response
             measures, each corrected for regression. Each
             point is the geometric mean of twenty responses.

The issue may be raised of the clinical utility of this procedure as a whole; what are the advantages over simple verbal descriptors and categorical scaling techniques that justify implementation of the method? Categorical pain scales result in interval scales sensitive to bias effects such as stimulus frequency, range, distribution and category end effects (Stevens & Galanter, 1957), and in fact may deliver at best ordinal data (Lodge & Tursky, 1979). In addition, a patient's verbal descriptions of clinical pain at present have to be accepted at face value in that there can be no comparison to independent measures of pain (i.e. varying stimulus intensities). Thus, the validity and reliability of a patient's self-report cannot be determined directly. This serves to detract from the diagnostic potential of the verbal report of pain.

One major advantage of the ratio-scaling procedure described is that it preserves the perceived ratios of the pain descriptors; that is, we can quantitatively evaluate one pain term as "twice as" or "ten times as" intense as another. Table 1 shows the calculated bias-free scale values for the descriptors in each of the three scales. If these words were given categorical values, the range of each set would be controlled by the number of terms in the set. Thus, the fifteen intensity words translate into a 15:1 ratio with equal intervals of intensity arbitrarily assigned to each successive pair of words. Examination of the obtained scale values shows that the true range of intensity words is 40:1 and that the intervals between adjacent pairs are not equal. The power of this method is further demonstrated upon comparison of the categorical and magnitude ratios for the thirteen descriptors in the sensory dimension. The categorical ratio would be 13:1, but due to the lack of an intensity dimension in these terms, the magnitude ratio is reduced to 7:1. These data clearly indicate the superiority of magnitude scaling over category scaling in the evaluation of pain perception descriptors.

This procedure would now allow the clinician or physician to more meaningfully interpret the verbal reports of their patients so that statements such as "the pain is intolerable but last night it was excruciating" can be evaluated and validated. In addition, these verbal scales would allow for more precise differentiation between dimensions common to pain syndromes. Such verbal scales would also permit comparisons among experimental, acute, and chronic pain responses. Within individuals, repeated administration of this procedure allows for a more powerful assessment of pharmacological and nonpharmacological pain-control techniques along several dimensions of the pain experience (i.e. intensity, reaction, and sensory).

Table 1. Relative scale values for three dimensions
          of pain

INTENSITY		REACTION		SENSATION	
*Excruciating	227	Agonizing	153	Piercing	113
Intolerable	167	Intolerable	145	Stabbing	109
Very Intense	154	Unbearable	128	Shooting	106
Extremely Strong	135	Awful	98	Burning	80
*Severe	132	Miserable	97	Grinding	79
Very Strong	129	Distressing	50	Throbbing	75
Intense	123	Unpleasant	43	Cramping	67
*Strong	101	Distracting	36	Aching	58
Uncomfortable	58	Uncomfortable	35	Stinging	50
*Moderate	50	Tolerable	23	Squeezing	46
*Mild	23	Bearable	23	Numbing	40
*Weak	15			Itching	25
Very Weak	10			Tingling	17
*Just Noticeable	8				
Extremely Weak	8				

*Scale values were derived from the following formula:

$$\psi(S) = (ME^{1/n_1}HG^{1/n_2}SP^{1/n_3})^{1/3}$$

    The utility of this device in the experimental and clinical
evaluation of pain control techniques and in the differentiation
of pain syndromes has been demonstrated in two recent
investigations.

    Elmore and Tursky (1981) conducted an investigation to
assess the relative effectiveness of two psychophysiological
approaches to the treatment of migraine: hand temperature
increase biofeedback and temporal pulse amplitude reduction
biofeedback. Both groups demonstrated significant learned
physiological control. Analysis of the clinical outcome data
was based on comparison of a novel pain diary, developed in our
laboratory, in which patients utilized previously
(psychophysically) scaled pain perception descriptors to
evaluate the intensity, unpleasantness, and sensation of each
pain episode for one month pre- and post-treatment (see Figure
1). Patients were also asked to directly scale these pain
episodes by providing a numerical evaluation of each episode.
Medication intake, sleep duration, and stressful events were
also recorded. While there were no significant differences

between groups during the pre-treatment periods, post-treatment
analysis indicated significant differences in reactivity,
sensation, and medication intake with each difference indicating
the superiority of temporal pulse amplitude reduction
biofeedback.

These results, which indicated a major alteration only in
the perceived unpleasantness and quality of the sensation rather
than an alteration in the reported intensity of the headache
pain, are consistent with the experimental literature (e.g.
Gracely et al., 1978b, 1980; Chapman & Feather, 1973), which
states that a number of pain modulators exhibit an effect mainly
due to their influences on the affective-motivational dimension
of the pain experience.

In another application of this method Blanchard, Andrasik,
Arena and Teders had the pain descriptors psychophysically
scaled by patients suffering from either tension, migraine,
combined tension and migraine, or cluster headaches and by a
group of non-headache sufferers. Blanchard et al. reported that
the reactivity descriptors consistently discriminated headache
from non-headache subjects, in that patients with a chronic pain
problem tended to reliably scale the reactivity descriptors
higher than non-pain patients. In addition, they report that
prior experience with differing types of pain did not affect
perceptions of the intensity dimension and that scaling of the
intensity descriptors did not differentiate between patients
with and without chronic pain disorders. A second interesting
finding of their study was the tendency for patients who had had
experiences with more intense pain (i.e. cluster headaches) to
exhibit differing response patterns than those patients whose
pain experiences are taken to have been less severe (i.e.
tension headache), hence demonstrating the diagnostic potential
of the described procedure.

A point to be made here is that the findings in our
laboratory and those reported by Blanchard et al. indicate that
either exposure to chronic pain or interventions aimed at
alleviation serve to reduce the intercept but not the slope of
the linear regressions of responses against the scale values of
the descriptor words. Gracely (1979) indicates that
traditionally, the focus has been on the slope and not the
intercept of magnitude scaling functions, and that intercept
changes observed after a pain manipulation may reflect
psychological processes. Further research in determining the
factors that influence intercept variability is required so that
we can properly assess the role that the intercept may have.

The values of these descriptors and the methods used to
scale them have also been confirmed by other investigators.

Richard Gracely and his colleagues (Gracely, McGrath & Dubner, 1978a, 1978b; Gracely, 1979; Heft, Gracely, Dubner & McGrath, 1980) have replicated and greatly extended the psychophysical scaling of and use of pain descriptors. Although utilizing a somewhat different set of descriptors, their research verifies the validity, reliability, and usefulness of this technique. In one study, Gracely et al. (1978b) had oral surgery subjects use cross-modality matching of handgrip force and verbal descriptor responses to rate the sensory (intensity) and unpleasantness (reactive) responses to noxious electrocutaneous stimulation before administering diazepam, a mild tranquilizer assumed to alter the unpleasantness but not the intensity of noxious stimulation. In a second study (Gracely, McGrath, Heft and Dubner, 1977), subjects were asked to make the same responses to painful electrical tooth stimulation before and after administration of fentanyl, a narcotic assumed to alter the intensity but not the unpleasantness of the stimuli. Their obtained results indicated that only the descriptor scales of unpleasantness were reduced following diazepam and only the sensory (intensity) scales were reduced following fentanyl. In addition, when subjects were asked to rate the sensory-intensity or unpleasantness of painful electrical tooth stimulation before and after a double-blind intravenous administration of fentanyl or saline, they found that sensory intensity responses were significantly reduced after fentanyl but remained unaltered after placebo (Gracely, McGrath and Dubner, 1979). Recently they have shifted their research toward demonstrating the validity of the use of quantified verbal descriptors for the assessment of both experimentally controlled noxious stimulation and the uncontrolled sensations of clinical pain (Heft, Gracely, Dubner and McGrath, 1980).

Behavioral Measures of Pain

The behavioral assessment and treatment of chronic pain has become increasingly common in pain clinics since the impressive reports by Fordyce et al. (1973, 1976) and others (e.g. Swanson, Swenson, Maruta & McPhee, 1976; Wooley, Blackwell & Winget, 1978). The basic elements of this approach are the examination and measurement of pain in terms of its behavioral expression, and the determination of factors in the individual's environment which are assumed to be exerting an influence on the occurrence of those pain behaviors (Fordyce, 1978).

The behavioral analysis of clinical pain typically consists of multiple interviews with the patient and his/her spouse, an MMPI, and a sample (usually two weeks) of diaries recorded by the patient to show the amount of time engaged in several behaviors such as sitting, standing or walking, and reclining, including a record of medications taken and hourly ratings of subjective pain. This information is then used in evaluating and identifying positive reinforcers and their relationship to pain behaviors, as well as identifying the negative reinforcers of pain (those aversive events that are avoided by pain behavior). For a detailed examination of the methods involved in behaviorally evaluating the pain patient see Fordyce (1976) pp. 103-146.

Another approach has been to engage chronic pain patients in prescribed exercises "until pain, weakness, or fatigue causes you to want to stop. You decide when to stop" (Fordyce, 1979, pp. 662). Various mechanical and electronic devices can calibrate the amount of effort expended, or time duration can be monitored. To evaluate change due to a particular manipulation, the amount performed during the first exercise session is assigned an arbitrary value of 1. Successive measurement sessions can then be calculated in terms of a ratio to the initial session.

Improvements in behavioral measures such as these have been correlated with an improvement of the subjective rating of pain (Sternbach, 1978). It is important to note that the behavioral assessment of pain focusses on clinical pain and has not been employed in experimental human studies.

Pain Perception Profile

Thus far, this paper has examined many of the more commonly employed measures of pain, measures which are differentially sensitive to the various dimensions of pain (sensory, affective, and behavioral). It seems reasonable that it should be of primary interest to the physician or clinician to be able to accurately assess all dimensions of their patients' pain experiences. It is indeed surprising to find that until very recently very little had been done by physicians and psychologists to develop a methodology to achieve this purpose. Unlike the physician's devotion to keep continuing clinical records for each of his patients, the recording of specific information related to the various dimensions of the patient's pain experience is almost non-existent. This is unexplainable since physicians do devote a great deal of effort to compiling both a normative and clinical physical and physiological record or health profile for each of their patients. This clinical information is utilized as a source of reference for the physician as well as for consulting physicians and hospital staff when there are complications in the patient's condition.

It seems paradoxical that so little comparable information about each patient's pain perception is recorded by the physician despite the fact that pain is often the basic

complaint that brings the patient to the physician and that pain
may be the primary symptom that the physician is trying to
alleviate.

No one would question the utility of a series of simple
assessment procedures that could establish for each individual a
personal pain perception profile. This profile could become
part of the individual's medical record and would provide vital
information about the individual's perception of and reaction to
each pain experience.

It also seems reasonable to believe that the physician
could accumulate normative pain perception information by
administering a similar series for pain perception tests as part
of the patient's routine physical examination.   In this
pain-free examination, the physician could establish a profile
of the qualitative and quantitative aspects of each patient's
reaction to a series of standard pain stimuli and the patient's
normal understanding and use of pain descriptors. Similar
information could be obtained when the patient is being examined
or treated for some specific complaint, thus giving the
physician the same comparative capabilities for assessing pain
symptoms as he/she has for clinical complaints. The basic pain
profile would provide the following information:

       1. an accurate measure of the individual's sensation
threshold;
       2. a measure of the level of controlled nociceptive
stimulation that each individual identifies as uncomfortable,
painful, and intolerable;
       3. psychophysical evaluations of controlled pain
stimuli to establish the individual's judgment function for
pain;
       4. psychophysical evaluation of a standard set of
pain descriptors to enable the physician to better understand
the patient's use of language to describe pain, and an
assessment of how the individual uses pain descriptive words to
describe controlled pain stimulation.

These items are not the only, nor necessarily the best,
variables that can be utilized to provide a comprehensive,
multidimensional pain profile; nor is this the only suggestion
of such an approach in the literature (Duncan, Gregg, & Ghia
(1978) describe a computer-based system to assess and quantify
the pathophysiologic, psychological, and behavioral aspects of
chronic pain, with the expectation of aiding the clinician to
find the most appropriate therapy. Rather, it is hoped that the
concept of a pain profile will serve as a seed for a
multidisciplinary approach to the problem of pain in which "the
anatomist,   neurophysiologist,   pharmacologist   and   other

laboratory workers join forces with clinicians and clinical
investigators to learn the nature of pain" (Bonica, 1953, pp.
24). And, as Bonica (1981) recently stated, "The ultimate
benefit will be improved care of patients with chronic pain,
which is, after all, the raison d'etre of the scientist and
clinician" (pp. ix).

REFERENCES

Beecher, H.K. Measurement of Subjective Responses. New York:
     Oxford, 1959.
Blanchard, E.B., Andrasik, F.A., Arena, J.G., & Teders, S.J.
     The effects of differing chronic pain experience on the
     psychophysical scaling of pain descriptors. Paper
     submitted for publication.
Bloedel, J.R., McCreery, D.B. & Erikson, D.L. Modification of
     subjective responses to thermal stimulation by electrical
     stimulation: An evaluation using signal detection theory.
     (Abstract) First World Congress on Pain, 1975, pp. 185.
Bonica, J.J. Foreward in N. Hendler, Diagnosis and Nonsurgical
     Management of Chronic Pain. New York: Raven Press, 1981.
Bonica, J.J. (ed.) Pain (Vol 58). Raven Press, 1980.
Bonica, J.J. The Management of Pain. Philadelphia: Lea and
     Febiger, 1953.
Chapman, C.R. Pain: The perception of noxious events. In R.A.
     Sternbach, (ed.) The Psychology of Pain. New York: Raven
     Press, 1978.
Chapman, C.R. Sensory decision theory methods in pain research:
     A reply to Rollman. Pain, 1977, 3, 295-305.
Chapman, C.R. Psychophysical evaluation of acupunctural
     analgesia: Some issues and considerations. Anesthesiology,
     1975, 43, 501-506.
Chapman, C.R. and Feather, B.W. Effects of diazepam on human
     pain tolerance and sensitivity. Psychosomatic Medicine,
     1973, 35, 330-340.
Chapman, C.R., Gehrig, J.D. & Wilson, M.E. Acupuncture, pain,
     and signal detection theory. Science, 1975, 189, 65.
Chapman, C.R., Murphy, J.M. & Butler, S.H. Analgesic strength of
     33 percent nitrous oxide: A signal detection theory
     evaluation. Science, 1973, 179, 1246-1248.
Chapman, C.R., Wilson, M.E. & Gehrig, J.D. Comparative effects
     of acupuncture and transcutaneous stimulation on the
     perception of painful dental stimuli. Pain, 1976, 2,
     265-283.
Clark, W.C. Pain sensitivity and the report of pain: An
     introduction to sensory decision theory. I M. Weisenberg,
     & B.Tursky, (eds.) Pain: New perspectives in therapy and
     research. New York: Plenum, 1976, pp. 195-222.

Clark, W.C. Sensory decision theory analysis of the placebo effect on the criterion for pain and thermal sensitivity (d'). Journal of Abnormal Psychology, 1969, 74, 363-371.

Clark, W.C. & Goodman, J.S. Effects of suggestion on d' and Lx for pain detection and pain tolerance. Journal of Abnormal Psychology, 1974, 83, 364-372.

Clark, W.C. & Mehl, L. Thermal pain: A sensory decision theory analysis of the effect of age and sex on d', various response criteria and 50 percent pain threshold. Journal of Abnormal Psychology, 1971, 78, 202-212.

Clark, W.C. & Yang, S.C. Acupunctural analgesia? Evaluation by signal detection theory. Science, 1974, 184, 1096-1098.

Craig, K.D. Social modeling influences on pain. IN Sternbach, R.A. (ed.) The Psychology of Pain. New York: Raven Press, 1978.

Craig, K.D. & Coren, S. Signal detection analysis of social modeling influences on pain expressions. Journal of Psychosomatic Research, 1975, 19, 105-112.

Cross, D.V., Tursky, B. & Lodge, M. The role of regression and range effects in determination of the power function for electrical shock. Perceptionand Psychophysics, 1975, 18, 9-14.

Degenaar, J.J. Some philosophical considerations of pain. Pain, 1979, 7, 281-304.

Duboisson, D. & Melzack, R. Classification of clinical pain descriptions by multiple group discriminant analysis. Experimental Neurology, 1976, 51, 480-487.

Duncan, G.H., Gregg, J.M. and Ghia, J.N. The pain profile: A computerized system for assessment of chronic pain. Pain, 1978, 5, 275-284.

Elmore, A. & Tursky, A. A comparison of the psychophysiological and clinical response to biofeedback for temporal pulse amplitude reduction and biofeedback for increases in hand temperature in the treatment of migraine. Headache, in press.

Fordyce, W.E. Learning processes in pain. IN Sternbach, R.A. (ed.) The Psychology of Pain. New York: Raven Press, 1978.

Fordyce, W.E. Behavioral methods in chronic pain and illness. St. Louis: C.V. Mosby Co., 1976.

Fordyce, W.E., Fowler, R.S., Lehmann, J.F., Delateur, B.J., Sand, P.L. & Trieschmann, R.B. Operant conditioning in the treatment of chronic pain. Archives of Physical Medicine Rehabilitation, 1973, 54, 399-408.

Gracely, R.H. Psychophysical assessment of human pain. IN Bonica, J.J. et al. (eds.) Advances in Pain Research and Therapy. New York: Raven Press, 1979.

Gracely, R.H., McGrath, P. & Dubner, R. Ratio scales of sensory and affective verbal pain descriptors. Pain, 1978a, 5, 5-18.

Gracely, R.H., McGrath, P. & Dubner, R. Validity and sensitivity of ratio scales of sensory and affective verbal pain descriptors: Manipulation of by diazepam. Pain, 1978b, 5, 19-29.

Gracely, R.H., McGrath, P.A., Heft, M.W., & Dubner, R. Narcotic analgesia: Fentanyl reduces the intensity but not the unpleasantness of painful tooth pulp sensations. Science, 1979, 203, 1261-1263.

Green, D.M. & Swets, J.A. Signal detection theory and psychophysics. New York: John Wiley, 1966.

Hardy, J.D., Wolff, H.G. & Goodell, H. Pain sensations and reactions. Baltimore: Williams and Wilkins, 1952.

Harkins, S.W. & Chapman, C.R. Detection and decision factors in pain perception in young and elderly men. Pain, 1976, 2, 253-264.

Heft, M.W., Gracely, R.H., Dubner, R. & McGrath, P.A. A validation model for verbal descriptor scaling of human clinical pain. Pain, 1980, 9, 363-373.

Hendler, N. Diagnosis and nonsurgical management of chronic pain. New York: Raven Press, 1981.

Huskisson, E.C. Measurement of pain. Lancet, 1974, 2, 1127-1131.

Lodge, M. & Tursky, B. Comparison between category and magnitude scaling of political opinion employing SRC/CPS items. American Political Science Review, 1979, 73, 50-66.

Marks, L.E. & Stevens, J.C. Perception and Psychophysics, 1966, 1, 17-24.

Melzack, R. Psychologic aspects of pain. IN Bonica, J.J. (ed.) Pain Vol 58. New York: Raven Press, 1980.

Melzack, R. The McGill Pain Questionnaire: Major properties and scoring methods. Pain, 1975, 1, 277-299.

Melzack, R. The Puzzle of Pain. New York: Basic Books, 1973.

Melzack, R. & Torgerson, W.S. On the language of pain. Anesthesiology, 1971, 34, 50-59.

Mersky, H. Some features of the history of the idea of pain. Pain, 1980, 9, 3-8.

Mersky, H. Pain and personality. IN Sternbach, R.A. (ed.) The Psychology of Pain. New York: Raven Press, 1978.

Mersky, H. IN M. Weisenberg, (ed.) Pain: Clinical and Experimental Perspectives. St. Louis: Mosby, 1975.

Ohnhaus, E.E. & Adler, R. Methodological problems in the measurement of pain: A comparison between the verbal rating scale and the visual analogue scale. Pain, 1975, 1, 379-384.

Rollman, G.B. Adaptation-level effects in the rating of acute pain. IN J.J. Bonica, et al. (eds.) Advances in Pain Research and Therapy, Vol 3. New York: Raven Press, 1979. pp. 825-829.

Rollman, G.B. Signal detection theory measurement of pain, a review and critique. Pain, 1977, 3, 187-211.

Scott, J. & Huskisson, E.C. Graphic representation of pain. Pain, 1976, 2, 175-184.

Sternbach, R.A. "Clinical aspects of pain". IN R.A.Sternbach, (ed.) The Psychology of Pain. New York: Raven Press, 1978.

Sternbach, R.A. Pain: A psychophysiological analysis. New York: Academic Press, 1968.

Sternbach, R.A. "Congenital Insensitivity to Pain: A Critique:". Psychological Bulletin, 1963, 60, 252-264.

Sternbach, R.A. & Tursky, B. Ethnic differences among housewives in psychophysical and skin potential response to electric shock. Psychophysiology, 1965, 1, 241-246.

Sternbach, R.A. & Tursky, B. On the psychophysical power function in electric shock. Psychonomic Science, 1964, 1, 217-218.

Stevens, J.C. & Mack, J.D. Scales of apparent force. Journal of Experimental Psychology, 1959, 58, 405-413.

Stevens, S.S. Psychophysics: Introduction to its Perceptual, Neural and Social Prospects. New York: John Wiley and Sons, 1975.

Stevens, S.S. "On the psychophysical law". Psychological Review, 1957, 64, 153-181.

Stevens, S.S. & Galanter, E. Ratio scales and category scales for a dozen perceptual continua. Journal of Experimental Psychology, 1957, 54, 377-411.

Swanson, D.W., Swenson, W.M., Maruta, T. & McPhee, M.C. Program for managing chronic pain. 1. Program description and characteristics of patients. Mayo Clinic Proceedings, 1976, 51, 401-411.

Timmermans, G. & Sternbach, R.A. Human chronic pain and personality: A canonical correlation analysis. In Advances in Pain Research and Therapy, Vol 1, edited by J.J. Bonica and D. Albe-Fessard. New York:Raven Press, 1976. pp. 307-310.

Timmermans, G. & Sternbach, R.A. Factors of human chronic pain: An analysis of personality and pain reaction variables. Science, 1974, 184, 806-808.

Tursky, B. Development of a pain perception profile: A psychophysical approach. In M. Weisenberg and B. Tursky (eds.) Pain: New Perspectives in Therapy and Research. New York: Plenum Press, 1976. pp. 171-194.

Tursky, B. & Sternbach, R.A. Further physiological correlates of ethnic differences in responses to shock. Psychophysiology, 1967, 4, 67-74.

Weisenberg, M. Pain and pain control. IN G.E. Schwartz, and D. Shapiro, (eds.) Consciousness and Self-Regulation: Advances in Research. New York: Plenum Press, 1976.

Wolff, B.B. Measurement of human pain. IN J.J. Bonica, (ed.) Pain, Vol 58. New York: Raven Press, 1980.

Wolff, B.B. Validity of different experimental pain response parameters for human analgesic assays. In J.J. Bonica, et al. (eds.) Advances in Pain Research and Therapy, Vol 3. New York: Raven Press, 1979.

Wolff, B.B. Behavioral measurement of human pain. In R.A.
    Sternbach, (ed.) The Psychology of Pain. New York: Raven
    Press, 1978.
Wolff, B.B. The role of laboratory pain induction methods in the
    systematic study of human pain. Journal of Abnormal
    Psychology, 1971, 78, 292-298.
Wolff, B.B. & Horland, A.A. Effect of suggestion upon
    experimental pain: A validation study.   Journal of
    Abnormal Psychology, 1967, 72, 402-407.
Wooley, S.C., Blackwell, B. & Winget, C. A learning theory model
    of chronic illness behavior: Theory, treatment and
    research. Psychosomatice Medicine, 1978, 40, 379-400.

BEHAVIORAL ANALYSIS OF CHRONIC PAIN

Wilbert E. Fordyce

Department of Rehabilitation Medicine & Pain Service
University Hospital
Seattle, Washington

Introduction

Chronic pain has long been one of the most prevalent, costly, and puzzling problems to health care delivery. In 1967-68, a case study was reported (Fordyce, Fowler & deLateur, 1968) in which chronic pain was treated as a behavior change rather than medical problem. In the ensuing years, this idea has proliferated to the point where nearly all of the comprehensive pain evaluation and treatment programs in the United States and Canada, as well as a number of other countries, base their procedures in significant part on behavioral concepts and behavioral methods. Because the problem of chronic pain is so prevalent, evaluation and management by behavioral methods and analysis of the problem in the framework of behavioral science is somewhat prototypic of the role of behavioral science in health care delivery. This paper will describe a conceptual and empirical foundation for viewing chronic pain in behavioral terms and will describe some implications for evaluation and treatment.

Before we can proceed, we must consider what is meant by "pain", for confusion abounds. The term is often used in two somewhat different ways. In one use, "pain" refers to a quality of sensory experience occurring in response to specific afferent stimuli. In that use, it is a neurophysiological phenomenon which also can be described as a stimulus-response phenomenon; i.e. given an adequate nociceptive stimulus, the organism responds by reporting "pain". That is the concept of pain which has been used in the voluminous research on pain thresholds, pain tolerance, and the like. It is the concept used in experimental pain and, traditionally, in health care in relation to clinical problems.

There is a second use of "pain" rather different from the first, though often the two are confounded. It is the report of "pain" by patients in clinical settings. In the second use, "pain" refers to patient reports of suffering an affective state, but which they label as "pain". This second use is broader in a number of ways. Perhaps the most critical difference between this second and the first use of "pain" is that in chronic pain there is usually no independently measurable and verifiable antecedent nociceptive stimulus. The data are not stimulus and response, but only "response"; i.e. the report by a person that what he/she labels as "pain" is being experienced.

These two seemingly similar but in fact somewhat different uses of the term pain present major problems to a proper understanding of chronic clinical pain. One problem is that the user is at risk to fall into the trap of assuming that use by a person of the term "pain" implies there is a peripheral nociceptive stimulus, when such an assumption is neither inherently necessary nor, as is often the case in chronic pain, valid. The second and related problem is the confounding of "pain" with "suffering". People may use "pain" to report "suffering". The clinician may then pursue a nociceptive stimulus which may not exist, and may fail to search for the other factors producing the "suffering" behavior.

We shall endeavor to be more strictly behavioral here. We shall use the term "pain" to mean visible or audible communications from a person which are labelled as "pain" or which are likely to cause observers to infer the person is suffering and using the language of pain to communicate that fact. however, the problem began. In short, the focus is on "pain behavior", whatever its origins. Another important distinction to make is between acute and chronic pain. That distinction is a time function. Acute refers to recent onset, chronic to a long-standing problem. In the clinical context, chronic pain can be seen as pain behaviors continuing past healing time of the initiating injury.

The Problem

Pain viewed in the "sensory pain" or "stimulus-response" sense uses a Disease Model perspective. The assumption is made that the symptoms of pain -- actually pain behaviors -- occur because of or are controlled by underlying body damage factors; i.e. physical findings or a peripheral nociceptive stimulus. That is a useful way of viewing recent onset or acute pain problems. It is the traditional way health care professionals, as well as consumers, view pain. Diagnosis seeks to identify the alleged underlying physical factors. Treatment seeks to eliminate it, whether by natural healing or with the help of surgery or some other treatment intervention. If the Disease Model strategy works, the pain problem is resolved and there is no problem. If it does not work and pain persists, typically the patient is recycled through the diagnostic and treatment process though still within the framework of a Disease Model conceptualization of the problem. Ultimately, after repeated failed treatment trials, the patient may be told the pain is "all in your head," or that it is untreatable and "...you will have to learn to live with it." Of course, neither of those propositions is a solution at all. In a sense, continuing reports of "pain" following either initial or repeated interventions are an index of the failures of such a system.

One common variant of this issue is eventually to term the problem "psychogenic." Like the term pain itself, "psychogenic" is also subject to surplus and confounding meaning. Usually it is taken to imply the patient has some kind of emotional, motivational, or personality problem. But the basis for such a conclusion is largely that the diagnostician finds, or seems to find, a discrepancy between physical findings and complaints of pain. That may be a problem of the observer deriving from limitations in the conceptual model and not a characteristic of the patient.

To summarize the statement of the problem, traditional views of pain are based largely on a Disease Model perspective. Diagnosis and treatment proceed from that. When treatment fails, there is a dearth of alternatives.

Loeser (1980) has proposed a different conceptualization of the domain of pain; one which appears eminently suited to deal with the critical issues. In Loeser's model, there are four components. These are illustrated in Figure 1 and defined as follows:

Nociception: Refers to thermal or mechanical stimuli impinging on peripheral receptors and which activate A-delta and C fibers;

"Pain": Refers to a sensory experience elicited by the perception of nociception. We are capable of perceiving nociception in its absence (as in Tic Douloureux) and of not experiencing "pain" in its presence (as in combat wounds not perceived until hours later). Thus the nociception-pain linkage is less than perfect. But it is not this "pain" that the pain patient manifests. The sensation of "pain" in turns activates

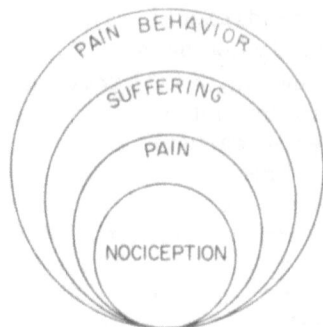

Figure 1.   Conceptual Component of Pain

higher nervous centers, leading to:

Suffering: Negative affective responses generated in
higher nervous centers by "pain" and other situations; e.g. loss
of loved objects, stress, anxiety, etc.   But it is also not
suffering that the clinician observes when he evaluates a
problem of chronic pain.   Suffering in turn generates:

Pain Behavior: Defined as all forms of behavior generated
by the individual commonly understood to reflect the presence of
nociception, including speech, facial expression, posture,
seeking health care attention, taking medications, refusing to
work.

As one progresses through each step from nociception to
"pain" to suffering to pain behavior, new factors enter in.   The
result is that the linkage between nociception and each
succeeding element may diminish.   Each element after nociception
may occur for reasons other than nociception and the probability
that they do occur for other reasons increases as one moves
through the conceptual sequence.

It is pain behavior which constitutes the bulk of the
observations about chronic pain.   Except in cases of ongoing
disease (e.g. rheumatoid arthritis) or clearly observed
structural defect, the so-called physical findings are often
evidence only that there has been a body damage factor at point
of onset of pain, supplemented by postural actions or reports
implying but not providing the continued existence of the
originating nociception.

A Data Base for the Behavioral Perspective

Empirical support for the proposition that chronic pain may be viewed in behavioral terms should show that pain behaviors may occur independently of nociception and that their occurrence is linked to conditioning effects.   Several studies will be cited in support of the proposition.

One study bearing on the matter was done by Fordyce, Caldwell and Hongladarom (1979).   A series of 77 chronic pain patients in treatment, including prescribed Physical Therapy exercises, were asked to exercise "...until pain, weakness, or fatigue cause you to stop. You decide when to stop." They were instructed to work to tolerance. A total of 442 exercise trials were examined to determine the last digit of the number of repetitions   completed   under   these   working-to-tolerance directions.   There should be an approximately equal probability subjects would quit on 11, 12, 13, for example, or 21, 22, 23, etc.   Instead, fully 50% of the time exercises were terminated on a multiple of 5. Clearly awareness of how much had been done and a readiness to quit on a "round number" influenced their exercise tolerance (as shown in Figure 2).

On   the   basis   of   those   findings,   a   second   study   was undertaken. Apparatus was set up to provide for a minimum of informational feedback to exercising patients as to how much work   had   been   done.   On   example   will   illustrate,   though exercises   were   prescribed   from   among   ten   different   exercise setups used in the study. For some patients, riding a fixed

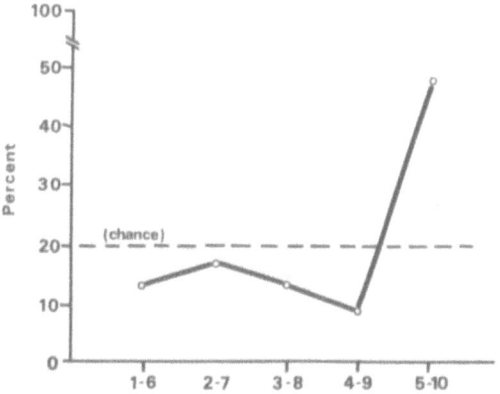

Figure 2.   Ending digit frequencies of exercise repetitions
             when exercising to tolerance by chronic pain
             patients.

bicycle was prescribed but with no speed or distance indicator at hand. Electronically controlled equipment changed the gear ratio of the bicycle on a random schedule. The drag on the axle also changed each few seconds randomly and asynchronous to gear ratio changes.

Two groups of subjects were studied. The Experimental or Pain Group consisted of chronic pain patients in initial evaluation. One experimenter (Hongladarom), following medical evaluation, prescribed exercises from the experimental set according to the nature of the pain problem. The second or Control Group consisted of miscellaneous subjects who volunteered to participate and who reported no present or past history of pain problems. All subjects performed the exercises in four sessions, once daily on approximately consecutive days. Because of the nature of the electronic programming, amount performed could be recorded only as distance into the electronic program before the subject indicated his/her tolerance had been reached. "Tolerance" was defined above: "Work until pain, weakness, or fatigue cause you to want to stop. You decide when to stop." The amount performed during the first session was assigned an arbitrary value of 1. Each successive session was calculated as a ratio of the first.

Results of this study are shown in Figure 3. Results indicate the Pain patients and the Control subjects are quite comparable in the relative amounts done across four sessions. Moreover, both groups show a consistent increase in amount done. The increase is probably essentially an accommodation factor to the apparatus and experimental conditions. Figure 3 also shows as a reference point the performance curve for a random sample

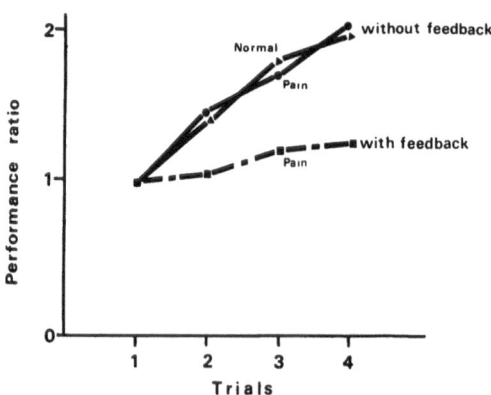

Figure 3.   Exercise tolerance ratios across sessions with and without performance feedback

of chronic pain patients exercising when they could readily count amount done. These data also support the notion that "tolerance" can hardly be understood as simply an extension of nociception and allegedly associated "pain".

In addition to these two studies, reports of treatment effects from these behavioral methods lend further support to the proposition that pain behaviors may have little or no link to current, ongoing nociception (Cairns, Thomas, Mooney & Pace, 1976; Fordyce, Fowler, Lehmann, deLateur, Sand & Trieschmann, 1973; Gottlieb, Strite, Koller, Madorsky, Hockersmith, Kleeman & Wagner, 1977; Greenhoot & Sternbach, 1974; Newman, Seres, Yospe & Garlington, 1978; Roberts & Reinhardt, 1980; Swanson, Maruta & Swenson, 1979). Those reports have shown that modifying pain behaviors without attacking alleged underlying nociception can produce significant and lasting improvement.

Evidence in support of the thesis that pain behaviors link to environmental contingencies can also be gained from those reports. There is more direct evidence, as well.

Block and others (Block, Kremer and Gaylor, 1980) studied a series of chronic pain patients to determine relationships between patient reports of pain intensity and spouse support/non-support toward pain behaviors. By interview of patient and spouse, patients were categorized as having spouses who responded to pain behaviors in a supportive or non-supportive way. Each subject then was taken through a standard interview in a room equipped with a one-way mirror. Subjects were told -- correctly -- that their spouses were observing during 10 of the 20 minutes of interview and that the spouse was gone but staff therapists were observing during the other 10 minutes. Midway in each 10-minute time block, subjects were asked to rate their present pain intensity on a 0-10 scale, with 10 being the more intense. As shown in Figure 4, patients

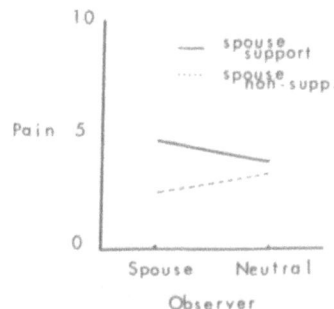

Figure 4.   Pain ratings with supportive and non-supportive observers.

having pain behavior supportive spouses reported higher levels
of pain when the spouse was observing and lower levels when
spouse was absent. Conversely, patients having non-supportive
spouses reported lower levels of pain when the spouse was
present and higher levels when spouse was absent. The ANOVA
interaction effects of present/absent - supportive/non-
supportive subgroup pain ratings were statistically significant.
Thus, we have direct evidence that what people say about their
pain intensity may vary systematically as a function of their
perception of the social milieu.

A recent study by Redd (1981) provides perhaps the
strongest evidence yet reported of the impact of social
contingencies on pain behaviors and, therefore, of the potential
independence of pain behaviors from nociception.

Redd worked in a hospice setting with a terminal cancer
patient suffering intense pain, leading to frequent screams and
cries loud enough to be audible two floors away. Hospice
procedures routinely provided for 24-hour surveillance. A nurse
sat in the room at all times. When the patient cried out, she
checked to see if anything could be done to ease suffering, did
it, and then resumed her seat.

First, full informed consent was obtained from all who were
involved. Next, baseline observations were obtained in which
the observer recorded each 10 minutes whether the patient was
crying, sleeping, talking, visiting, or lying quietly. After
124 hours of continuous observation, the intervention began.
The nurse continued as before, including freedom to interact
with the patient as needed. When crying/screaming began, she
first checked to see what help might be given and gave it. Then
she repositioned herself in a chair in the corridor just outside
the open room door, continuing to maintain surveillance but more
remotely. Two minutes after crying stopped, she returned to her
seat inside the room. Results are shown in Figure 5. Data are
recorded in 124-hour blocks or phases, each based on data
recorded each 10 minutes around the clock. The third or last
phase was but 25 hours long because this dying patient then
lapsed into coma and expired soon thereafter. The results show
graphically that this simple modification of social feedback,
contingent on crying/non-crying, had a remarkable effect.

The Conditioning of Pain Behaviors

The long-observed tendency for different cultures to show
differences in the intensity or form of pain behaviors indicates
the influence of modeling or imitation learning on acquisition
of pain behaviors. Similarly, intrafamily similarities in pain
behaviors presumably occur through modeling effects by which

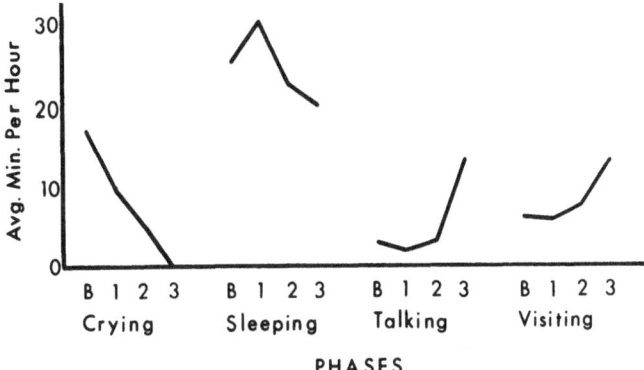

Figure 5.   Influence of Nurse proximity on pain behavior in
terminal cancer patient.

parents inadvertently influence or condition the pain behavior
repertoire of their children.

In addition to modeling, we now conclude that there are at
least   two,   and   perhaps   three,   ways   in   which
learning/conditioning appear to influence pain behaviors.   One
is through direct positive reinforcement.   There   are   many
illustrations   of   potentially   reinforcing   events   occurring
contingent upon the emission of pain behaviors.   A few will be
described.

Traditionally, physicians have prescribed analgesics on a
prn or "take only as needed" basis.   The prn regimen has the
effect of requiring the person to emit pain behaviors in order
to   receive   sanction   to   take   the   analgesic.   That   is,   the
analgesic   is   pain   behavior   contingent.   For   many   patients
sedation,   or   whatever   the   effect   of   the   medication,   is   a
reinforcing consequence.   Therefore, a prn regimen is also an
operant conditioning regimen.   The operant pain behaviors are
being   systematically   reinforced;   a   state   of   affairs well
designed to increase or strengthen pain behaviors leading to the
sanctioned use of the analgesic.

A second illustration is to be found in the oft-encountered
prescription   by   physicians   to   pain   patients   to
"work-to-tolerance.   Let pain be your guide.   If it begins to
hurt,   stop."   That   arrangement   makes   rest   pain   behavior

contingent. One must emit pain behaviors in order to gain
sanctioned rest. Working to tolerance is probably a prudent
prescription early in the life of a trauma-induced pain problem
but after healing time, such a plan risks systematically
reinforcing pain behaviors and punishing efforts toward
resumption of activity.

It should be noted that these two illustrations are common
to medical practice. They encourage iatrogenic persistence of
pain behaviors beyond healing time. It should also be noted
that in neither case is it necessary to postulate a personality,
motivational, or attitude problem in the patient to account for
persistence of pain behaviors.

Family members also can play a role, as indicated by the
Block study cited above. Spouses and other family members may
let specialized attention be pain behavior contingent. When
that pattern develops, there is risk that pain behaviors may be
encouraged or conditioned to persist past healing time.

The direct positive reinforcement of pain behavior can be
paraphrased as, "When I hurt (emit pain behaviors), 'good'
things happen which otherwise would not."

Indirect Positive Reinforcement (Avoidance Learning)

Pain behaviors which serve to limit activity may thereby
lessen or postpone nociception particularly during the early
history of a trauma- induced pain problem. They are therefore
potentially reinforcing, and in nearly all cases will be so.
Limiting activities to lessen, postpone, or avoid nociception
and ensuing pain also means the activities themselves are
avoided. If those activities have aversive qualities
independent of the issue of nociception, the activity-limiting
pain behaviors are reinforced in yet another way; namely, by
"time out" from aversive activities. To illustrate, the
impotent husband who pleads inability to participate in
intercourse because of back pain may receive reinforcement for
those limiting behaviors, both from postponing or avoiding
nociception emanating from his injured low back and from the
aversive social consequences of displaying yet again his
impotence. This kind of avoidance conditioning corresponds
approximately to the psychodynamic concept of secondary gain.
Examples of potential aversive activities avoided by pain
behaviors are legion and should need no further elaboration
here. This concept of avoidance learning can be paraphrased as,
"When I hurt (emit pain behaviors) 'bad' things don't happen
which otherwise would."

The kind of avoidance learning phenomenon just described is often, though not always, associated with personality problems or, more operationally stated, with fewer effective well behaviors. Clinically, it is very frequently encountered in chronic pain patients.

There is a second kind of avoidance learning phenomenon which should receive some consideration. It will be recalled that stimuli associated with aversive stimuli can ultimately take on aversive properties themselves. Redd, (personal communication), for example, describes an illustration from a series of breast cancer patients receiving twice monthly chemotherapy on an outpatient basis. The chemotherapy was extremely noxious. Virtually all of the patients displayed nausea and intense vomiting for several hours following each chemotherapy dose. After but a few sessions, several of the patients reported they would experience severe nausea and would vomit simply on driving into the hospital parking lot for their next chemotherapy session. Similarly, Fordyce and Shelton (1981) report a case in which a young man who came to be confined to a wheelchair because of severe pain and dizziness when attempting to walk, was deconditioned from his guarding pain behaviors by a walking regimen in which rest (avoidance of walking) was made contingent on meeting a speed of walking quota. What appeared to be happening in this case was that cues originally associated with nociception became capable of eliciting nociception or, at least, guarding behaviors against it. It is as if the person displays pain behaviors in anticipation of nociception, just as the breast cancer patient displayed vomiting in anticipation of the chemtherapy agent.

Implication of a Behavioral Perspective of Chronic Pain on Patient Evaluation

Chronic pain patients should undergo a two-part evaluation. First, there should be a careful assessment of neurophysiological factors; i.e. a medical workup. The intent of that workup is to assess the role nociception or body damage factors, whether from the original injury or from subsequent treatments, may currently be playing in the pain behaviors observed and reported. The second part is a behavioral analysis. The objective of that component is to assess the viability of an alternative to understanding the pain problem in Disease Model terms; namely a learning/conditioning-based explanation.

The behavioral analysis of a problem of chronic pain seeks to study in detail pain and "well" behaviors and their relationship to physical and social environmental events. The objectives are to assess the closeness of the linkage between inferred nociception or physical load factors and pain behaviors, on the one hand, and between pain behaviors and environmental events, on the other.

There are different ways to gather the requisite data for a behavioral analysis of a problem of chronic pain. Our experience has led us to focus on obtaining from patients several days of completed diary forms prior to seeing them. On the diary forms are recorded by hours of the day and night when the patient is sitting, standing/walking, or reclining; pain ratings on a 0-10 scale; and medications consumed. In addition, patients complete an MMPI at time of first visit. Finally, we will not agree to see a patient for evaluation unless the spouse comes along. Data from the spouse is not obtained so much as a reliability check on data from the patient but because the perspective of the spouse often adds crucial information. Each is interviewed separately. Those interviews, along with diary forms and the MMPI, constitute the data of the behavioral analysis.

The focus of the interviews is to determine as fully as possible what cues or consequences in the patient's environment show systematic relationships to pain behaviors and to unrestrained activity or well behaviors. Though there is an almost infinite variety of cues or consequences–pain/activity behavior relationships one may encounter, one case example will be used to illustrate. More detailed interview guidelines can be found in Fordyce (1976a,b) and Heaton, Getto, Lehman, Fordyce, Brauer and Groban (1981). For example, should it be found that a given body position, when maintained for significant periods, systematically leads to suffering and pain behaviors and shifting away from that body position leads to easing of suffering and pain behaviors, that would be evidence consistent with there being a nociceptive stimulus actively involved in maintaining the pain problem. If, however, it could be shown that body positions of equivalent physical load do not produce pain behaviors, and further, that the exceptions relate to apparently reinforcing environmental consequences, there would be evidence of a probable conditioning component.

The illustrative case is of a middle-aged white-collar managerial-level worker. He came with a complaint of low back and hip pain sufficiently severe to require frequent interruptions of his demanding work schedule and occasional trips to the Emergency Room of a hospital for narcotic injections. Interviews with patient and spouse, however, revealed an interesting pattern. The patient reported and his wife confirmed that working at his desk for long intervals was the principal activity leading to severe episodes of suffering. Periodically, these episodes would reach such intensity as to require an immediate narcotic injection. However, it also came out that the patient spent many hours on weekends sitting in the cockpit of his small sailboat, sailing alone. Here then was an indication that body position physical load factors had differential effects according to whether it was in an office or the cockpit of a sailboat. Additional interview data indicated that developments on the job corresponding in time roughly to onset of the pain problem had led a previously very rewarding position to become somewhat aversive. Concomitantly, it also came out that the patient did not find parenting an easy and enjoyable activity. The solitude of sailing alone both avoided his employment and also what was for him the rigors of trying to raise a number of children. Moreover, his very faithful and dutiful wife was quite ready to assume the extra load of parenting and to provide emotional support to her husband in his time of need.

This example illustrates how evidence may emerge supporting that the nociception - pain behavior link may be questionable and that the pain behaviors may be linked systematically to potentially reinforcing environmental consequences; in this case both avoidance of work and parenting burdens and achievement of special attention and support from his wife.

Heaton and colleagues (Heaton et al., 1981) have developed an empirically weighted multiple-choice format of a behavioral analysis interview based on the model described here. Their form provides "scores" calibrated against a criterion of medical consensus as to physical findings, in a set of chronic pain patients. While those data are still limited, they offer considerable promise as a way of testing further precise predictors or descriptors of pain behaviors which have come under control of conditioning effects.

Implications of a Behavioral Perspective on Treatment of Chronic Pain

A number of papers have been cited here which report results of treatment of chronic pain via various behavioral methods. The variations used, all reporting significant levels of success, make clear that no one best way has yet been documented. That is not surprising when one considers the rich variety of chronic pain problems encountered. Certain generalizations seem warranted, however, and will be discussed briefly. For more detailed accounts of procedures used, one should consult Fordyce (1976a).

One general conceptual point should be made before proceeding. In chronic pain, as in any chronic illness, treatment programs must recognize the probable need for dealing with several issues. One is the reduction of "sick" or pain behaviors. With appropriately selected patients and a well prepared treatment environment, that usually proves to be relatively easy to do. However, people who have been limited in function for a long time do not find it easy to resume pre-onset activities. It is usually necessary also to provide a systematic activation program. In addition, if their pain behaviors were being sustained in part because they permitted the person to avoid aversive activities, it would of course be essential to help the person to remediate the problems which have limited his/her ability to attain satisfaction or success in the activities. Otherwise, following treatment the person would be going back into an environment which helped produce or sustain the pain problem in the first place. This is, in addition to the contingency management program for diminishing pain behaviors, the principal contribution of Psychology to management of chronic pain. The diversity of problems and of possible methods for dealing with them is, however, too great to permit dealing with them in any detail here.

The principal underlying point to a behavioral approach to treatment of chronic pain is to recognize that the target of treatment is not some alleged or inferred underlying "cause" of the pain behaviors, but the pain behaviors themselves. The objective is to reduce pain behaviors and to increase their counterpart, effective activity and well behavior. The criterion of change or success/failure is a change in what the person does; i.e. his/her pain behaviors. This, of course, presupposes that evaluation has indicated pain behaviors are substantially controlled by conditioning effects and not by nociception.

The behavioral tactics used generally focus either on contingency management or on cognitive behavioral methods. It is beyond the scope of this paper to deal with the characteristics and differences between these two approaches in detail. For the purposes here, two general propositions will be offered. One is that contingency management methods make a direct attack on the social/environmental contingencies to sick and well behavior. Cognitive behavioral methods, in contrast, seek to modify the person's experience of pain/suffering and of non-pain/suffering. This may be done by working at altering expectancies, by seeking to develop alternative cognitive meanings for pain/non-pain experiences, or by working on the person's sense of self-mastery over events. The second proposition is that neither contingency management nor cognitive behavioral methods have demonstrated meaningful and enduring successes except where the approaches have been multi-modal, involving such diverse components as reduction of analgesic consumption, reduction of communications to others about pain/suffering, increases via systematic exercise programs of activity level, and systematic efforts through vocational and social counseling components to help the person to again become re-engaged into his/her environment and into sustaining activities.

So much for the methods. Let us now examine the implications of a behavioral approach by identifying the more likely specific behavioral targets needing change.

Reduction of pain behaviors

1. Decrease or elimination of visible or audible signals to others that pain is being experienced.

2. Decrease or elimination of specific guarding behaviors, such as a limp, or the avoidance of certain body positions or movements which are essential to the way of life projected following treatment.

3. Decrease or elimination of analgesic, psychoactive, muscle relaxants, sedative hypnotic or other pain-related medication ingestion.

4. Decrease of health care utilization rate related to the pain problem bringing the person for treatment.

Increase in activity or well behaviors

1. For those patients who have been limiting activity because of a pain problem, a systematic exercise program to restore physical condition, to restore or renew access to the activities of daily living essential to functioning following treatment, and to communicate to the patient and others that the patient can perform. Clinical experience suggests this component is the most significant single element in a multi-modal pain management program.

2. Increase in involvement and rehearsal with specific activities making up a significant part of the person's post-treatment way of life; e.g. return to employment, resumption of homemaking activities, resumption of recreational activities, resumption of sexual activity, resumption of family chores and responsibilities.

Modification of social contingencies to pain behavior and well behavior

1. Decrease and elimination of spouse, family member, or significant others' encouragement, support, or special

attention when pain behaviors are emitted. This component
usually means involving the family in the treatment
program.

2. Decrease or elimination of spouse, family member, or
significant others' protective admonitions to the patient
to avoid activities productive of pain behaviors.

3. Increase in the systematic reinforcement and support by
those around the patient for activity and for patient
efforts to move ahead against previously constraining
activity barriers.

REFERENCES

Block, A.R., Kremer, E.F. & Gaylor, M. Behavioral treatment of
    chronic pain: The spouse as a discriminative cue for pain
    behavior. Pain, 1980, 9, 243-252.
Cairns, D., Thomas, L., Mooney, V. & Pace, J. A comprehensive
    treatment approach to chronic low back pain. Pain, 1976, 2,
    301-308.
Fordyce, W., Fowler, R. & deLateur, B. An application of
    behavior modification techniques to a problem of chronic
    pain. Behavior Research and Therapy, 1968, 6, 105-107.
Fordyce, W., Fowler, R. Lehmann, J., deLateur, B., Sand, P. &
    Trieschmann, R. Operant conditioning in the treatment of
    chronic   pain.   Archives   of   Physical   Medicine   and
    Rehabilitation, 1973, 54, 399-408.
Fordyce, W. Behavioral methods in chronic pain and illness. St
    Louis: C.V. Mosby, Company, 1976a.
Fordyce, W. Behavioral concepts in chronic pain and illness.
    In P.O. Davidson (ed.) Behavioral management of anxiety,
    depression and pain. New York: Brunner/Mazel Press, 1976b,
    pp. 147-188.
Fordyce, W., Caldwell, L. & Hongladarom, T. Effects of
    performance feedback on exercise tolerance in chronic pain.
    Unpublished manuscript, University of Washington, 1979.
Fordyce, W., Shelton, J. & Dundore, D. The modification of
    avoidance learning in pain behaviors. Journal of Behavior
    Medicine, 1981.(in press)
Gottlieb, H., Strite, L., Koller, R., Madorsky, A., Hockersmith,
    V., Kleeman, M. & Wagner, J. Comprehensive rehabilitation
    of patients having chronic low back pain. Archives of
    Physical Medicine Rehabilitation, 1977, 58, 101-108.
Greenhoot, J. & Sternbach, R. Conjoint treatment of chronic
    pain. Advances in Neurology, 1974, 4, 595-603.
Heaton, R., Getto, C., Lehman, R., Fordyce, W., Brauer, E. &
    Groban, S. A standardized evaluation of psychosocial
    factors in chronic pain. Pain, in press, 1981.
Newman,   R.,   Seres,   J.,   Yospe,   L.   &   Garlington, B.

Multidisciplinary treatment of chronic pain: Long-term follow up of low back pain patients. Pain, 1978, 283-292.

Redd, W. Treatment of excessive crying in a terminal cancer patient: A time series analysis. Journal of Behavior Medicine, 1981, in press.

Roberts, A. and Reinhardt, L. The behavioral management of chronic pain: Long term follow up with comparison groups. Pain, 1980, 8(2), 151-162.

Swanson, D., Maruta, T. & Swenson, W. Results of behavior modification in the treatment of chronic pain. Psychosomatic Medicine, 1979, 41, 55-61.

BEHAVIORAL TREATMENT OF PAIN

Francis J. Keefe - Dept. of Psychiatry, Duke University
Medical Center, Durham, North Carolina

Andrew R. Block - Dept. of Psychology, Perdue University
Indianapolis, Indiana

Recently, there has been a great deal of interest in the
use of behavior therapy and behavior modification techniques for
treating chronic pain patients. This interest has been
generated by several factors. First, medical and surgical
approaches known to be effective for acute pain have met with
only limited success when used with chronic pain patients
(Aitken, 1959; White, 1966). Second, there is a growing
awareness among physicians that behavioral and psychological
factors are important in the etiology and maintenance of chronic
pain problems (Bonica, 1977; Holden, 1978). Third, there is
evidence that treatment programs incorporating behavioral
techniques are successful in treating many chronic pain patients
who have failed to respond to more conventional treatment
efforts.

Behavioral treatment approaches to chronic pain can be
grouped into two major categories: 1) operant conditioning
methods, and 2) self-management techniques. The operant
conditioning approach, developed and refined by Fordyce over the
past decade, is well summarized in his preceding paper. This
approach has made major contributions to the field. It has
placed the behavior of the chronic pain patient in a theoretical
context, led to the development of systematic behavioral
intervention procedures and has proven to be effective in
helping chronic pain patients increase activity level and
decrease medication intake.

Self-management techniques, in contrast, have only recently
been applied to the treatment of chronic pain. Initially
developed in the 1960s by clinicians working with behavioral

disorders such as smoking and obesity, self-management
strategies were extended in the 1970s to the treatment of
medical disorders such as hypertension, asthma and peripheral
vascular disease. Chronic pain patients have been taught to
control maladaptive behavior patterns using self-management
techniques such as electromyographic (EMG) biofeedback,
progressive relaxation and cognitive behavior therapy methods.
While successful applications of these techniques to chronic
pain, either alone or in combination, have been reported, a
comprehensive review of this area is lacking.

The purpose of this paper is to provide a critical review
of behavioral self-management approaches to chronic pain. In
this review, we 1) describe the basic concepts and procedures
of this approach, 2) review outcome data, and 3) consider
relevant theoretical, methodological and clinical issues.

BASIC CONCEPTS AND PROCEDURES

Techniques in which the patient assumes some part
of the therapist's role and attempts to modify his own
behavior are not an innovation of behavior therapy. The
importance of self-regulation in achieving happiness has
been stressed by ancient philosophers and moralists and
is encountered in many religions. Behavior therapists
differ not in application of these general techniques, but
in their efforts to incorporate them into a theoretically
consistent model of behavior modification and to provide
the patient with supplementary behaviors that make
self-regulation easier.
                                      Kanfer & Phillips, 1970

Theoretical conceptualizations of how self-management
techniques work have been advanced by prominent behaviorists
(Skinner, 1953; Ferster, 1965; Kanfer & Phillips, 1970).
According to this viewpoint, self-regulatory behaviors are
normally acquired during childhood socialization training.
Under certain circumstances, however, such as trauma or
prolonged stress, an individual's normal ability to regulate
his/her own behavior may become impaired. Thus, when faced with
severe and chronic pain, patients' attempts to regulate activity
and medication intake may become inconsistent. Further,
expressions of inability to control behavior may be socially
reinforced by sympathetic family members. Eventually, patients
may relinquish all attempts at self-control, as powerful
environmental contingencies (narcotic medication, financial
compensation, avoidance of work and other responsibilities) come
to maintain and control behavior. The behavioral perspective
maintains that self-management techniques work because they both
strengthen such impaired self-regulatory skills and also teach
individuals new methods of self-regulation.

Training programs to enhance self-control skills involve several basic elements. These are self-observation, training in an alternative response, intervention early in the chain of learned responses, and generalization and maintenance training.

Self-observation is begun before patients are actually trained in self-management techniques. Self-observation is used to help patients become more aware of 1) their own behavior, 2) environmental factors that may be important, and 3) the efficacy of self-control strategies that they may already be using. Patients are typically asked to keep records of duration, frequency and severity of behavioral problems. In our laboratory we ask patients to keep records of subjective tension, medication intake and pain. Self-observation enables patients to recognize relationships between problem behaviors. For example, patients may learn that they take pain medication when subjective tension is high or in anticipation of pain increases, rather than in response to increases in pain.

Self-observation is a method of assessment continued throughout treatment. Simple graphs are often used to display data gathered during pre-treatment and treatment stages of training. This helps make patients more aware of improvements that occur. Self-observation assists patients in making a much more objective and accurate analysis of their own behavior. When chronic pain patients are asked to describe their own behavior, they usually portray themselves in a negative light either because of cognitive errors (Lefebvre, in press) or because they fail to systematically attend to their own levels of well behavior (Kremer, Block & Gaylor, 1981). Self-observation, particularly when coupled with periodic feedback on performance by external observers, can help patients correct misperceptions of their behavior. Perhaps this is one of the reasons that self-observation has been found to be an effective means of reducing maladaptive behaviors, even in the absence of the therapeutic interventions (Nelson, 1977).

Following a period of self-observation, patients are trained to use a self-control method as an alternative to typical pain behavior patterns. Chronic pain patients engage in behaviors such as intake of narcotic agents or resting in bed, which have reinforcing short-term consequences. The long-term consequences of these behavior patterns, however, are clearly deleterious for the patient. The self-management approach trains patients to engage in an alternative behavior, such as relaxation, that is beneficial on both a short and long-term basis. Training sessions are carried out by a therapist and involve a demonstration of the alternative response, guided practice, performance feedback, and specific instructions as to home practice regimens.

Patterns of pain behavior involve chains of related behaviors in which one patient activity serves as a stimulus for the next. For example, a twinge of pain may lead to an immediate reduction of activity and request for pain medication. Learning to re-exert control over pain behavior sequences involves breaking an almost automatic chain of events by substituting an adaptive coping behavior. Once patients have mastered the specifics of a coping technique, they are trained to apply the technique early in the chain of events that ordinarily would lead up to maladaptive pain behavior.

As patients experience success with self-management techniques, the focus of training is shifted to teaching generalization and maintenance skills. These skills enable patients to transfer learned control over pain behavior to other settings and to maintain initial gains over long time periods. Training techniques used at this stage of treatment include overlearning the target behavior, training under stressful conditions, extensive home practice assignments, booster sessions and fading the frequency of therapist contact (Lynn & Freedman, 1979).

Outcome Data

Published research studies have examined the efficacy of four self-management approaches to chronic pain. These are: 1) EMG biofeedback, 2) EMG-assisted relaxation training, 3) cognitive-behavioral interventions, and 4) multi-modal approaches.

EMG Biofeedback

A major psychophysiologic mechanism believed to be responsible for the maintenance of chronic pain is the pain-tension-pain cycle (Bonica, 1977). In many chronic pain patients, pain produces reflex skeletal muscle spasm. The resultant vasospasm in turn leads to the release of pain-producing substances. These substances lower the threshold of peripheral nociceptors, thereby increasing pain. Research clearly indicates that chronic pain patients, in fact, do often show excessive muscle activity in specific muscles affected by pain. For example, Kravitz (Kravitz, Moore & Glaros, 1981) examined EMG activity in the paraspinal musculature in chronic low back pain patients and normals. When asked to contract muscles other than those in the low back, chronic low back pain patients responded with increased paraspinal muscle activity, while normals did not. Wolf (Note 1) also reports abnormal patterns of paraspinal EMG activity in low back pain patients. When compared to a normative sample (Wolf, Basmajian, Russe, & Kutner, 1979), low back pain patients showed EMG hyperactivity on one side of the spine and EMG hypoactivity on the other. This pattern is evident at rest, during anterior flexion and during dynamic activity.

Electromyographic biofeedback training has been primarily used to help chronic low back pain pateints gain voluntary control over muscles contributing to their pain. In training, surface electrodes are placed over targeted muscles. The EMG activity recorded controls an audio and/or visual feedback display. The patient is instructed to decrease EMG activity using the feedback provided.

Nouwen and Solinger (1979) trained patients suffering from chronic low back pain to reduce EMG activity from electrodes placed bilaterally on the erector spinae muscles at L2. During training, patients were in a static prone position on an examining table. Eighteen patients were given EMG biofeedback training and seven served as waiting list controls. Significant reductions in EMG activity and pain were obtained by subjects in the EMG feedback group, while subjects in the control condition showed no change. Follow-up conducted three months following treatment revealed that patients treated with biofeedback continued to report substantially lowered pain. However, EMG levels recorded at follow-up had returned to initially high levels.

Freeman, Caslyn, Page & Halar (1980) reported success in treating eight chronic low back pain patients with EMG feedback from the paraspinal musculature. Patients were trained to reduce EMG activity to a criterion of 50% baseline. Four patients were able to reach this criterion and all four reported pain relief and increases in activity. An additional two patients also evidenced behavioral improvement.

One study has reported treatment failure with EMG biofeedback training from the lumbar paraspinal muscles. Peck and Kraft (1977) found that patients with chronic upper and low back pain were able to reduce muscle activity but reported no concomitant reductions in pain.

Wolf and his associates (Jones & Wolf, 1980; Wolf, Nacht & Kelly, in press) have recently developed an EMG biofeedback training program for low back pain that has been more consistently effective. In this program, patients are given dual-channel EMG feedback to teach them to equalize EMG activity from two lumbar paraspinal muscle placements. Results indicate that as patients learn to equalize the EMG activity, decreases in pain also occur. An important feature of this work is that training takes place while patients are engaged in both static and dynamic activities which they report are painful for them.

This procedure may facilitate generalization of self-control over muscle activity from laboratory tasks to activities of daily living.

To summarize, EMG biofeedback has been used to teach chronic pain patients to reduce excessive tension in muscles affected by pain. Applications have been limited primarily to low back pain and results are variable. The relationship between changes in EMG activity occurring over treatment and pain report is inconsistent. Most patients who reduce EMG activity report pain relief but some do not. Recording EMG from a single muscle site only during a static activity may be problematic. Readings may be unrepresentative of muscle activity during more strenuous activity. The most consistent and positive results have been reported by Wolt, who uses an intensive training program that teaches patients to control patterns of muscle activity during static and dynamic activity.

The EMG biofeedback approach addresses a limited target behavior, i.e. muscle tension. As pain becomes chronic, muscle tension problems may be compounded by other well-entrenched behavior patterns, such as excessive medication dependence and low activity levels. Thus, for more chronic patients, EMG biofeedback may not be enough. Along these lines, it is interesting to note that EMG feedback has been most successful when used by investigators who have screened out patients with evidence of spinal dysfunction (Nouwen & Solinger, 1979) or who have had prior surgery (Wolf, et al., in press). In fact, the utility of this approach may be minimal for the great number of chronic pain patients have had numerous surgical procedures in which the paraspinal musculature has been cut. Electromyographic activity recorded from surface electrodes placed over these muscles is invalidated because of denervation potentials (Wolf, et al., in press).

In conclusion, EMG biofeedback has been shown to be effective in treating chronic pain patients. While this approach is appropriate for only a small percentage of chronic patients (i.e. those who have never had surgery), it may prove to be an effective preventative training method for patients at risk of developing chronic pain.

EMG-Assisted Relaxation Training

The rationale for EMG-assisted relaxation training is based on two major problems displayed by many chronic pain patients. First, chronic pain patients often show generalized and sustained increases in muscle activity (Holmes & Wolff, 1952). Hyperactivity in muscles from the primary site of pain may exacerbate pain, may lead to other pain problems (headaches, bruxism), and may increase fatigue. Muscle overactivity becomes apparent to those around the patient as the patient's movements take on a stiff, interrupted and guarded quality. Recent data indicate that both the patient's own report of pain and observers' judgments of pain correlate with the frequency of such stiff and guarded movements (Keefe & Block, in press).

A second major problem characteristic of many chronic pain patients is that they often report anxiety symptoms. Experimental studies indicate that anxiety tends to heighten the intensity of pain and that anxiety reduction techniques can lower pain intensity (Weisenberg, 1977).

In EMG-assisted relaxation training, patients are initially instructed in a variant of progressive relaxation (Jacobson, 1938; Wolpe & Lazarus, 1966; Bernstein & Borkovec, 1973) and then given feedback from a number of muscle groups to facilitate the learning of relaxation skills. EMG biofeedback from the frontalis, upper trapezius and paraspinal musculature has been used to assist relaxation training. Three studies have investigated the efficacy of EMG-assisted relaxation training.

The first such study was conducted by Hendler, Derogatis, Avella & Long (1977). In this study, results obtained in 13 chronic low back pain patients who had undergone a five-session course of frontalis EMG-assisted relaxation were examined. Changes in EMG activity, pain, and in psychiatric symptoms were monitored over treatment. Of 13 patients treated, six reported decreases in pain on four out of five days during training. For these patients, maintenance of treatment gains was reported at one-month follow-up. Patients reporting pain relief had significant reductions in psychiatric symptoms as measured by the SCL-90. They showed reductions in somatization, obsessive compulsiveness, interpersonal sensitivity and hostility.

A recent study conducted in our laboratory (Keefe, Schapira, Brown, Williams & Surwit, 1981a) evaluated the efficacy of frontalis EMG-assisted relaxation training in the management of 18 chronic low back pain patients. All patients were given a minimum of six laboratory training sessions and were given extensive generalization training involving home practice with cassette tapes, and a prompting procedure to cue frequent practice. Significant decreases in EMG activity were obtained both within and across the sessions. A highly significant decrease in pain also occurred during the training sessions. Patients reported success in using relaxation to decrease pain in home practice settings. One year follow-up data were collected on 13 of the 18 patients. Seventy percent said they had maintained or extended gains achieved during treatment. This group of patients reported substantial improvements in their ability to control pain, in their mood, in function, and a high degree of compliance with treatment instructions.

Sherman, Gall & Gormly (1979) used EMG-assisted relaxation training to treat phantom limb pain in a group of 16 amputees. Biofeedback was provided from frontalis and stump muscle placements. The results of training were impressive. Highly significant reductions in pain and anxiety were obtained in 14 out of 16 patients. Eight patients were actually pain-free at one to three year follow-ups. Changes in pain correlated significantly with changes in anxiety (p = .69), supporting the contention that for this population pain and anxiety are related. Changes in a variety of anxiety symptoms, including sleeping problems, and major reductions in narcotic intake also were reported.

In summary, EMG-assisted relaxation training has been used with patients who have very long histories of continuous pain, multiple surgical procedures, and psychological deficits. This training appears to not only help patients reduce the relevant target behaviors (anxiety and muscle tension), but to also help them reduce medication intake and alleviate symptoms of mood disorder, as well as often relieving pain sensations. However, as with EMG feedback alone, the results are variable. In addition, none of the studies reviewed has compared the effects of EMG-assisted relaxation to other common treatment approaches or to appropriate control conditions. However, the long history of treatment failure certainly provides a baseline control against which the initial and long-term positive results obtained can be compared. Further controlled research is obviously needed in this area.

Cognitive-Behavioral Intervention

Research conducted in naturalistic settings has demonstrated that patients normally develop strategies to cope with pain. Copp (1974) interviewed hospitalized patients and found a wide variety of pain coping strategies. Many patients used cognitive manipulations involving distraction or reinterpretation of pain. Research in our laboratory (Rosenstiel & Keefe, Note 1) has shown that chronic low back pain patients also utilize similar strategies and that the type of strategies employed relates to adjustment. Patients who used active coping strategies (such as behavioral distraction) were found to be more active, less depressed and anxious; whereas those who tended to use passive techniques (such as hoping or praying) were significantly more impaired by their pain.

There has been great interest in the possibility that chronic pain patients might benefit from training in cognitive coping skills. Such interest is based on studies which have found that normals exposed to laboratory pain stressors have increased pain threshold or pain tolerance level if they have been taught active cognitive coping techniques (Weisenberg, 1977). Cognitive coping skills training for chronic pain involves many treatment techniques such as imagining pleasant events, focusing on other things, dissociating oneself from pain, imagining an affected area as numb, and concentrating on sensations other than pain (Scott & Barber, 1977). These are usually combined in a treatment package (e.g. "stress inoculation training", Meichenbaum & Turk, 1976).

A number of case studies have assessed the effects of such cognitive interventions. Cautela (1977) treated a woman suffering from chronic arthritic pain with a number of cognitive strategies, including thought stopping, thought replacement, and covert reinforcement for pain relief. After three weeks of treatment, the patient no longer had pain in her hands or feet, gains which were apparently maintained at eight month follow-up. Similarly, Varni (1981) used a cognitive strategy termed "guided imagery" for hand warming, combined with progressive relaxation, in treating three patients suffering from chronic arthritic pain secondary to hemophilia. Systematic treatment was carried out over long time periods (from seven to fourteen months). Patients reported substantial decreases in pain, and in general improvement as reflected by improved mobility and ability to sleep. Grzesiak (1977) also combined progressive relaxation with guided imagery to successfully treat four spinal cord injured patients who complained of chronic pain.

Two controlled group outcome studies have compared cognitive strategies to various control conditions. In the first study (Rybstein-Blinchik, 1980), 44 patients suffering from mixed chronic pain syndromes were assigned to either an experimental group or one of three control groups. The experimental group was instructed to utilize a cognitive strategy in which they attempted to omit the word pain from their thinking and to reinterpret their pain experience as a different sensation such as numbness or general arousal. Two control groups were instructed to engage in cognitive strategies which were considered less relevant, e.g. to focus on important life events. The remaining group was a waiting list control. Subjects in the experimental group had significant decreases in pain and in overt pain behaviors displayed during brief standardized observation periods conducted before and following treatment. Subjects in the control groups showed no improvement.

In the second study (Turner, Note 2), patients suffering from chronic low back pain were randomly assigned to one of three conditions: 1) progressive relaxation training alone, 2) a cognitive-behavioral coping skills program modeled after Meichenbaum and Turk (1976), and 3) a waiting list control condition. Patients in both relaxation and cognitive training groups reported significant decreases in pain over treatment as compared to the waiting list control. At one-month follow-up, however, patients in the cognitive pain management group were able to maintain improvements, whereas, patients in the relaxation group had reverted to pre-treatment levels. Patients in the cognitive group reported significantly lower levels of pain and anxiety, and an enhanced ability to tolerate pain and to participate in routine daily activities.

The results of studies that have evaluated the use of cognitive coping strategies with chronic pain patients are quite interesting. Early applications to single patients or small groups of patients have yielded encouraging results. Subsequent controlled studies have suggested that patients trained in cognitive strategies show improvements in measures of pain and adjustment relative to patients in control conditions. One feature of studies using cognitive strategies that differentiates them from studies in which EMG feedback or EMG-assisted relaxation training has been used is the number of different treatment strategies employed. Under the rubric of cognitive strategies investigators have placed relaxation training, distraction techniques, reinterpretation strategies, expectancy manipulations, covert reinforcement, guided imagery, thought stopping, etc. The distinction between these various techniques is not clear. What is apparent, however, is that these multiple strategies do appear to help patients. Given the complex nature of chronic pain (Fordyce, 1976), it may be that more multimodal approaches are helpful because they address the broad spectrum of behavioral problems exhibited by patients.

Multimodal Approaches

The vast majority of studies examining behavioral techniques for chronic pain have involved combined treatment approaches. There are two major reasons for this. First, ethical concerns are raised when one randomly assigns chronic pain patients so that some are in control conditions in which they receive no treatment. Second, many chronic pain patients present with a complex picture of psychiatric, behavioral and physical symptoms. Comprehensive evaluation is usually considered a prerequisite for dealing with these patients. An interdisciplinary evaluation approach naturally leads to application of multiple treatment modalities, each aimed at different target problems.

Gottlieb, Strite, Kollar, Madorsky, Hockersmith, Kleeman & Wagner (1977) described results achieved with 72 chronic low back pain patients who were exposed to a multidisciplinary treatment program emphasizing behavioral self-management techniques. Behavioral techniques included EMG-assisted relaxation training, self-paced medication reduction, assertion training, and self-directed physical exercise programs. Patients were given a central role in treatment and were even involved in staff meetings to help determine their own treatment plans. Staff members rated patients throughout treatment on a number of four-point scales. Ten scales were used to assess functional improvements and four scales were used to measure improvements in pain and mood. Of 72 patients treated, 50 completed the program and were rated by clinicians as showing significantly less pain behavior and increases in functional activity. At six-month follow-up, 82% of patients contacted were either employed or in job training.

Keefe, Block, Williams and Surwit (1981b) described results achieved in a multidisciplinary treatment program for 111 chronic low back pain patients. Treatment procedures consisted of EMG-assisted relaxation training, self-paced medication reduction, physical therapy and psychotropic medication. Significant decreases in subjective tension and muscle tension were reported. Also, significant improvements occurred in medication intake and general physical activity. Individual differences in patient characteristics associated with pain relief were also examined. The 28 patients reporting the largest decreases in pain were compared to 28 patients showing the least relief in pain. The patients in the worst outcome group were found to have longer histories of continuous pain, to be more likely to have had multiple surgeries, and more likely to be on disability than patients in the best outcome group.

A recent study by Herman and Baptiste (1981) described outcome data obtained in treatment of 75 patients reporting mixed chronic pain complaints. Self-management strategies used included cognitive techniques and EMG-assisted relaxation training. Training in self-management was carried out in a group format. Although the results of this study are difficult to interpret because of the multitude of paper-and-pencil tests of questionable relevance to chronic pain, the authors report 79% of patients achieved marked to moderate success. Treatment appeared to be most successful in reducing patients' levels of depression and pain, and in moving patients towards a more internal locus of control (Rotter I-E Scale).

Khatami and Rush (1978) have reported on one of the few outpatient treatment programs. In this study, cognitive strategies were combined with EMG biofeedback and social systems intervention to treat five patients suffering from chronic pain. Treatment was intensive and carried out systematically over a 12-month period. Large and dramatic reductions in pain were obtained over the course of 12 months' treatment. Patients treated reported significant decreases in pain, medication intake and depression.

Multidisciplinary approaches combining self-management and other treatment techniques have yielded promising initial results. These programs are particularly impressive in that the population of pain patients treated in these studies is extremely chronic and has demonstrated an unresponsiveness to previous treatment efforts. However, since these studies do use multiple treatment procedures, little information is available on the effectiveness of any specific treatment component. Data that are available clearly indicate that these techniques do not work for all chronic pain patients. This appears to be true even though treatment combinations are typically tailored in these programs to the patient's apparent needs. Given the promising initial results achieved with these techniques, further research is needed both to evaluate specific treatment components and to identify behavioral patterns that may pre-dispose patients to respond well or poorly.

Critical Issues

We now consider issues that are relevant whenever self-management techniques are used with chronic pain patients.

Theoretical Issues

Although the diverse self-regulation techniques reviewed may appear to be loosely related, all are based on conditioning and learning theory. Basic to this theory is the notion that behavior is modulated by environmental events. Behaviors followed by relatively positive consequences increase in probability; whereas those followed by aversive consequences decrease in probability (Skinner, 1953). As Fordyce (1976) has indicated, chronic pain behaviors may have tremendously positive consequences. Whereas the operant approach attempts to arrange the patient's environment so that the patient no longer receives such powerful reinforcers for pain behavior, the self-management approach teaches the patient to recognize the long-term maladaptive consequences of these patterns and to substitute a response that is more adaptive on both a short- and long-term basis. A number of theoretical questions concerning reinforcement available for self-regulation strategies remain unanswered. First, are self-management techniques rewarding for patients because they reduce the severity of behavioral problems such as chronic muscle tension, anxiety, and depression?

Second, does an enhanced sense of control over pain behaviors serve as an effective reward in its own right? Third, is it necessary for self-management techniques to produce an actual decrease in pain in order for them to be reinforcing? Finally, can any of the potential rewards for exerting self-control over pain behavior outweigh the powerful naturally occurring environmental consequences of pain behaviors?

Training in self-management techniques, as we have seen, is not typically carried out by the patient alone. A therapist's active guidance and encouragement during the early stages is probably crucial to treatment success. As treatment progresses, the patient gradually assumes more responsibility. Simple instruction in self-management via books or cassette tapes generally has not been found to be an effective treatment for behavioral disorders. Thus, the term "self-management" is used to emphasize the relative contribution made by the patient to treatment. A continuum of patient involvement exists in behavioral treatment programs. On one end of the continuum are radically operant conditioning programs in which no attempt is made to instruct or elicit patient cooperation; on the other extreme are naturally-occurring self-regulation techniques developed and used by patients on their own. Most behavioral programs that have been described in the literature fall somewhere between these two extremes. For example, a number of self-management techniques (e.g. self-observation) are evident in the operant conditioning approach described by Fordyce (Fordyce, Fowler, Lehmann, DeLateur, Sand and Trieschmann, 1973). The eventual goal of any behavioral program is to help the patient improve enough so that self-generated or intrinsic reinforcements maintain behavior.

Methodological Issues

Research on self-management techniques for chronic pain is only in its early stages. The studies reviewed primarily consist of uncontrolled, single-case reports or descriptions of outcome in a series of similar patients. There are few controlled outcome studies comparing the efficacy of various behavioral self-management techniques, and no outcome studies comparing these techniques to more traditional treatment procedures. Although this lack of sophistication in experimental design is common in the preliminary stages of research in any area (Kuhn, 1962), the absence of data from controlled studies makes it difficult if not impossible to evaluate treatment outcome. A number of variables that have been uncontrolled may confound the results obtained, e.g. the application of simultaneous treatments, collection of data by clinicians involved in the patient's treatment, imprecision in the description of treatment modality, and the relative absence of follow-up data.

Several recommendations might be given for future research. First, experimental designs that provide an opportunity to definitively evaluate treatment outcome, while simultaneously not denying patients access to treatment need to be utilized. For example, in the analysis of single cases, a multiple baseline approach in which treatments are sequentially administered across subjects may be utilized (Hersen & Barlow, 1976). In the case of group designs, counterbalanced designs in which groups of subjects are sequentially exposed to different orders of treatment can be utilized (Campbell & Stanley, 1963, Design #11).

Second, a broader spectrum of behavioral assessment techniques need to be utilized. Chronic pain patients are a heterogeneous group presenting a complex set of behavioral responses. Development of more reliable and valid behavioral measures for this population is sorely needed (Keefe, Brown, Scott and Ziesat, 1982). Routine assessment across multiple response systems - overt behavioral, cognitive-verbal and physiological - is needed both to improve our understanding of pain behavior and to help assess treatment outcome.

Clinical Issues

The selection of chronic pain patients appropriate for self-management training is an important issue. Not all patients who have behavioral problems are appropriate. Patients must have the capacity for self-control. Variability in response to treatment programs is apparent. Our work (Keefe, Block, Williams and Surwit, 1981b) has suggested that chronicity is an important variable. Very chronic patients may have firmly entrenched behavior patterns and may only respond when the environmental contingencies for pain behavior are altered. We recommend that clinicians attempt to identify patients at a pre-chronic stage (Keefe & Brown, 1982) and utilize self-management training to prevent the development of further behavioral problems.

Another clinical issue is the need to utilize a systematic approach when training patients. Self-management techniques are not a bag of therapeutic tricks. Research has shown that the patient's ability to comprehend the rationale for training, and his willingness to keep records and comply with treatment instructions is critical to the ultimate success of treatment. Careful attention to the delivery and follow-up of treatment instructions is essential (Shelton & Ackerman, 1974).

Clinically, self-management strategies appear to have a number of advantages. They are inexpensive and may help maintain and extend treatment gains over time and across settings. At present these strategies appear to offer a promising treatment alternative to clinicians working with chronic pain patients. A great deal more controlled clinical research is needed, however, before that promise can be fulfilled.

REFERENCE NOTES

1. Rosenstiel, A.R. and Keefe, F.J. Development of a questionnaire to assess cognitive coping strategies in chronic pain patients. Paper presented at the Annual Meeting of the Association for the Advancement of Behavior Therapy, New York, NY.
2. Turner, J. A comparison of two behavioral treatment procedures for low back pain: Progressive relaxation vs. stress inoculation training. Unpublished manuscript, University of Washington Medical Center, Seattle, WA.

REFERENCES

Aitken, A. T. The present status of intervertebral disc surgery. Michigan State Medical Society, 1959, 58, 1121-1127.

Bernstein, D. A., and Borkovec, T.D. Progressive relaxation training: A Manual for the Helping Professions. Champaign, Illinois: Research Press, 1973.

Bonica, J.J. Neurophysiologic and pathologic aspects of acute and chronic pain. Archives of Surgery, 1977, 112, 750-761.

Campbell, D. T., and Stanley, J.C. Experimental and quasi-experimental designs for research. New York: Rand McNally, 1963.

Cautela, J.R. The use of covert conditioning in modifying pain behavior. Journal of Behavior Therapy & Experimental Psychiatry, 1977, 8, 45-52.

Copp, L. The spectrum of suffering. Americen Journal of Nursing, 1974, 74, 491-495.

Ferster, C.B. Essentials of a science of behavior. IN J. I. Nurnberger, C. B. Ferster, and J.P.Brady (Eds.) An introduction to the science of human behavior. New York: Appleton Century Crafts, 1963.

Hersen, M and Barlow, D.H. Single case experimental designs: Strategies for studying behavior change. New York: Pergamon, 1976.

Holden, C. Pain, dying and the health care system. Science, 1978, 203, 984-985.

Holmes, T. and Wolff, H.G. Life situations, emotions and backache. Psychosomatic Medicine, 1952, 14, 18-33.

Jacobson, E. Progressive Relaxation. Chicago: University of
    Chicago Press, 1938.

Jones, A.L. and Wolf, S.L. Treating chronic low back pain: EMG
    biofeedback training during dynamic movement. Physical
    Therapy, 1980, 60, 58-63.

Kanfer, F. and Phillips, J. Learning foundations of behavior
    therapy. New York: John Wiley and Sons, 1970.

Keefe, F. J. and Brown, C. Behavioral treatment of chronic pain
    syndromes. IN P. Boudewyns and F. Keefe (Eds.) Behavioral
    medicine in general medical practice. Menlo Park, CA:
    Addison-Wesley, 1980.

Keefe, F. J., Brown, C., Scott, D. and Ziesat, H. Behavioral
    assessment of chronic pain. IN F. Keefe and J. Blumenthal
    (Eds.) Assessment Strategies in Behavioral Medicine. New
    York: Grune & Stratton, 1981.

Keefe, F.J., Block, A.R., Williams, R.B. and Surwit, R.S.
    Behavioral treatment of chronic low back pain: Clinical
    outcome and individual differences in pain relief. Pain,
    1981a, 11, 221-231.

Keefe, F.J., Schapira, B., Brown, C., Williams, R. and Surwit,
    R.S. EMG-assisted relaxation training in the management of
    chronic low back pain. American Journal of Clinical
    Biofeedback, 1981b, 4, 93-103.

Khatami, M. and Rush, A.J. A pilot study of the treatment of
    outpatients with chronic pain: Symptom control, stimulus
    stimulus control and social system intervention. Pain,
    1978, 5, 163-172.

Kravitz, E., Moore, M.E. and Glaros, A. Paralumbar muscle
    activity in chronic low back pain. Archives of Physical
    Medicine & Rehabilitation, 1981, 62, 172-176.

Kremer, E.F., Block, A.R. and Gaylor, M.S. Behavioral
    approaches to treatment of chronic pain: The inaccuracy of
    patient self-report measures. Archives of Physical
    Medicine and Rehabilitation, 1981, 62, 188-191.

Kuhn, T.S. The Structure of Scientific Revolutions. Chicago:
    University of Chicago Press, 1962.

Lefebvre, M. Cognitive errors in depressed psychiatric and low
    back pain patients. Journal of Consulting & Clinical
    Psychiatry, 1981, 49, 517-525.

Lynn, S.J. and Freedman, R.F. Transfer and evaluation of
    biofeedback treatment. IN A. Goldstein and F.H. Kanfer
    (Eds.) Maximizing treatment gains: Transfer enhancement in
    psychotherapy. New York: Academic Press, 1979.

Meichenbaum, D.H. and Turk, D.C. The cognitive-behavioral
    management of anxiety, anger, and pain. IN P. O. Davidson
    (Ed.) The behavioral management of anxiety, anger, and
    pain. New York: Brunner/Mazel, 1976.

Nelson, R.O. Methodological issues in assessment via
    self-monitoring. IN J. D. Cone and R.P. Hawkins (Eds.)
    Behavioral assessment: New direction in clinical
    psychology. New York: Brunner/Mazel, 1977.
Nouwen, A. and Solinger, J. The effectiveness of EMG
    biofeedback in low back pain. Biofeedback &
    Self-Regulation, 1979, 4, 103-112.
Peck, C.L. and Kraft, G.H. Electromyographic biofeedback for
    pain related to muscle tension. Archives of Surgery, 1977,
    112, 889-895.
Rybstein-Blinchik, E. Effects of different cognitive strategies
    in the chronic pain experience. Journal of Behavior
    Medicine, 1979, 2, 93-102.
Scott, D.S. and Barber, T.Y. Cognitive control of pain: Four
    serendipitous results. Perceptual & Motor Skills, 1977, 44,
    569-570.
Sherman, R.A., Gall, N. and Gormley, J. Treatment of phantom
    limb pain with muscular relaxation training to disrupt the
    pain-anxiety-tension cycle. Pain, 1979, 6, 47-55.
Shelton, J. L. and Ackerman, J.M. Homework in Counseling and
    Psychotherapy. Springfield, Illinois: C. Thomas, 1974.
Skinner, B.F. Science and human behavior. New York: Macmillan,
    1953.
Varni, J.W. Self-regulation techniques in the management of
    chronic arthritic pain in hemphilia. Behavior Therapy,
    1981, 12, 185-194.
Weisenberg, M. Pain and pain control. Psychological Bulletin,
    1977, 84, 1008-1044.
White, A.W., Low back pain in men receiving workmen's
    compensation. Canadian Medical Association Journal, 1966,
    95, 50-56.
Wolf, S.L., Basmajian, J. V., Russe,T.C. and Kutner, M.
    Normative data on low back mobility and activity levels:
    Implications for neuromuscular re-education. American
    Journal of Physical Medicine, 1979, 58, 217-229.
Wolf, S., Nacht, M. and Kelly, J.L. EMG feedback training
    during dynamic movement for low back pain. Behavior
    Therapy, (in press).
Wolpe, J. and Lazarus, A.A. Behavior therapy techniques: A
    guide to the treatment of neuroses. New York: Pergamon,
    1966.

MILITARY APPLICATIONS OF BEHAVIORAL MEDICINE

Robert J. Biersner

National Medical Research Center

Bethesda, Maryland

The North Atlantic Treaty Organization (NATO), under the auspices of which this conference on behavioral medicine was held, represents a group of nations that have united together to form a common determent and defense against military invasion from Warsaw Pact nations. Much of the research described at this conference would be useful in treating victims of combat between NATO and Warsaw Pact forces, most notably those victims who have developed physical signs and symptoms that are mediated directly by anxiety accruing from poor adjustment to combat environments, or physical illnesses and injuries that are aggravated by impaired psychological adjustment to combat stress. Combat casualties of psychogenic origin have become substantially more prevalent recently compared to earlier periods of armed conflict. Official records show that during World War II, such casualties accounted for about one per cent of total casualties among Navy and Marine Corps personnel, while during the Vietnam Conflict, over seven per cent of the total Navy and Marine Corps casualties involved some form of psychological disturbance (Hoiberg & Gunderson, Note 1). Recent findings show that resistance to infectious diseases may also be impaired by psychological stress (Keller et al., 1981), indicating that these previous casualty statistics for psychogenically-mediated disorders may be inordinately conservative (i.e.,the data of Hoiberg and Gunderson show that over nine per cent of the total Navy and Marine Corps casualties during the Vietnam Conflict were attributed directly to infectious diseases).

Many of the behavioral medicine techniques described in previous chapters of this book are currently and routinely used in major military hospitals throughout the United States and other NATO countries. As familiarity is gained with these techniques, especially with long-term effectiveness and the extent to which skilled technicians can be used to provide these services, treatment of casualties at field health care facilities (e.g., division or capital ship level) becomes a distinct possibility. While these techniques hold promise for enhancing military effectiveness by substantially improving remission rates for psychogenic disorders, those behavioral medicine techniques described above, which are used typically in clinical medicine settings (even during combat), do not contribute directly toward enhancing the performance effectiveness of normal (non-casualty) military personnel. However, a number of military and quasi-military R&D funding agencies, including the Office of Naval Research, the Defense Advanced Research Projects Agency, and the National Aeronautics and Space Administration, have supported research and development efforts using behavioral medicine techniques to improve directly performance in military situations. These applications include augmentation and/or suppression of (a) brain waves (alpha and theta rhythms), (b) cardiovascular and muscle activity (including hand/digit temperature), (c) vestibular dysfunction, and (d) visual accommodation. These applications involve improvement of performance tasks that are unique to military settings, or from which the military would be a principal benefactor. Each of these applications will be discussed separately below, as well as some implications from previous research indicating that the principles and techniques of behavioral medicine may be useful in understanding and improving the effectiveness with which military personnel endure and cope with a variety of environmental and physical stressors.

Prior to this discussion, note should be made of the appropriateness of retaining these applications within the purview of medical psychology instead of personnel psychology, the latter traditionally having responsibility for matters involving performance enhancement (at least in most military organizations). While the stimulus and response characteristics of these applications may be largely behavioral (unlike conventional behavioral medicine applications that use physiological responses as criterion measures), the intervening variables are undoubtedly physiological, and as such involve the health and viability of those on whom these techniques are being used. Additionally, while few of these military applications have dealt directly with interactions between drugs and behavioral medicine techniques, the evidence from other work cited in previous chapters in this volume (e.g., the chapter by Shapiro) indicates that combinations of these two procedures may be more effective in attaining the criterion response than either procedure separately. The use of behavioral medicine as an adjunct to drug therapy may permit reduced drug dosages to be used, thereby preserving performance integrity and enhancing the overall effectiveness and safety of those undergoing treatment. If future research should show similar interactive effects for military applications, then such applications would be most conveniently and effectively pursued in conjunction with the pharmacological expertise and legal safeguards provided typically within the medical community.

PREVIOUS RESEARCH AND APPLICATIONS

Brain Waves. Research on the effectiveness of augmentation and suppression of brain waves as a means of improving performance has been reviewed recently and comprehensively by Johnson (1977) and Lawrence and Johnson (1977). Only the highlights of these reviews will be discussed in this chapter. The two types of brain waves most commonly featured in this work are alpha and theta rhythms. Enhancement of alpha has been analyzed in association with recuperation from sleep loss effects, cognitive (verbal and numerical) performance (both learning and memory), complex discrimination tasks, and relief of pain. In military settings, research on these topics would have substantial utility because (a) sleep loss or fragmented sleep is a frequent occurrence because of unusual watchstanding schedules or extended military operations, (b) verbal/numerical skills and complex discrimination responses are necessary in aviation and fire control (gunnery) situations, and (c) control of pain would benefit the combat victim who may have to continue to defend against enemy action until medical aid arrives. After noting that learned control of alpha unmediated by other physiological events (especially muscular activity) has yet to be demonstrated, Johnson (1977) concludes that "Enhanced alpha activity does not prevent sleep-loss effects or substitute for sleep, is not related to memory or choice-reaction performance, does not provide a recuperative break period, and is incompatible with cognitive tasks requiring effort." Johnson also describes as equivocal the work of Melzack and Perry (1975) and Lehmann et al. (1976) in which learned alpha control was used to reduce chronic and headache pain. While Melzack and Perry (1975) obtained positive effects using alpha control in combination with hypnosis, neither Melzack and Perry nor Lehmann et al. (1976) obtained positive effects with alpha control alone. Johnson attributes the positive effects found by Melzack and Perry using alpha used conjointly with hypnosis to the general relaxation effects normally associated with maximum alpha activity, and further states that alpha feedback does not appear to have "... unique recuperative or therapeutic powers."

Findings involving control of theta rhythms appear to be more equivocal than the negative data described above for alpha rhythm control. Vigilance, which is important to many military situations including sonar and radar operations, has been the skill most often analyzed in association with theta control. Early data obtained by Beatty et al. (1974) showed that college students trained to suppress theta were more accurate in detecting visual targets during a two-hour monitoring task than a control group, while students taught to enhance theta performed more poorly on this task than controls. However, these results on target detection accuracy could not be replicated on a similar task of one-hour duration using college students trained to suppress theta, although both this student group and a subsequent group of naval personnel trained as radar observers could detect targets more rapidly than comparison groups not trained in theta suppression (Beatty & O'Hanlon, Note 2). Using a similar task of three-hours duration on a highly experienced group of Navy air traffic controllers, Beatty and O'Hanlon (1975) failed to demonstrate any performance effect for those who were trained to suppress theta, while Wilson et al. (Note 3) could not show an effect for inexperienced Navy enlisted personnel who were trained to suppress theta and who performed a three-hour sonar task. Questioning the reliability of the early theta suppression findings, Johnson (1977) states that if these results do prove to be robust, the effect is most likely the consequence of an elevated arousal state and is not specific to theta. This interpretation is supported by the failure of these researchers to limit the physiological effects only to theta, as well as earlier findings by Williams et al.(1962) showing that those deprived of sleep perform more effectively during periods of theta activity. Further substantiation of an arousal effect was demonstrated later by Johnson (Note 4). Effective performance among a group of Navy volunteers on an alpha-numeric vigilance task of three-hours duration was correlated principally with elevated activity in EEG frequencies of 8 Hz and 15-20 Hz. The pattern of significant frequency elevations across individual group members was, however, highly variable and idiosyncratic. Johnson interprets these results as showing that EEGs (or specific frequencies within the EEG) are an "... index of arousal (that) may correlate with cognitive functioning. EEG is probably the best physiological variable to measure, but EEG frequency per se is not the most relevant factor. Rather, changes in EEG frequency intensity appear to be predictive of performance only when they are secondary changes in general arousal which, in turn, affect performance."

Cardiovascular and Muscle Activity.   While   research described in other chapters of this book shows that relaxation (as evidenced by reduced cardiovascular and muscle activity) appears to benefit health, research on the performance effects of such relaxation is equivocal. Using a measure of vigilance (the Mackworth Clock Vigilance task) and a reaction-time task, Stephens et al. (1972), Harris et al. (Note 5), and Stephens et al. (1975) showed that under normal, non-stressful conditions, self-regulated elevations and reductions in heart rate did not result in significant performance differences across conditions, nor did performance interfere with self-regulated elevations and reductions in heart rate. However, on another measure of vigilance (the Continuous Performance Task), self-regulated elevations in heart rate were found to shorten the response time significantly compared to self-regulated reductions in heart rate. Response times during non-regulatory periods were at intermediate levels. Accuracy did not differ across any of the conditions. Under stressful conditions in which a random schedule of click-shock combinations were imposed on the Continuous Performance Task, self-regulated reductions in heart rate were eliminated, while self-regulated elevations in heart rate fell from 13 bpm to 7 bpm above normal heart rate levels. Under these stressful conditions, accuracy fell to 73% of normal (non-stressful) levels during rest periods, and to 60% of normal levels during self-regulated reductions of heart rate. Performance remained at normal levels under conditions of self-regulated elevations in heart rate. Whether or not the beneficial performance effects of elevated heart rate (and, conversely, the detrimental performance effects of lowered heart rate) are specific to the cardiovascular system or are representative of a more general arousal effect is unknown, as is the effectiveness of this technique for improving other forms of learning, memory and vigilance.

Use of muscle relaxation techniques as a means of coping with stress and consequently to improve performance was developed by Jacobson (1938), and may well be the precursor of current biofeedback technology. The techniques developed by Jacobson were used by pilots during World War II to reduce the stresses associated with repeated, highly dangerous sorties. However, data establishing the effectiveness of this application were never published. Interestingly, these same stressors were found by Brictson et al. (Note 6) to be associated with the performance of U.S. Navy aviators during the Vietnam Conflict. Despite the advancements made in biofeedback technology during the intervening years, applications were not made to this highly stressful combat setting. Perhaps this is just as well because the recent data on the performance effectiveness of muscle relaxation have been mixed. Stoyva and Budzynski (Note 7), using a loud, distracting noise as a stressor, failed to show a difference between self-induced muscle relaxation and control (recovery) conditions for performance involving eye-hand steadiness, visuomotor coordination (pursuit rotor), simple intelligence (serial sevens task), and complex perceptual reasoning (figure completion). Smith (Note 8) successfully trained a small group of participants to relax by using biofeedback to reduce heart rate, blood pressure and muscle tension. In order to determine if this technique was effective in reducing the performance effects stress, the performance of this group on an arithmetic task, a verbal intelligence test, and a measure of vigilance (i.e., monitoring gauges and responding to specific deviations beyond an established limit) was compared to the performance of various control groups (i.e., a group provided with simulated relaxation training, a group that was asked to relax only, and a control group that was not provided with any special instructions). The stressful condition consisted of exposure to a hyperbaric (i.e., high pressure) chamber for a period of one hour. Before pressurizing the chamber, the dangers of hyperbaric exposure were explained to the groups, including barotrauma (e.g., ruptured eardrums, sinus hemorrhage and lung compression that results in hemorrhage and pain), carbon dioxide poisoning resulting from inadequate ventilation of the chamber, spontaneous combustion and oxygen fires, power and communication failures, decompression sickness (or "bends"), and over-pressurization of the lungs (leading to air embolism or interstitial emphysema). The experimental and control groups described above were assigned either to this experimental condition (hyperbaric chamber exposure to a pressure equivalent to 30 feet of sea water) or to one of two control conditions--chamber only without pressurization or simulated pressurization (i.e., movement of gauge dials and pressurization to an equivalent depth of 30 feet of sea water, followed by a reduction of pressure to one atmosphere absolute--the pressure of air under normal surface conditions). Under both the experimental and simulated pressurization conditions, the gauges were adjusted to indicate a pressure equivalent of 100 feet of sea water, and conditions were such as to convince most members of the pressurized and simulation groups that this was the true chamber pressure. Despite these elaborate preparations and procedures, as well as an admonition to use previously trained relaxation techniques, the experimental group failed to demonstrate physiological responses indicative of relaxation while under pressure, and showed performance decrements similar to the other control groups while under the pressurization condition. While none of the differences between the groups were significant under the hyperbaric condition, the trend of these differences was consistently in the unexpected direction--control groups showed lower physiological responses and better performance than the experimental group. The implication of this finding is that biofeedback training may in some way have interfered with the normal orienting response typical of initial exposure to unfamiliar (and stressful) conditions, thereby permitting high levels of arousal to disrupt adaptive behavior. Another implication of these findings is that biofeedback is highly contingent on the conditions (both environmental and physiological) under which the training occurs, and does not readily transfer to other conditions. Biofeedback may therefore be constrained by variables and conditions typical of those found in the state-dependent learning paradigm. Burish and Schwartz (1980) have demonstrated that even under conditions in which self-induced reductions in EMG do transfer to a stressful situation, such reductions are not accompanied by lower states of general arousal as measured by skin temperature, pulse rate and finger pulse volume, thereby bringing into question the usefulness of such effects.

The work of Burish and Schwartz (1980) may have some bearing on results obtained by Tebbs et al. (Note 9) who trained U.S. Air Force cadets to relax using EMG feedback techniques. The data were again mixed, with the first experimental group failing to show an advantage compared to untrained controls during a check flight in a T-41 aircraft simulator, while a second experimental group was judged by instructors as significantly better than untrained controls. EMG training was provided several months before the check flight, and EMG status during the check flight was not monitored. These data indicate that sampling differences may be accounting for the results (i.e., only select individuals are responsive to biofeedback training or capable of retaining this learning for long periods). Work on self-regulation of hand and digit temperature has demonstrated that such effects are feasible using biofeedback techniques, but concomitant performance effects remain to be demonstrated (Taub, 1976; Taub, Note 10). In addition to advantages that such effects may have for those in the military who must perform dexterous tasks in cold weather (such as aiming a rifle or maintaining/repairing equipment), other possible military applications include reduction of edema and pain following tissue damage, control of diffuse bleeding that cannot readily be staunched, induction of sleep by lowering core temperature (useful during combat operations that prohibit sleep during normal periods), promotion of wound-healing, and suppression of local infection through phagocyte concentration and tissue oxygenation (the latter two applications involving vasodilatation and the remaining applications involving vasoconstriction). In a cool ($55\frac{1}{4}F$) environment, self-regulated hand temperatures averaging $6-12\frac{1}{4}F$ above the non-regulated hand were found (with a difference of nearly $24\frac{1}{4}F$ being obtained in one case). While such differences would be of little practical significance in extremely cold environments, if similar effects could be obtained in ambient temperatures of $20-40\frac{1}{4}F$ (the likely temperature range in a northern European or sub-Arctic combat environment), then the benefits to performance and health (including avoidance of frostbite) would be substantial. While this research provided much useful information on methodology and effects (e.g., long-term retention, specificity, transfer, discriminability of feedback, relationship to hypnotic susceptibility, and maintenance of self-regulation in the presence of distracting activities), the usefulness of this technology, especially the benefits that may accrue for performance or the management of vascular casualties, remains to be demonstrated.

Vestibular Dysfunction. Perhaps the most promising application of biofeedback technology for military purposes has been in the prevention and treatment of motion sickness. Motion (air) sickness has been shown to afflict over eighty percent of aviator candidates in the U.S. Navy, and is a serious problem even among highly experienced aviators (Hixson et al. (Note 11). Motion sickness on board ships in moderate to high seas can also be a serious problem (Hershey, 1980). Typically, these applications have involved some form of relaxation activity combined with physiological monitoring to provide appropriate feedback. Levy et al. (1981) trained a group of 20 aircrew personnel who had a history of chronic, disabling air sickness to regulate autonomic responses (using EMG, GSR and skin temperature feedback). The members of this group had been screened prior to treatment for an absence of other physical and psychological complications, as well as for a high level of motivation for flying. During biofeedback training, the group was exposed to progressively faster rotations (combined with 40¼ rolls) in a modified Barany chair. Following the treatment 16 (80%) were returned successfully to full flying duties.

Cowings et al. (1977) used autogenic exercises combined with visual (CRT display and meters), verbal, and auditory (pure tone) biofeedback of autonomic responses (heart rate, respiration rate, and blood volume pulse of the face and hands) to effect self-regulated reductions of physiological responses to motion stress. Motion stress was induced by using a rotating chair that varied progressively in speed from 6 to 30 rpms (or until severe facial pallor, sweating, and epigastric discomfort were observed). During each 5-minute interval at a constant rotational velocity, the head was moved every two seconds to a 45¼ angle in each of four quadrants. The combination of head movements and whole-body rotation resulted in a Coriolis acceleration effect. Two and one-half hours of biofeedback training occurred between test trials 2 and 3 and 3 and 4 or between test trials 4 and 5 and 5 and 6 (experimental groups I and II respectively). Training involved rotation of the chair only without head movement. Results showed that for both experimental groups, motion sickness effects occurred at much higher velocities and were reduced in severity and duration compared to the untrained control group. Four members of the two experimental groups (N=16) attained maximum acceleration (30 rpm) without developing symptoms. Using essentially the same procedures, Toscano and Cowings (1978) showed that the beneficial effects of biofeedback training in one rotational direction would transfer to rotations in the opposite direction, and that those who are highly susceptible to motion sickness benefited as much from the training as those who were average or low in susceptibility. In order to determine if the effects of biofeedback training and chair rotation would transfer to other components of vestibular function, Stewart et al. (1978) assessed experimental and control groups using six measures of vestibular function before and after training intervention. These measures included two ataxia tests, two measures of thresholds for rotary acceleration, cupulogram slopes, and power exponents for high level accelerations. Findings showed that this particular form of biofeedback training did not result in significant differences between pre- and post-test scores for these six measures. These data were interpreted as evidence that the vestibular mechanisms involved in the biofeedback/rotation procedures were separate from those vestibular mechanisms involved in performance of these six tests.

While the Cowings group has stated that individual variability in response to this intervention procedure remains a problem, the data cited above provide evidence that this technique is one of the most promising and reliable applications of biofeedback developed to date for use in military settings.

Visual Accommodation Training. While flying either military or commercial aircraft[1], pilots often enter visual environments in which detail is absent because of low illumination or homogeneous visual fields. Such conditions arise because of clouds, fog, bright light or flashes, dust, gases or darkness. These conditions can be extremely detrimental to air-to-air search (e.g., locating another aircraft or missile) and landing operations. Not only are normal visual cues occluded or degraded, the situation is made worse by the typical visual response of pilots under these conditions, a response commonly referred to as "empty field myopia". In viewing a homogeneous visual field, the accommodation system normally stabilizes at one diopter and fluctuates about this level. The eye is therefore myopic and is focused for targets at a distance of about one meter. A more

[1]While the work cited in this section emphasizes aviator performance, the data would be equally applicable to any military or civilian group (such as tank crews or lookouts on board ships) that is required to detect distant, moving objects in an occluded visual field.

advantageous response would be to maintain focus at zero
diopters or infinity. Such a response would keep the eye
focused outside the cockpit, a position from which the pilot
could better detect distant objects for more effective
missile/aircraft avoidance, targeting, and aircraft control than
is possible if the eyes are focused on a finite point within the
cockpit.

The accommodative response is under the control of the
ciliary muscle system, a group of smooth muscles governed by the
autonomic nervous system. This response was generally assumed
to be reflexive and outside of the control of higher centers
within the central nervous system. However, Randle (Note 12)
was successful in bringing the response under voluntary control
using conventional biofeedback techniques. The pure tones used
for biofeedback ranged in frequency from 800 to 3200 Hz and
provided feedback to accommodative responses in the range of
zero to six diopters. Variations in the accommodative response
were detected using an automatic infrared optometer (described
in detail by Cornsweet & Crane, Note 13)). Electrical input
from the optometer was used to regulate pure tone biofeedback.
Inasmuch as the optometer is a monocular device, only the
accommodative responses of the left eye were trained (the right
eye was occluded). Using this device, Randle (Note 12) trained
six participants to voluntarily maintain accommodation as close
to the zero diopter level as possible in a darkened visual
field. A patterned target was used initially to stimulate the
accommodation response over the full dioptric range in order for
the participant to learn the association between auditory
feedback and response conditions. Later, the participant was
trained to maintain accommodation at a single criterion level
while the target was moved above and below this level. The
criterion level was indicated by a constant auditory signal
provided to one ear, while variable auditory signals in the
other ear provided feedback information regarding the present
level of accommodation. (The objective was to match the two
sounds.) Following this training procedure, the participant was
asked to vary the accommodative response while viewing a dark,
empty field (using auditory feedback). During the final phase
of training, the participant viewed a moving target at three
diopters and was asked to accommodate to zero diopters after the
target was extinguished (i.e., viewing a dark, empty field).
Training was completed after the participant effected a zero
diopter response to this latter condition without the aid of
auditory feedback. Findings showed that participants could be
trained successfully to maintain dioptric levels above and below
those levels attendant with the patterned target, as well as
vary the accommodative response with shifts in auditory feedback
while viewing an empty visual field. Training on the criterion
task (maintaining zero diopter accommodation in an empty field

without auditory feedback on extinction of a three diopter
target) was moderately successful, with the average response
being about one-third of a diopter. The mean for an untrained
control group was about one diopter. Results also showed that
this trained response could be retained for at least ten days.
In later work, Cornsweet and Crane (1973) showed that
self-regulation of the accommodative response using auditory
feedback could be readily transferred to visual biofeedback
conditions (i.e., matching horizonal lines on an oscilloscope).

Two serious problems exist with this form of biofeedback
training. The first problem is individual variability to
training, while the second is the absence of data showing
whether or not this technique would be useful under either
operational or simulated flying conditions. These problems must
be adequately resolved before this extremely promising technique
can be applied in the settings.

DISCUSSION AND SUMMARY

None of the biofeedback techniques described above has been
proven ready for application, either for commercial or military
purposes. Results to date have been limited to basic research,
and testing under operational or field conditions is many years
away. Of these techniques, those that appear to be most
applicable to military problems are those used to prevent
vestibular dysfunction and to adjust the accommodative response
to infinity while viewing a homogeneous visual field. While
self-regulation of hand and digit temperature has been shown to
be feasible, the usefulness of this technique to improve
performance or to treat vascular wounds remains to be
determined. (Treatment of vascular wounds using biofeedback
techniques under combat conditions is likely to prove
impractical, if not infeasible.) While self-regulation of heart
rate and use of biofeedback to induce muscle relaxation have
been amply demonstrated, neither of these techniques appears to
benefit performance among normal, healthy individuals. This
conclusion is even more definite for self-regulation of brain
waves--despite overwhelming evidence that alpha and theta
rhythms can be voluntarily controlled, the enhancement of either
frequency does not improve performance effectiveness among
normal individuals. The primary applications of these
techniques may well lie in regulating sleep and rest under
conditions of sustained wakefulness or fragmented sleep, or in
reducing performance impairments among sleep-deprived
individuals. Both of these applications, however, may be useful
under sustained combat conditions.

In addition to the need for more performance effectiveness
data, as well as a demonstration of the operational usefulness

of these techniques, several other problems must also be
addressed. More knowledge is required about optimum feedback
strategies, not only to obtain maximum effectiveness but also to
establish standard procedures acceptable to practitioners. In
this regard, variability of effectiveness across different
conditions, especially transfer of effectiveness from
non-stressful to stressful conditions or from one state of
consciousness to another, must be better understood.

Another major problem is individual variability in response
to biofeedback. Little is known about the personality
characteristics or traits of those who respond readily to
biofeedback and those who do not. Cowings (1977) found that
sympathetic/parasympathetic responsiveness differentiated
between individual biofeedback learning rates, as did hypnotic
susceptibility and autonomic perception. This is an important
area remaining to be explored, and may even contribute
significantly to selection of military personnel for physically
and psychologically stressful combat operations. That
perception of physiological states may be important to
successful performance has been demonstrated in previous
research. Morgan and Pollack (1977) have shown that successful
marathon runners are conscious of physiological processes
(including pain), and appear to use these cues to regulate
energy expenditure so as to avoid wasteful use of these limited
resources during the early and middle portions of the race.
Rahe et al. (1976) provided evidence that Underwater Demolition
Team (UDT) trainees who passed this physically stressful program
showed significant correlations between blood factors (serum
uric acid and cholesterol) and emotional states during those
phases of training known to be especially stressful, while those
who failed did not show such correlations. These data could be
interpreted as indicating that the successful UDT group was more
sensitive to physiological states than was the failure group,
and that this trait was critical to completion of the program.
Whether this trait (either among marathon runners or UDT
trainees) is inherited or learned remains an unanswered
question. If trainable, then perhaps biofeedback technology
could be used to identify those who possess this characteristic,
thereby improving current selection procedures. Another
possibility is that biofeedback techniques could be used to
improve the probability of success among those who would
normally fail by enhancing physiological perceptiveness among
this group. As this discussion shows, much research remains to
be done and the usefulness of biofeedback technology for
enhancing military performance has only been superficially
explored.

REFERENCE NOTES

1. Gunderson, E.K.E., & Hoiberg, A. Navy epidemiological
    research in the 1980's (Staff paper). San Diego,
    California: Naval Health Research Center, March 1980.
2. Beatty, J., & O'Hanlon, J.F. EEG theta regulation and radar
    monitoring performance of experienced radar operators and
    air traffic controllers (Tech. Rep.). Los Angeles:
    University of California, March 1975.
3. Wilson, C., Hord, D., Townsend, R., & Johnson, L. Lack
    of^{a}theta suppression effect on performance during a complex
    visual sonar vigilance task. Paper presented at the
    meeting of the Biofeedback Research Society, Colorado
    Springs, February 1976.
4. Johnson, L. Use of physiological measures to monitor
    operator state. Paper presented at the meeting of the
    Cybernetics Technology Office of the Defense Advanced
    Research Projects Agency, Chicago, April 1978.
5. Harris, A.H., Stephens, J., & Brady, J.V. Self-regulation
    of performance-related physiological processes (Annual Rep.
    under Contract N00014-70-C-0350). Baltimore: Johns Hopkins
    University, 1973-1974.
6. Brictson, C.A., McHugh, W.B., & Naitoh, P. Prediction of
    pilot performance: Biochemical and sleep-mood correlates
    under heavy workload conditions. Paper presented at the
    meeting of the Advisory Group for Aerospace Research and
    Development, Oslo, April 1974.
7. Stoyva, J., & Budzynski, T. Biofeedback training in the
    self-induction of sleep (Annual Rep. under contract
    N0014-70-C-0350). Denver: University of Colorado Medical
    Center, June 1973.
8. Smith, R.W. Self-regulation as an aid to human
    effectiveness (Final Rep. under contract
    N00014-70-C-0350). Coral Gables: Applied Science
    Associates, Inc., June 1975.
9. Tebbs, R., Eggleston, R., Prather, D., Simondi, T., &
    Jarboe, T. Stress management through scientific muscle
    relaxation training and its relation to simulated and
    actual flying training (Final Rep. under ARPA Order 2409).
    Colorado Springs: U.S. Air Force Academy, November 1974.
10. Taub, E. Self-regulation of vasomotor tone in peripheral
    vascular beds (final Rep. under Contract N00014-70-C-0350).
    Silver Spring: Institute for Behavioral Research, October
    1975.
11. Hixson, W.C., Guedry, F.E., Holtzman, G.L., Lentz, J.M., &
    O'Connell, P.F. Airsickness during naval flight officer
    training: Basic squadron VT-10 (new syllabus) (Report No.
    NAMRL-1275). Pensacola, Florida: Naval Aerospace
    Medical Research Laboratory, March 1981.
12. Randle, R.J. Volitional control of visual accommodation.

Paper presented at the meeting of the Advisory Group for
Aerospace Research and Development, Garmisch-Partenkirchen,
September 1970.
13. Cornsweet, T.N. & Crane, H.D. Experimental study of visual
accommodation (Rep. under Contract NAS 2-5097). Menlo
Park: Stanford Research Institute, March 1972.

REFERENCES

Beatty, J., Greenberg, A., Deibler, W.P., & O'Hanlon, J.F.
Operant control of occipital theta rhythm affects
performance in a radar monitoring task. Science, 1974,
183, 871-873.
Burish, T.J. & Schwartz, D.P. EMG biofeedback training,
transfer of training, and coping with stress. Journal of
Psychosomatic Research, 1980, 24, 85-96.
Cornsweet, T.N., & Crane, H.D. Training the visual
accommodation system. Vision Research, 1973, 13, 713-715.
Cowings, P.S. Observed differences in learning ability of heart
rate self-regulation as a function of hypnotic
susceptibility. Theory and Therapy in Psychosomatic
Medicine, 1977, 4, 221-226.
Cowings, P.S., Billingham, J., & Toscano, B.W. Learned control
of multiple autonomic responses to compensate for the
debilitating effects of motion sickness. Theory and
Therapy in Psychosomatic Medicine, 1977, 4, 318-323.
Hershey, R. Sea sickness in the sexes. Naval Institute
Proceedings, 1980, 106, 95-97.
Jacobson, E. Progressive relaxation. Chicago: University of
Chicago Press, 1938.
Johnson, L.C. Learned control of brain wave activity. In J.
Beatty & H. Legewie (Eds.), Biofeedback and behavior. New
York: Plenum Press, 1977.
Keller, S.E., Weiss, J.M., Schleifer, S.J., Miller, N.E. &
Stein, M. Suppression of immunity by stress: Effect of
graded series of stressors on lymphocyte stimulation in the
rat. Science, 1981, 213, 1397-1400.
Lawrence, G.H. & Johnson, C.C. Biofeedback and performance. In
G.E. Schwartz & J. Beatty (Eds.), Biofeedback: Theory and
research. New York: Academic Press, 1977. Lehmann, D.,
Lang, W. & Debruyne, P. Kontrolliertes
EEG-alpha-feedback-training bei gesunden und
kapfschmerzpatientinnen. Archiv fur Psychiatrie
und Nervenkrankheiten, 1976, 221, 331-343.
Levy, R.A., Jones, D.R. & Carlson, E.H. Biofeedback
rehabilitation of airsick aircrew. Aviation, Space, and
Environmental Medicine, 1981, 52, 118-121.
Melzack, R. & Perry, C. Self-regulation of pain: The use of
alpha-feedback and hypnotic training for the control of
chronic pain. Experimental Neurology, 1975, 46, 452-469.

Morgan, W.R. & Pollock, M.L. Psychologic characterization of
     the elite distance runner. In P. Milvy (Ed.), The
     marathon: Physiological, medical, epidemiological, and
     psychological studies. New York: New York Academy of
     Sciences, 1977.
Rahe, R.H., Ryman, D.H. & Biersner, R.J. Srum uric acid,
     cholesterol, and psychological moods throughout stressful
     naval training. Aviation, Space, and Environmental
     Medicine, 1976, 47, 883-888.
Stephens, J.H., Harris, A.H. & Brady, J.V. Large magnitude
     heart rate changes in subjects instructed to change their
     heart rates and given exteroceptive feedback.
     Psychophysiology, 1972, 9, 283-285.
Stephens, J.H., Harris, A.H., Brady, J.V. & Shaffer, J.W.
     Psychological and physiological variables associated with
     large magnitude voluntary heart rate changes.
     Psychophysiology, 1975, 12, 381-387.
Stewart, J.D., Clark, B., Cowings, P.S. & Toscano, W.B.
     Learned regulation of autonomic responses to control
     coriolis motion sickness: Its effects on other vestibular
     functions. Proceedings of the 49th Annual Meeting of the
     Aerospace Medical Association, 1978, 134-135. Taub, E.,
Feedback aided self-regulation of skin temperature
     with a single feedback locus: I. Acquisition and reversal
     training. Biofeedback and Self-Regulation, 1976, 1,
     147-168.
Toscano, W.B. & Cowings, P.S. Transference of learned autonomic
     control of symptom suppression across opposite directions
     of coriolis acceleration. Proceedings of the 49th Annual
     Meeting of the Aerospace Medical Association, 1978,
     132-133.
Williams, H.L., Granda, A.M., Jones, R.C., Lubin, A. &
     Armington, J.C. EEG frequency and finger pulse volume as
     predictors of reaction time during sleep loss. Electro-
     encephalography & Clinical Neurophysiology, 1962, 14,
     64-70.

THE CONTROL OF BEHAVIOR: THEORETICAL CONSIDERATIONS FOR BEHAVIORAL MEDICINE

Arthur J. Bachrach

Naval Medical Research Institute

Bethesda, Maryland

In any discussion of the experimental analysis of behavior and in particular the techniques of operant conditioning, it is not uncommon to find that there is a general assumption that theory has no place in a rigorous operant conditioning approach. This is simply not the case. Operant conditioners have never been truly atheoretical. The concern over theory was simply a matter of believing that theories, when they dictated experiments to prove or disprove, interfered with research. There was a time when behavioral engineers, armed with M & M candy pellets or their equivalent for reinforcers, set out to change the world. Witness, for example, the bold words of McConnell (1970), "I foresee the day when we could convert a worst criminal into a decent respectable citizen in the matter of a few months--or perhaps even less time than that."

I am using the term "engineer" to describe some of these enthusiasts in the sense that an engineer applies the carefully developed principles and functional relationships that physicists have worked out in research laboratories. It is probable that the engineer or technician who applies principles without understanding fully the theoretical background underlying them will run into difficulties. Suppose that reinforcement doesn't work and the agent of reinforcement, sitting there with a bagful of M & Ms, needs to know something about stimulus control?

This very zeal that characterized the behavior technicians some years back led us (Bachrach and Quigley, 1966) to express concern that the field was being ruined by amateurs. Actually, the observation we made was ruined by a timid editor: in the original version we had said that behavior modification was in a

405

condition similar to what George Bernard Shaw had said about prostitution--it was a field in danger of being ruined by amateurs. Our greatest concern was the lack of a sound foundation of theory upon which to base the application. That concern is valid to this day.

There is another type of amateur with whom you are all familiar, a figure I like to refer to as the Piltdown Man.[1] You will recall that just before World War I an enthusiastic amateur geologist and archeologist discovered what was claimed to be the missing link between ape and man, joining the existing 3/4 ape and 1/4 man with the 1/4 ape-3/4 man. The half-and-half was hailed as a major scientific find, although, to be sure, there were doubters. Almost a half century later, the Piltdown Man was revealed as a forgery, a hoax with a human skull fragment but with the jaw of an Orang-Utan. The professional Piltdown Man may become an equally ill-fitting hybrid--the person who is trying to be half of each profession--1/2 physiologist-1/2 behaviorist. To properly function in behavioral medicine, each professional must be expert in his or her own field and knowledgeable in the fields of colleagues. Interdisciplinary research is not a chorus line in which each dancer is doing the same steps--it's more like a minuet where each partner does his or her own dance and meets to touch only when required.

Procedural Bases of Theoretical Questions

One question that arises with some frequency in discussions of particular behavior techniques is whether or not they derive from or depart from procedures as defined in a laboratory context. In a very real sense this is an important question inasmuch as one can hardly determine the efficacy, validity, and reliability of a particular method if the applications vary widely. Fads among psychologists who indulge in bandwagon-leaping behavior are notorious. Witness, for example, the surge of interest in programmed instruction a few years back or, of even greater intensity, the enthusiastic (if uncritical) adoption of verbal conditioning so that every clinician seemed to have been born with a Greenspoon in his mouth. Procedures derived from the laboratory in these applications were so drastically altered as they were applied that they were barely recognizable at times and sometimes, in World War I submarine parlance were Spurlos gesunken, sunk without a trace. The serious consequence of this approach was that individuals who did not properly apply the procedures were then free to declaim they didn't work.

[1] A recent series of articles on the Piltdown forgery appeared in the New Scientist beginning with L.H. Matthews' article in the April 30, 1981 issue.

Such circumstances have made me a purist regarding the application of techniques. To discuss operant procedures and to assess their applicability, we should first agree upon what these procedures are. To that end, let me review basics, using as a focus Goldiamond's operant paradigm (1962) slightly modified (Bachrach, 1980a) (Figure 1). Goldiamond stated that Dollard and Miller (1950) listed the four variables of learned behavior as drive, response, cue, and reinforcement. He further observed that these variables were, in effect, identical to the variables subsumed under operant procedures as depicted in his paradigm.

Presenting a discriminative stimulus (S_D) in the presence of other constant stimuli (SS_C) is expected to occasion a response (R). Whether this R occurs or not is contingent upon the consequences (S^r) of that R under these specific conditions as well as the state variables (usually called "needs" and "motives" as part of the subject's reinforcement history) that make the consequences of the response effective in controlling it. State variables are difficult to determine and are "always inferred from the subject's history."

The S_Ds are those stimuli to which the experimenter wishes to have the subject respond, discriminating these stimuli from the world of constant stimuli (SS_C). Those stimuli to which the discriminating subject is not to respond are classed under a general rubric of S (S delta).

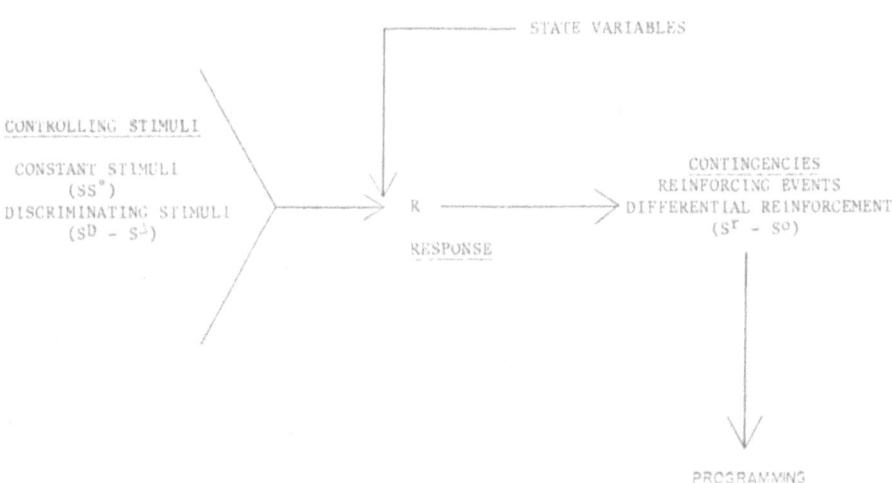

Figure 1. Goldiamond's (1962) operant paradigm modified by Bachrach (reproduced with permission from Comprehensive Textbook of Psychiatry, III, edited by H. Kaplan, A. Freedman, and B. Sadock. Baltimore: Williams & Wilkins Co., 1980)

Although these observations have centered around operant procedures, they reflect one behavioral approach which relates to the statement of Ullmann and Krasner (1965, p. 24) that, despite differences in approaches and techniques, all behavior modification can be viewed as "procedures utilizing systematic environmental contingencies to alter the subject's response to stimuli."

Given such controlled conditions and procedures, operant conditioning and other rigorous behavior modification techniques are generally highly successful. The wide range of productive efforts in the treatment of disease processes by behavioral means is clearly demonstrated in this Symposium.

GENERALIZATION

In the operant paradigm we have discussed, the assumption is made that the conditions, the stimulus complex, and the reinforcing contingencies are arranged by the experimenter, hence, the controlled condition. Skinner (1953, p. 238) used that very phrase: "arranged a sequence of events in which certain behavior has been followed by a reinforcing event." The transfer of these arrangements to a nonexperimental, nonclinical environment presupposes the role of generalization from one set of conditions to another. Even with excellent results in a controlled condition, the transfer to another situation, such as the home, is at best difficult under many if not most circumstances.

This problem of generalizing the methods and results from one stimulus condition to another is illustrated by an observation made by Keefe, Surwit and Pilon (1979) in a follow-up study of Raynaud's patients treated with behavior therapy. Success in training these patients to maintain digital temperature during a cold stress challenge was high, but a year later these patients had reverted to a baseline level. These investigators observe, "A major factor that may have been responsible for the subjects' failure to maintain the ability to deal with cold stress is that they tended to stop practicing once the initial study was completed," adding that this failure to practice behavioral techniques outside the controlled condition "is probably not surprising to any practicing clinician." (p. 390) Neither, I might add, is it surprising to the experimental scientist for whom generalization is still a methodological question.

Such a transfer of responsibility to the subject has been referred to clinically as "patient compliance," and it is a source of wonderment and dismay to the practicing clinician to see patients "who should know better" fail to observe techniques that would clearly benefit them, such as the control of peripheral vasodilation in Raynaud's patients.

When we speak of generalizing the control of the laboratory or clinic to the condition of, let us say, the home, what are we attempting to transfer? The two major components are environmental contingencies (reinforcement) and stimulus control. Here, it is especially crucial to be aware of specific conditions. The use of reinforcement, transferring the reinforcement to a self-controlled subject definition, presents problems. Stimulus control may also present problems if we are trying to replicate a stimulus complex in the home where the constant stimuli are probably greater in number, intensity, and reinforcement history. In his work with obesity, Ferster used a purple tablecloth in the home as a clear S_D to be associated with eating behavior, a technique several of us adopted with a home transfer of eating behavior control in an anorexic (Bachrach, Erwin and Mohr, 1965). Attention to stimulus control, recognizing the problems, should be a primary focus in analyzing conditions. It appears, in any consideration of literature in behavior modification, that the emphasis is usually on reinforcing contingencies. We know that reinforcement is most effective in establishing behavior, but it is probable that stimulus control is more effective in maintaining behavior.

Behavior established under certain conditions has a higher probability of recurrence under these same conditions or conditions generalized from the stimulus complex. Stimulus generalization is, indeed, basic to learning and performance. Early classification of all quadrupeds by a child as "doggies" is a generalization which, with learning, evolves to a differentiation of quadrupeds into "doggies" and "not-doggies" (cows, horses). Ultimately, there is an added discrimination among doggies into beagles, bassetts, and boxers and, if it seems important and reinforcing, into discriminating, on the basis of breeding points, good boxers from poor boxers. Learning is always a balance between generalization and discrimination. Stimulus generalization allows us the freedom to assume that two names juxtaposed on curbs, poles, or building walls are street signs when we enter a strange city; discrimination allows us to ascertain where we are (Bachrach, 1980a).

Stimulus control exercises much more control over our behavior than we may recognize. Let me present in Figure 2 a very rudimentary example of stimulus control:

You are in Room A and move to Room B to find an object you desire but, when you arrive in Room B, you have found that you have forgotten what it was you went to find so you return to Room A, where the object is not but where the stimuli that evoked the seeking behavior are (Bachrach, 1980b).

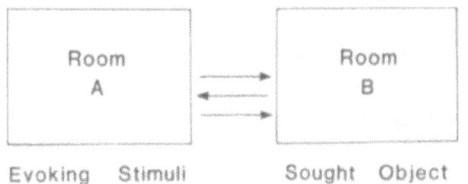

Figure 2.   Example of stimulus control

The programming of stimulus control can be an important tool in behavior modification involving, as it does, the use of stimulus complexes to aid in establishing, evoking, controlling, and maintaining behavior. It is, we have seen, basic to stimulus generalization--behavior learned under certain stimulus conditions may be more readily generalized to stimulus conditions having topographical or behavioral relationships to the original. An area that has developed in recent years is that of architectural or environmental psychology, which has an awareness of stimulus conditions so that hospital rooms, classrooms, therapy rooms, space vessels, and underwater habitats have all come under scrutiny.

The area of greatest difficulty in transferring control from the laboratory or clinic to the noncontrolled environment such as the home is the transfer of reinforcement. We have seen the term "arrange" used by Skinner to describe the establishment of behavior and its reinforcement. How is reinforcement arranged in the absence of the experimenter?

Self Control and "Self-Reinforcement"

At this point I wish to distinguish between "self-control" and "self-reinforcement." I have no doubt that there is such a behavior as self-control. It would be specious to question the fact that individuals edit or modify their own behavior. It is valid that there is a reciprocal relationship of the individual operating on the environment, which, in turn, alters the individual's behavior. The control by one's self of one's behavior is hardly free of environmental influence. Kanfer (1977, p. 4) referred to this reciprocal relationship as the major direction of self-management therapy. "A person is the product of his environment. His behavior, in turn, shapes the environment and, thus, the individual is able to modify the conditions under which he lives."

What is in question, however, is whether there is such an event as "self-reinforcement." In another comment Kanfer (1977, p.2) stated that "Self-management therapy ...involves the use of

the same person as subject and therapist," thus implying that
there is no question that self-reinforcement is possible.

The argument regarding self-reinforcement continues to be
heard, with rumblings remaining of the cannon fire between Catania
(1975) and Bandura (1976) and the rejoinders exchanged among
Mahoney (1976), Thoresen and Wilbur (1976), and Goldiamond (1976b)
related to Goldiamond's paper (1976a).

To those among you who are wondering why I sound so
compulsive about terminology, let me reaffirm my belief in the
linguistic principle expounded by Benjamin Lee Whorf--that what
you name an event determines how it will be perceived (Whorf,
1956). Thus, to use the term "self-reinforcement" appears to give
the same status to the consequential events arranged by the
subject as those arranged by the experimenter and, in the words of
that philosopher from Porgy and Bess, Sportin' Life, "it ain't
necessarily so."

Skinner expressed the question and the confusion surrounding
it:

> The place of operant reinforcement
> in self-control is not clear. In
> one sense, all reinforcements are
> self-administered since a response
> may be regarded as 'producing' its
> reinforcement. (1953, p. 237)

Another question derives from the standpoint of schedules of
reinforcement. While continuous reinforcement is usual in first
establishing behavior, the likelihood of moving toward
intermittent schedules of reinforcement is high so that the
organism's response requirement may be altered (arranged) by the
experimenter to be, for example, a fixed ratio of 10, whereby ten
responses are required to produce the reinforcement. The
topography of response-producing reinforcement is not simple.

Skinner further commented about operant behavior and self-
reinforcement:

> Self-reinforcement of operant behavior
> presupposes that the individual has it in
> his power to obtain reinforcement but does
> not do so until a particular response has been
> emitted. This might be the case if a man
> denied himself all social contacts until he
> had finished a particular job. Something
> of this sort unquestionably happens, but is
> it operant reinforcement? It is certainly
> roughly parallel to the procedure in con-
> ditioning the behavior of another person.

> But it must be remembered that the individual
> may at any moment drop the work in hand
> and obtain the reinforcement. (1953, pp.
> 237-239)

Goldiamond, with characteristic pith, observed regarding Skinner's comment about dropping the work and getting the reinforcement anyway, "Stated otherwise, he can cheat." (1976a, p. 510)

As Goldiamond[2] further notes, cheating may also happen in the laboratory but the experimenter attempts to arrange it otherwise.[3]

Skinner also spoke about the possibility of cheating; he suggested that the failure of the individual to cheat may derive from the consequential control of others, to which Goldiamond remarked, "if such consequential control is punitive, the cheating may then be accompanied by the experience of guilt." (1976a, p. 510) Guilt is an interesting behavioral event in and of itself. Reports are found of patients expressing guilt over having failed to comply with a regimen of medication or self-management. Guilt, itself, is under a measure of behavioral self-control and, to a large degree, our conditioning history suggests that, following sincere expression of guilt, expiation is very likely.   Once again, the consequential control of others plays a role.

The failure to adhere to a program when positive reinforcements (such as social contacts as reinforcers for accomplishment) are used is magnified considerably when aversive consequences are programmed. This is a finding reported by Azrin and Powell (1968), who arranged for an electric shock to be delivered by an apparatus each time a subject reached for a cigarette. The subjects consented to the experiment but, in most cases, failed to activate the apparatus they wore or did not wear it at all. Dropouts from the program were high. In this regard two considerations are not only obvious but obviously important: one is what is always loosely referred to as "motivation"; as

[2] It is apparent that Israel Goldiamond has been a major influence on my thinking over the years. Goldiamond is very much like John Hughlings Jackson--you can learn more from reading his footnotes than from reading most authors' articles.

[3] You will recall the ingenuity of rats who learned to place their feet on alternate bars of a shock grid, standing on bars of the same polarity, thereby reducing or eliminating shock delivery. This taxed the ingenuity of the experimenters who responded by developing a grid scrambler with shifting polarities.

Morris observed (1974), "such successes as have been achieved in aversive conditioning programs are to be found particularly in situations where the subjects are profoundly anxious to change the patterns of their lives." To that comment I would add that this anxiety to change the life pattern and, accordingly, comply with a behavioral program is not limited to aversive conditioning alone. The second consideration, also obvious, also important, is that of risk-taking. I believe it would be useful for behavior modification specialists to look carefully into the risk literature for, in effect, each decision made to comply or not comply in the behavioral treatment of disease is a risk-assessment by the subject. The overweight person with hypertension and the compulsive smoker assess risks. Risk-taking in our society appears to be on the rise in many ways. The Yankelovich Monitor (1979) presented a survey of trends in recreation and found that high-risk sports such as sky-diving, hang-gliding, high-speed auto racing, and scuba diving were increasing in popularity and participation. Also of interest is the finding that shoplifting among middle and upper middle class individuals seems to be on the rise as a high-risk sport. (also see Bachrach, 1978 and in press)

Durkheim (1897) stated that only when there is absolutely no chance of an individual emerging alive from an event, with the person being fully aware of that fact, can an act be defined as a suicide. The risks assumed by overeaters, smokers, drinkers, and drug users are balanced against the reinforcement received from these activities and the chance is taken. Pitted against control of such behaviors are the cultural reinforcement of excess and the delayed aversive consequences of the action.

It is apparent that in a noncontrolled situation the responding subject may also become the agent of reinforcement, determining whether or not the response requirement has been met and the reinforcement ready for delivery. Herein lies the crux of the linguistic problem of the term self-reinforcement, for the term itself implies that the agent/subject is the delivering vehicle. What is a problem and what differentiates the self-reinforcement from operant reinforcement lies "in who evaluates whether or not the response requirement for the delivery of the consequence has been met" (Goldiamond, 1976a, p. 511). Laboratory procedures are automated so that the required pressure on a microswitch (key, bar, et cetera) activates a recording device (such as a counter), which is programmed to set a response requirement and thereby activates a food hopper or other vehicle of reinforcement. The programming can be exquisitely complex but still automated. The equipment defines the response requirement. "It is this independent definition of a response as a requirement for delivery of a consequence, in a specifiable relation, that defines an operant contingency. And operant reinforcement is a contingency." (Goldiamond, 1976a, p. 512)

Equipment definition of response requirements is basic to laboratory procedures. It is also basic to self-programmed teaching machines and to biofeedback (where, for example, a digital thermometer informs the subject of successfully fulfilling the response requirement of temperature change). It is obviously possible that subjects, perceiving that they have met the response requirement as defined by the equipment, may award themselves a contingency upon the occasion of the success--but this is not, properly speaking, self-reinforcement. For many purposes, I like the term proposed by Goldiamond--self- congratulation.

Throughout this discussion, the elements of confusion and uncertainty must stand out. This, in large measure, reflects the field itself for we are, I believe, not much closer to understanding generalized or secondary reinforcement now than we were 30 years ago when Newman (1951, p. 419) observed that "one has the suspicion that if we completely understood secondary reinforcement we should understand the whole mechanism !of reinforcement!." We still don't fully understand secondary reinforcement and yet we have a general feeling that secondary reinforcement works, that approval, social contact, affection, and the like are very powerful controllers of behavior. To be sure, when we talk as Skinner did about the consequential control by others in controlling behavior, we are invoking a so-called secondary reinforcer. One has the suspicion that the behavior modification model is very dependent on the consequential control by others, in this case the experimenter/therapist who appears to have excellent success in the laboratory, a success that dims in many instances when this control is transferred to the subject.

At this point one can ask if we are any clearer on primary reinforcers and the answer is, of course, no. During SEALAB II, the Navy's underwater habitat experiment, there were some marvelous trained dolphins used in the underwater work. One of them, a curmudgeon named "Tuffy," would perform his work and swim, usually several hundred feet, to receive his fish from the diver. What was striking about this was that Tuffy swam by hundreds or thousands of equally delectable fish on his way to the diver and could have stopped to snack had hunger been a factor. At least this is what we assumed. What made that particular fish (a primary reinforcer) more "desirable" than the others was, also an assumption, that it was the interaction with the diver. Perhaps we should have done a more intensive study of the behavior at that time, the statistics were ready-made--after all, we did have a Poisson distribution.

Throughout this discussion I have tried to concentrate on the centrality of specification of procedures, in particular environmental conditions, response topography and reinforcement. These procedures must have their base in a thorough understanding

of theory, in particular learning theory, in order to be most
meaningful and effective. It is my strong belief that
practitioners who lack a firm theoretical understanding of their
procedures become technicians. Let me conclude with some specific
comments on effective procedures. (Is it not true that the
discriminative stimulus found in the phrase "let me conclude..."
sets the occasion for marked relief behavior on your part?)

I would like to leave you with yet another observation of
Goldiamond's, which deals with a semantic analysis of behavioral
interaction (Goldiamond, 1976c). Goldiamond states that so much
of the behavior we analyze is expressed in <u>intransitive</u> terms:
!the <u>child</u> learns,1
thus the responsibility appears to be placed on the child. "The
child is not learning," "the patient is resistant to therapy."
If we then add environmental variables we can transform the
event into a <u>transitive</u> one:
!The <u>teacher</u> instructs the <u>child</u>1 Another
transformation is needed, for now the responsibility is focused on
the teacher.
!The teacher's <u>procedures</u> instruct the child.1
"If the child is not learning and the teacher is not
teaching, we should look to the <u>procedures</u> used in the transaction
and change them to produce the effects desired." (Goldiamond,
1976c, p. 125) There is, however, a third transformation needed.
This transformation makes programming more specific:
!The <u>procedures</u> of the teacher program the <u>target
behaviors</u>
of the child1

In this analysis of programming procedures Goldiamond uses
the term "target behaviors" of the subject. In programmed
self-instruction, it is necessary to specify some target
performance toward which the program is directed. Buying a
program to learn a language, for example, specifies the target
performance to be increased fluency in that particular tongue. He
notes that a program states a positive goal of accomplishment and
does not begin with a <u>deficit</u> orientation.

The target, then, becomes the outcome behavior of the program
and should be specified and agreed upon by the teacher and the
student or by the experimenter/therapist and the subject. The
term "therapeutic contract" has gained currency in characterizing
such an agreement in which the behaviors of both parties,
therapist and subject, are stated specified as <u>measurable</u>
outcomes, not in some vague fashion such as "improvement."
Specific weight loss can be measured or, perhaps, more
reinforcing, is the ability to get into a specific size bikini.

Specifying target behavior necessitates an assessment of what Goldiamond calls the <u>current relevant repertoire</u>, what skills or behaviors currently available to the subject may be brought into responding. This current relevant repertoire, is an <u>entry</u> behavior into the program and defines, through the use of the cataloguing technique common in behavior modification, what responses are within the subject's repertoire.[4]

The next focus of the programming procedures is on the <u>sequence of change steps</u>, steps which will be programmed to mediate between the <u>current relevant repertoire</u> and the <u>target</u> behavior. A step is a specified, defined behavioral requirement—a response. As Goldiamond notes, "The step itself consists of a behavioral requirement which either differs somewhat from the requirement at the preceding step (shaping) or is identical to the preceding requirement under different stimulus control (fading), or both" (Goldiamond, 1974, p. 27).

Following this analysis and programming comes the stage of <u>response-contingent consequences</u> or <u>progression-maintaining consequences</u>. Completing a step is part of a response chain, providing a reinforcing contingency for the step accomplished (S^{r+}) as well as a discriminative stimulus (S_D) for the step to follow. Holland and Skinner (1961) in the introduction to their programmed text quote from the volume of Thorndike and Gates published thirty years before:

> If, by a miracle of mechanical ingenuity
> a book could be so arranged that only to
> him who had done what was directed on page
> one would page two become visible, and so
> on, much that now requires personal
> instructions could be managed by print.

Certainly, teaching machines and programmed texts attempt to provide the response-contingent visual situation of which Thorndike and Gates spoke.

Chaining as a behavioral event is effective and real. It is, in effect, smoothing a series of stimulus-response steps into a fluid sequence of performance. Steps in addressing a golf-ball characterize the initial awkwardness of movement in

---

[4]The term "repertoire" was selected by Goldiamond very wisely for as he notes (Goldiamond, 1974, p. 6) learning can affect present and future behavior, as in learning to read. Moreover, <u>repertoire</u> includes many behaviors not apparent, as in the repertoire of a stage company which has a number of plays possibly to be performed but not actively engaged engaged in at that moment. What an individual can <u>draw upon</u> is behavioral repertoire to be catalogued.

learning each step: holding the head properly, adjusting the arms, eye movement, and all the stages with which many of you are familiar. Ultimately, the swing is fluid, telescoping all the S-R steps into one perceived motion. The progression of the program, properly specified in measurable steps, should be reinforcing and should provide a maintaining consequence for advancement. We found this to be so in the case of the anorexic I mentioned (Bachrach et al., 1965) in which the recovery of hair, no longer brittle and now capable of being combed, was verbalized as a crucial sign of progress by the subject patient.

Procedures, then, are in need of the most careful attention in planning and executing a program in behavior modification, and these procedures must be based in sound theoretical understanding.

REFERENCES

Azrin, N. H., & Powell, J. Behavioral engineering: The reduction of smoking behavior by a conditioning apparatus and procedure. Journal   of Applied Behavior Analysis, 1968, 1, 193-200.

Bachrach, A. J. (Ed.). Experimental foundations of clinical psychology. New York: Basic Books, 1962.

Bachrach, A. J. Psychophysiological factors in diving. Weekly Update: Hyperbaric and Undersea Medicine, 1978, 1:2-7, Princeton, N.J.: Biomedia, Inc., 1978.

Bachrach, A. J. Learning theory. In H. I. Kaplan, A. M. Freedman & B. J. Sadock (Eds.), Comprehensive Textbook of Psychiatry, III. Baltimore: Williams & Wilkins, 1980a, p. 384.

Bachrach, A. J. Psychological research: an introduction (4th ed.). New York: Random House, 1980b.

Bachrach, A. J. The human in extreme environments. In A. Baum & J. E. Singer (Eds.), Advances in environmental psychology. New York: Lawrence Erlbaum Associates, in press.

Bachrach, A. J., Erwin, W. J., & Mohr, J. P. The control of eating behavior in an anorexic by operant conditioning techniques. In L. Ullmann & L. Krasner (Eds.), Case studies in behavior modification. New York: Holt, Rinehart & Winston, 1965, pp. 153-163.

Bachrach, A. J., & Quigley, W. A. Direct methods of treatment. In I. A. Berg & L. A. Pennington (Eds.), Introduction to clinical psychology (3rd ed.). New York: Ronald Press, 1966. Bandura, A. Self-reinforcement: theoretical and methodological considerations. Behaviorism, Fall, 1976, 4(2), 135-155.

Catania, A. C. The myth of self-reinforcement. Behaviorism, 1975, 3, 192-199.

Dollard, J. E., & Miller, N. E. Personality and psychotherapy. New York: McGraw Hill, 1950.

Durkheim, E. Le suicide. Paris: Alcan, 1897. Translated by
    J. A. Spaulding & G. Simpson, Glencoe, Ill: Free Press,
    1951.
Goldiamond, I. Perception. In A. J. Bachrach (Ed.),
    Experimental foundations of clinical psychology. New York:
    Basic Books, 1962, p. 295.
Goldiamond, I. Toward a constructional approach to social
    problems. Behaviorism, 1974, 2(1), 1-84.
Goldiamond, I. Self-reinforcement. Journal of Behavioral
    Analysis, 1976a, 9, 509-514.
Goldiamond, I. Fables, armadyllics and self-reinforcement.
    Journal of Applied Behavior Analysis, 1976b, 9, 521-525.
Goldiamond, I. Coping and adaptive behaviors of the disabled.
    In G. L. Albrecht (Ed.), The sociology of physical
    disability and rehabilitation (Chapter 5). Pittsburgh:
    University of Pittsburgh, 1976c, pp. 97-138.
Holland, J. G., & Skinner, B. F. The analysis of behavior. New
    York: McGraw Hill, 1961.
Kanfer, F. H. The many faces of self-control, or behavior
    modification changes its focus. In R. B. Stuart (Ed.),
    Behavioral self-management. New York: Brunner/Mazel,
    1977.
Keefe, F. J., Surwit, R. S., & Pilon, R. N. A 1-year follow-up
    of Raynaud's patients treated with behavioral therapy
    techniques. Journal of Behavioral Medicine, 1979, 2(4),
    85-391.
Mahoney, M. J. Terminal terminology: a self-regulated response
    to Goldiamond. Journal of Applied Behavior Analysis, 1976,
    9, 515-517.
Matthews, L. H. Piltdown man: The missing links. New
    Scientist, 1981,90, 280-282.
McConnell, J. V. Stimulus/response: criminals can be
    brainwashed--now. Psychology Today, 1970, 3, 14-18.
Morris, as cited in Goldiamond, I. Toward a constructional
    approach to social problems. Behaviorism, 1974, 2(1),
    1-84.
Newman, E. Learning. In H. Helson (Ed.), Theoretical
    foundations of psychology. New York: Van-Nostrand, 1951.
Skinner, B. F. Science and human behavior. New York:
    Macmillan, 1953.
The Yankelovich Monitor. Trend No. 44, Flirtation with danger.
    New York: Yankelovich, Skelly & White, Inc., 1979.
Thoresen, C. E., & Wilbur, C. S. Some encouraging thoughts
    about self-reinforcement. Journal of Applied Behavioral
    Analysis, 1976, 9, 518-520.
Ullmann, L., & Krasner, L. (Eds.). Case studies in behavior
    modification. New York: Holt, Rinehart & Winston, 1965.
Whorf, B. L. The relation of habitual thought and behavior to
    language. In J.B. Carroll (Ed.), Language, thought, and
    reality.   Selected   writings of Benjamin Lee Whorf.

Cambridge, MA: The Technology Press, Massachusetts
Institute of Technology, 1956, pp. 134-159.

ACKNOWLEDGEMENTS

The opinions and assertions contained herein are the private
ones of the writer and are not to be construed as official or
reflecting the views of the Navy Department or the Naval Service
at large.

The author expresses his appreciation to Mary M. Matzen,
Doris N. Auer and Regina E. Hunt for their editorial help.

BEHAVIORAL MEDICINE: THE BIOBEHAVIORAL PERSPECTIVE

Stephen M. Weiss
National Institutes of Health
Bethesda, Maryland
Gary E. Schwartz
Departments of Psychology and Psychiatry
Yale University
New Haven, Connecticut

As chronic disease has become the major health concern for Americans, the need to develop more relevant research and treatment models for disease prevention and control has led to multidisciplinary investigations of what appear to be multifactorial problems. Such explorations have led to consideration of biobehavioral research designs which emphasize pooling of the unique talents of the biomedical and behavioral research communities, resulting in a "whole which is greater than the sum of its parts." Implications of this biobehavioral approach to theory, methods, practice and training are discussed.

I. Introduction

The advances in biomedicine against infectious diseases over the past forty years in the United States and the current prevalence of chronic diseases have resulted in a major shift in health research emphasis. Cardiovascular disease and cancer, in particular, have become the major foci in the biomedical research arena, with accompanying reallocations of resources and manpower. The search for the pathogenic agent, so successful in combatting the acute infectious diseases, has failed to explain such health problems as high blood pressure, coronary heart disease, and cardiac arrhythmias. The traditional risk factors for cardiovascular disease (smoking, high blood pressure, obesity, diabetes, elevated serum cholesterol) have accounted for less than half of the variance associated with myocardial infarction (Keys, Aravanis, Blackburn, VanBuchen, Buzina,

Djordjevic, Fidanza, Karugnen, Mennotti and Taylor (1972).
Multifactorial hypotheses are therefore being explored;
environmental and personality variables are being considered
along with the more traditional physiological components.

The impetus for a more comprehensive approach to disease
prevention and control has come from several sources. The HEW
Forward Plan for Health (1975) and the Canadian "LaLonde Report"
(LaLonde, 1974) have both identified the need for expanded
research on the role of behavioral and lifestyle factors in the
prevention and control of chronic disease. The conditions in
which seminal theory and receptive environments combine to
foster a major paradigm shift (Kuhn, 1962) have been exemplified
in the emergence of behavioral medicine.

The founding of behavioral medicine in a formal sense can
be dated to the Yale Conference on Behavioral Medicine held in
February, 1977 (Schwartz & Weiss, 1978; Schwartz & Weiss, 1978).
The evolution of behavioral medicine, however, spans the entire
spectrum of the debate concerning the relationship of mind and
body, beginning with the writings of Homer, Plato and Aristotle
and continuing with Descartes, Spinoza, Sir William Osler,
Claude Bernard and Walter Cannon into more modern times. The
most recent 20th century effort to provide yet another
perspective to the mind-body linkage emerged from psychoanalytic thinking in the form of psychosomatic medicine some
forty years ago (Leigh & Reiser, 1977).

This extensive history notwithstanding, the greatest
deterrents to establishing meaningful conceptual, theoretical
and ideological linkages between the behavioral and biomedical
sciences have been encountered in the paucity of research
methodology and communication models acceptable to all relevant
scientific disciplines. The recent emergence of multifactorial
approaches to the pathogenesis of chronic disease has
unquestionably facilitated conceptual development in behavioral
medicine. Understanding the role of environmental and
behavioral factors as synergistic, catalytic, instigative,
modulating, and mediating agents in the complex physiologic and
biochemical reactions that result over time in biological system
dysfunction has become a significant challenge to the biomedical
and behavioral research communities. Efforts to broaden the
attack on chronic disease have identified many nontraditional
paths for potentially fruitful exploration. For example, diet,
exercise, stress reduction, weight control, smoking behavior,
and strategies for compliance and adherence have all recently
emerged as legitimate areas of research in the prevention and
control of chronic disease.

Most attempts by biomedical and behavioral researchers to independently investigate these common problems have not been gratifying. Too often, in both the behavioral and biomedical fields, scientists have made serious, disqualifying errors or omissions resulting from an insufficient knowledge of concepts outside of the traditional frameworks of their own disciplines. It has become increasingly obvious that successful research paradigms for chronic disease states require collaborative relationships among professionals with the collective expertise necessary to understand all of the multiple dimensions inherent in such research. Familiarity with one another's terminology, concepts, and perspectives of the many biomedical and behavioral disciplines involved has become a prerequisite to successful interactive efforts to model the combination of biobehavioral circumstances responsible for the development of chronic disease.

II. Multidisciplinary Investigations of Multifactorial Problems

A. Integrative Efforts in the Biomedical and Behavioral Sciences. Within the biomedical sciences themselves, new fields have emerged which have successfully integrated traditional disciplines such as biology, chemistry, neurology, physiology, and endocrinology; such fields as biochemistry, neurochemistry, and neurophysiology are a few examples. Through interactive efforts, these fields have integrated methods and theories to produce unique approaches to problems which could not be attempted by any one field by itself.

A similar trend has occurred within the behavioral sciences, although, due to the relative infancy of these disciplines, the illustrations are fewer. Separate disciplines such as psychology, anthropology, and political science have merged to create new fields such as psychological anthropology and political psychology. It is important to realize that even within a single discipline such as psychology, sub-specialties that were once seen as unrelated, even antithetical, are now being joined. Of special relevance to behavioral medicine, the new area of cognitive behavior modification is but one example of two different traditions being meaningfully integrated and advanced (Mahoney, 1977). The more classic integrative areas of physiological psychology, psychophysiology, and psychosomatic medicine have been augmented by new journals and research societies interested in biological psychology, neuropsychology, behavioral genetics, psychopharmacology, health psychology, biological psychiatry, and behavioral neurology. Wilson (1975) has written forcefully of the need for a "sociobiology," while more complex terms, such as "biopsychosocial," are now used by researchers such as Engel (Engel, 1977).

B. <u>Definition of Behavioral Medicine</u>. The field of behavioral medicine is maturing within this broad effort to better integrate the behavioral and biomedical sciences. As an illustration of the field's potential for dynamic growth, the definition of behavioral medicine developed by the participants at the Yale Conference (Schwartz & Weiss, 1978) has already been refined by a "second generation" definition proposed and adopted at the organizational meeting of the Academy of Behavioral Medicine Research in April, 1978 (See the journal <u>Psychoneuroendocrinology</u>).

<u>Yale Conference Definition</u>: Behavioral Medicine is the field concerned with the development of behavioral-science knowledge and techniques relevant to the understanding of physical health and illness and the application of that knowledge and these techniques to prevention, diagnosis, treatment and rehabilitation. Psychosis, neurosis, and substance abuse are included only insofar as they contribute to physical disorders as an end point (Schwartz & Weiss, 1978; Schwartz & Weiss, 1978).

<u>Academy of Behavioral Medicine Research Definition</u>: Behavioral Medicine is the interdisciplinary field concerned with the development and integration of behavioral and biomedical science knowledge and techniques relevant to the understanding of health and illness and the application of this knowledge and these techniques to prevention, diagnosis, treatment and rehabilitation (Schwartz & Weiss, 1978).

In comparing the two definitions, it is apparent that the emphasis has shifted from the contribution of the behavioral sciences to biomedicine to the integration of the biomedical and behavioral sciences -- a pooling of talent and diverse perspectives to capitalize on the interactive potential inherent in collaborative research efforts. Such integration may include (but should not be limited to) contributions from many disciplines such as

anthropology	health education
physiology	sociology
psychology	psychiatry
clinical and preventive medicine	neurobiology
(including cardiology,	biochemistry
nephrology, oncology,	epidemiology
neurology, family	
medicine, etc.)	

Recognizing this multidisciplinary mix of disciplines as the essential ingredient to developing "biobehavioral" paradigms, it is obvious that behavioral medicine itself is a "field of endeavor" rather than a "discipline" in its own right. One might think of behavioral medicine as a conceptual "crucible" which provides a forum for the many disciplines involved to undertake the process of jointly exploring problems of common concern. Thus, the challenge facing behavioral medicine is to establish the value of these "interactive" research hypotheses, designs, and methods in addressing central problems of health and illness. Such models already exist, but their significance is not widely appreciated.

C. Interactive Statistical Models. The traditional analysis of variance model (Winer, 1962) demonstrates mathematically how two independent variables (e.g. A and B) may each seem to have little effect on a given dependent variables (e.g. X) when examined independently, yet they may have significant synergistic, or sometimes even opposite, effects when examined concurrently (See Figure 1). Thus, specific combinations or patterns of A and B determine the behavior of X. Clearly, studies of only a single factor (e.g. A) with random variance occurring in the other (B), can lead to simplistic and erroneous conclusions regarding the role that each plays in the conduct of X. Although analyses of variance are widely used in the behavioral and biomedical sciences, the variance due to interaction is often viewed as an annoyance rather than as meaningful data.

Although one might ideally wish to find simple "one-to-one" relationships between independent and dependent variables, the systems under study are typically integrated and very complex. With each arithmetic increase in the number of subsystems composing the system under study, there is a corresponding geometric increase in the number of possible interactions among each of the subsystems that can contribute to the functioning of the system as a whole. The concept that the "whole is greater than the sum of its parts, yet is dependent upon the interactions among the parts for its unique properties" is a basic tenet of systems theory (Schwartz, 1979). This principle implies that it is not possible to examine subsystems in isolation if the complex behavior of the system as a whole is to be understood.

Applied to behavioral medicine, it follows that approaches examining the interactive effects of biomedical and behavioral variables will account for greater proportions of the variance related to understanding chronic disease than independent investigations by biomedical and behavioral scientists. Advanced multivariate analyses combined with highly sophisticated information processing systems permit us to quantitate the potential synergism inherent in such interactive efforts, and thus demonstrate the unique capability of the "behavioral medicine" model.

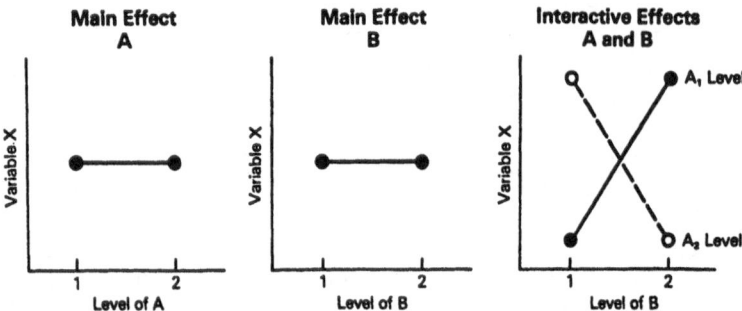

Figure 1.   Two independent variables, A and B, each of two
            levels, are examined for their effects on
            dependent variable X. The particular example is
            is selected to illustrate how in certain sets of
            data, it is possible for low levels in both A
            and B. (A1B1 combination) or high levels in both
            A and B (A2B2 combination) to be associated with
            low levels of X. However, if A is low but B is
            high (A1B2) or A is high but B is low (A2B1),
            high levels of X may occur. This sort of inter-
            action (a form of the inverted U-shaped function
            for two variables) is not uncommon in behavioral
            or biomedical data. These interactive effects
            of A and B on X can be uncovered by graphing the
            effects on X as a function of both levels of A
            and B (the interactive effects graph). The
            important point in this example, illustrated in
            the graphs depicting the main effects of A and
            B alone, is that research designs which manipu-
            lating only A (ignoring B) or manipulated only B
            (ignoring A) could come to the erroneous conclu-
            sion that A had no effect on X, or B had no
            effect on X. Interactive biobehavioral research
            designs that manipulate behavioral and
            biomedical variables concurrently have the
            potential to uncover important biobehavioral
            interactive effects.

D. <u>Interactive Research in Behavioral Medicine</u>. The potential fruits of interactive biobehavioral research designs are beginning to appear in the behavioral medicine literature. As one example, studies of blood pressure regulation have received extensive biobehavioral consideration (e.g. Weiner, 1977), and recent examples (e.g. Schwartz et al., 1979) illustrate the application of systems theory toward integration of behavioral and biomedical approaches to this multifactorial problem. In the past, such studies were typically conducted without regard to either the genetic or diet history of the subjects. However, recent studies in animals (Friedman & Iwai, 1977) indicate that the pressor effects of psychosocial stress are potentiated if the subjects are (a) genetically predisposed to respond to sodium in the diet, and (b) sodium is present in the diet. The presence of these interactive effects suggests that while the genetic and salt mechanisms may be different from the psychosocial mechanisms, the interaction of the different mechanisms may produce unique, emergent properties. A similar finding for caffeine has been observed by Henry (Henry & Stephens, 1980).

We predict that many sources of apparent confusion and random variability in health research will be clarified and resolved as interactive theories, methods, and designs are developed. Such interactive models may also clarify existing confusion and contradictory data noted in the research on personality correlates of essential hypertension. Esler, Julius, Zweifler, Randall, Harburg, Gardiner and DeQuattro (1977) proposed that essential hypertension as currently diagnosed is not a single disorder reflecting a single mechanism, but rather it represents a heterogeneous population consisting of many multifactorial disorders. In light of the tremendous complexity of the cardiovascular system, it should not be surprising to discover that numerous processes, in interaction, may be differentially involved in what may appear at the periphery to be a single problem or symptom (Schwartz et al., 1979).

Esler et al., for example, divided patients with essential hypertension into five groups according to the extent to which neurogenic factors were found to be important to the maintenance of their high blood pressure. When personality measures of aggression were examined, it was found that <u>only</u> the neurogenic hypertensives showed the predicted association between blood pressure and aggression. Since personality factors presumably operate through the central nervous system, it follows that personality correlates of high blood pressure should express themselves through the neurogenic mechanisms involved in raising blood pressure. Clearly, behavioral scientists interested in the contribution of personality to hypertension should be able to profit from considering new distinctions among different biological mechanisms underlying high blood pressure. In fact, it may be of questionable value to continue administering personality tests to hypertensive patients without considering the more subtle interactions with different biological mechanisms.

This same lesson applies to biomedical researchers interested in applying state-of-the-art knowledge in personality research. In the Esler et al. data, personality measures of aggression were associated with high blood pressure in neurogenic hypertensives, whereas personality measures of anxiety were not. Should one conclude, therefore, that only aggression is important to neurogenic hypertension?

Recent personality research indicates that the reasons (mechanisms) underlying subjects reporting of low scores on anxiety scales may not necessarily be the same. Using a second personality scale as a moderator variable (the second scale measuring defensiveness), subjects reporting low anxiety and low defensiveness were labelled by Weinberger, Schwartz and Davidson (1979) as "true low anxious," while subjects reporting low anxiety and high defensiveness were labelled "repressors." Note that in the same way the neurogenic factor was used to divide patients with high blood pressure into hypertensive subgroups, the defensive factor was used to divide patients reporting low anxiety into anxiety subgroups. Using this procedure, Weinberger et al. documented that in response to a moderate laboratory stress, subjects scoring low in both anxiety and defensiveness (true low anxious) showed less sympathetic arousal than subjects scoring high in anxiety (high anxious). Those subjects scoring low in anxiety but high in defensiveness (repressors) showed levels of sympathetic arousal equal to, if not higher than, the subjects scoring high in anxiety!

At this time, one can only speculate whether future studies interpreting the new personality differentiations with the new hypertension differentiations will result in better understanding of their interrelationships (e.g. do neurogenic hypertensives who report low anxiety also report high defensiveness, whereas non-neurogenic hypertensives who report low anxiety report low defensiveness?). The challenge for future research is to apply state-of-the-art concepts in the relevant fields in such a way as to promote the discovery of interactive effects. These discoveries will more likely occur if specialists in the relevant disciplines work together to develop multifactorial research designs capable of identifying such effects.

The above research in hypertension is but one illustration of the need for interactive models of biobehavioral research. Current research linking coronary-prone behavior with coronary heart disease is coming to a similar conclusion. For example, recent work by Jenkins, Zyzanski and Rosenman (1978) suggests that different subsets of Type A behavior are related to different types of coronary heart disease. Again, such findings can only emerge if multifactorial research designs and analyses account for both the behavioral and biomedical aspects of the problem. This approach has important implications for theory, methods, practice and training in all areas of behavioral medicine.

E. <u>Developing an Interactive Perspective</u>. How can the theoretical perspectives of the various behavioral and biomedical disciplines be brought together into interactive, biobehavioral models? It seems likely that only through close working relationships on research problems of common concern can such understandings take place. Problems currently exist in the use of terminology where shades of meaning differ unknowingly between behavioral and biomedical scientists. For example, differences in terminology used in psychology and pharmacology are apparently causing inadvertent controversy and confusion (Schwartz et al., 1979). Psychologists might state that operant conditioning had "direct effects" on blood pressure, meaning that the effects were relatively specific to blood pressure when compared with the other ANS-mediated responses. However, for the pharmacologist, the term "direct" has a specific meaning referring to the site of action of the drug (or agent) on the tissue in question. By this definition, operant conditioning of blood pressure must involve a complex set of "indirect effects." Furthermore, by pharmacologic definition, the word "effect" implies that the response is indirect. Hence, a pharmacologist would never use the words "direct effect" together.

The term "placebo" is another example. To a behavioral scientist, the term "placebo" generally refers to certain nonspecified components of a complex psychological process (involving implied "expectancy" and "set" on the part of the experimenter and subject). Using this terminology in the biofeedback situation, the subject's expectancies might be labelled "placebo factors," while the more specific aspects linked to the contingency of the biofeedback would be labelled "active ingredients." However, to a pharmacologist, the term "placebo" specifically refers to all factors involved in the administration of a drug that are not due to the direct action of the drug itself. From this definition, <u>all</u> behavioral aspects fall under the category of "placebo effects." Hence, if behavioral and biomedical scientists argue whether or not a specific behavioral factor is a "placebo" effect, the problem may lie in part in the different implicit meanings attached to the term "placebo."

Clearly, there is a need for behavioral and biomedical scientists to learn the precise meanings of one another's terminology if new understanding and models are to be developed. For example, in pharmacology there are a series of terms used to refer to different kinds of "interactions" between combinations of drugs or between drugs and organisms (See Schwartz et al., 1979). Pharmacologists describe, for example, "synergistic" effects of drugs, which are effects produced when the drugs are given in combination, and which are qualitatively greater than the simple sum effects produced by drugs taken individually. Behavioral scientists can profit from learning how such concepts can provide a framework for new models that include interactions of behavioral and pharmacologic agents. In other words, the pharmacologic model is not "antibehavioral" per se, but rather it may be expanded by appropriate concepts from the behavioral sciences towards a more comprehensive, biobehavioral analysis.

Such collaboration may reduce the likelihood of perpetuating mind/body or behavior/biology dichotomies traditionally characteristic of the behavioral and biomedical sciences. For example, it was recently demonstrated that a subgroup of patients responding to a "placebo" drug were found to show changes in endorphins in the brain (Levine, Gordon & Fields, 1978). Based on these findings, the authors concluded that such effects were not "placebo" effects, but rather they were neurophysiological! This conclusion is unfortunate because it fails to recognize that "psychological" (e.g. placebo) variables produce peripheral "physiological" effects via the central nervous system. One might more accurately describe the phenomenon in biobehavioral terms, where behavioral inputs elicit neuropsychological processes that modulate peripheral physiological responses.

These issues apply to all aspects of health and illness, from etiology and pathogenesis to treatment and prevention. As noted by Shapiro, Schwartz, Ferguson, Redmond and Weiss (1977), the concept of "phases" in clinical pharmacology research can be readily translated into more generic terms which apply equally to behavioral and biomedical interventions. With common terminology, it becomes easier to compare and contrast the relative progress made to date in understanding behavioral and biomedical intervention strategies.

F. Treatment Research Issues. The interactive approach of behavioral medicine suggests certain side benefits directed toward improving the quality of care and the promotion of health. In the treatment of high blood pressure, for example, preliminary research (Patel, 1977) suggests that the combination of drugs with relaxation training results in significant decreases in blood pressure with reduction or elimination of medication. Two interpretations are possible: (a) the drugs potentiate the relaxation effects; (b) the relaxation training potentiates the drug effects. In either case, this combination provides more effective therapy than could be provided by either biomedical or behavioral treatment alone.

Treatment procedures such as "biofeedback" or "relaxation training" do not at present adhere to a standardized protocol or consensually agreed-upon guidelines, which makes comparison between studies somewhat tenuous. For example, the considerable variation in the length and number of training sessions across biofeedback research studies concerning blood pressure reduction (Green, Green & Norris, 1979) almost certainly overpowers specific treatment effects, particularly when one assumes that the development of high blood pressure takes an extended period of years. It is naive to assume such long standing patterns can be reversed in the course of five to twelve one- hour weekly sessions (average training time for most biofeedback research studies). The most successful program to date has employed goal-limited rather than time-limited training, which has provided additional insights into both "person-treatment" compatibility and differences in responsivity across individuals to a given program (Green, Green & Norris, 1980).

The interaction between patient and health care provider may also have a profound effect on the efficacy of treatment, which is still not widely appreciated. Studies investigating the effects of a particular drug or procedure attempt to "hold constant" all sources of independent variation, except for the variable under study. The interactive model stipulates that the combinations of independent variables may produce a synergistic (or inhibiting) effect which cannot be identified by studying one variable at a time, all others being held constant. Such multivariate analyses must also include systematic variation of the "provider" variable, so the true complexities of the "patient-provider-treatment" paradigm can ultimately be better understood in terms of enhancing treatment efficacy.

G. Methods. Statistical and practical considerations are raised by recognition of the advantages of multifactorial, interactive designs. Typically one thinks of large numbers of groups and subjects, followed by the collection of masses of data requiring complex interpretations. Fortunately, new multivariate statistical techniques including multiple regression and multivariate analyses of variance (Kerlinger & Pedhazur, 1973) are being developed to reduce the necessity of doing complete factorial experiments. Progress is being made in

developing other procedures which might be fruitfully extended
to the problems inherent in behavioral medicine, including
modern techniques of pattern recognition and profile analysis.
Technological advances in instrumentation for measuring
biobehavioral responses have reached new levels of
sophistication. The challenge is to bring these new techniques
to bear on the design and interpretation of behavioral medicine
research.

H. <u>Training</u>. The range of knowledge required to
comprehend the theories, concepts, and technology associated
with the many disciplines contributing to the interdisciplinary
field of behavioral medicine is too broad to be found in any one
individual. Therefore, training experiences should have the
goal of bridging conceptual and communication gaps so that
effective collaboration is possible. Trainees from the
biomedical and behavioral disciplines should become conversant
with the principles, theories, concepts, and terminology of one
another's fields. In the course of such experiences, increased
familiarity within the behavioral and the biomedical sciences
will also facilitate the goals of biobehavioral understanding,
e.g. psychologists will become more familiar with the
perspectives of anthropology and sociology as well as with the
tenets of pharmacology and physiology.

Such research training programs are already underway at
various medical schools and universities across the country
(Weiss, 1978). The development of complementary programs in
clinical training will also require clinicians to become
conversant with other fields related to behavioral medicine.
The "clinical team" concept is hardly new to medicine. The
expansion of this team to include relevant biobehavioral
expertise will have direct implications for the expansion of the
prevention as well as the treatment capabilities of the health
care system as we know it today.

III. <u>Conclusion</u>

How is behavioral medicine different from other fields such
as psychosomatic medicine and holistic medicine? Perhaps the
most cogent difference is the commitment to develop integrative
models consistent with the best of biomedical and behavioral
research designs at the chosen level of investigation. Review
processes will demand excellence from both biomedical and
behavioral perspectives; joint expertise in research design and
implementation will be the <u>sine qua non</u> of behavioral medicine
research. Evaluation of the outcomes of such endeavors will be
subject to rigorous appraisal. In essence, the necessity for
satisfying the criteria established for all disciplines will

encourage a more ready acceptance of findings by the involved research communities and ultimately by the health care system.

Because behavioral medicine research has been developed by the key source for the nation's biomedical research effort - the National Institutes of Health - the scientific community can be assured of adequate and competent biomedical as well as behavioral scientific review. The NIH is also committed to facilitating the transfer of successful research efforts to the relevant sectors of the health care system. The final judgement concerning the value of behavioral medicine will depend upon the ability of those adopting the biobehavioral model to effectively contribute toward solutions to the highly complex issues surrounding the prevention and control of chronic disease.

REFERENCES

Engel, G. L. The need for a new medical model: A challenge for biomedicine. Science, 1977, 196, 129-136.
Esler, M., Julius, S., Zweifler, A., Randall, O., Harburg, E., Gardiner, H. & DeQuattro, U. Mild high-renin essential hypertension. New England Journal of Medicine, 1977, 296, 405-411.
Friedman, R., & Iwai, J. Dietary sodium, psychic stress and genetic predisposition to experimental hypertension. Proceedings of the Society for Experimental Biology and Medicine, 1977, 155(3), 449-452.
Forward Plan for Health, FY 1977-81. Publication No. (Os 76-50024, Department of Health, Education and Welfare, Washington, D.C., 1975.
Green, E. E., Green, A. M., & Norris, P. A. Preliminary observations on a new non-drug method for control of hypertension. Journal of the South Carolina Medical Association, 1979, 75, 575-582.
Green E. E., Green, A. M., & Norris, P. A. Self regulation training for control of hypertension: An experimental method for restoring or maintaining normal blood pressure. Primary Cardiology, 1980, 6, 126-127.
Henry, J. P., & Stephens, P. M. Caffeine as an intensifier of stress induced hormonal and pathophysiologic changes in mice. Pharmacology, Biochemistry and Behavior, 1980, 13, 719-727.
Jenkins, C. D., Zyzanski, S. J., & Rosenman, R. H. Coronary prone behavior: One pattern or several. Psychosomatic Medicine, 1978, 40, 25-43.
Kerlinger, F. N., & Pedhazur, E. J. Multiple Regression in Behavioral Research. New York: Holt, Rinehart, & Winston, 1973.
Keys, A., Aravanis, C., Blackburn, M., VanBuchen, F. S. P., Buzina, R., Djordjevic, B. S., Fidanza, F., Karugnen, M.,

Menotti, A., Puddu, V., & Taylor, H. L. Probability of
middle-aged men developing coronary
heart disease in five years. Circulation, 1972, 45,
815-828.

Kuhn, T. S. The Structure of Scientific Revolutions. Chicago:
University of Chicago Press, 1962.

LaLonde, M. A New Perspective on the Health of Canadians: A
Working Document (Ministry of Health and Welfare, Ottawa,
1974).

Leigh, H., & Reiser, M. F. Major trends in psychosomatic
medicine - Psychiatrists evolving role in medicine. Annals
of Internal Medicine, 1977, 87(2), 233-239.

Levine, J. D., Gordon, N. C., & Fields, H. L. The mechanisms of
placebo analgesia. Lancet, 1978, 2, 654-

Mahoney, M. J. Reflections on the cognitive learning trend in
psychotherapy. American Psychologist, 1977, 32, 5-13.

Meichenbaum, D. Cognitive Behavior Modification. New York:
Plenum Press, 1977.

Patel, C. H. Biofeedback-aided relaxation and meditation in the
management   of   hypertension.   Biofeedback   and
Self-Regulation, 1977, 2, 1-41.

Schwartz, G. E. Disregulation and system theory: A
biobehavioral framework for biofeedback and behavioral
medicine. In N. Birbaumer & H. D. Kimmel (Eds.),
Biofeedback and Self-Regulation. Hillsdale, New Jersey:
Erlbaum, 1979.

Schwartz, G. E., Shapiro, A. P., Redmond, D. P., Ferguson, D.C.
Ragland, D. R., & Weiss, S. M. Behavioral medicine
approaches to hypertension: An integrative analysis of
theory and research. Journal of Behavioral Medicine, 1979,
2, 311-363.

Schwartz, G. E., & Weiss, S. M. Behavioral medicine revisited:
An amended definition. Journal of Behavioral Medicine,
1978, 1, 249-251.

Schwartz, G. E., & Weiss, S. M. Proceedings of the Yale
Conference on Behavioral Medicine Publication No. (NIH)
78-1424, Department of Health, Education and Welfare,
Washington, D.C., 1978.

Schwartz, G. E., & Weiss, S. M. Yale conference on behavioral
medicine: A proposed definition and statement of goals.
Journal of Behavioral Medicine, 1978, 1, 3-12.

See the Journal Psychoneuroendocrinology.

Shapiro, A. P., Schwartz, G. E., Ferguson, D. C. E., Redmond, D.
P., & Weiss, S. M. Behavioral methods in the treatment of
hypertension: A review of their clinical status. Annals of
Internal Medicine, 1977, 86, 626-636.

Weinberger, D. S., Schwartz, G. E., & Davidson, R. J.
Low-anxious, high-anxious, and repressive coping styles:

Psychometric patterns and behavioral and physiological responses to stress. Journal of Abnormal Psychology, 1979, 88, 369-380.

Weiner, H. Psychobiology and Human Disease. New York: Elsevier, 1977.

Weiss, S. M. News and developments: Research training in behavioral medicine. Journal of Behavioral Medicine, 1978, 1, 241-247.

Wilson, E. O. Sociobiology. Cambridge, Mass.: Harvard University Press, 1977.

Winer, B. J. Statistical Principles in Experimental Design. New York: McGraw-Hill, 1962.

BEHAVIORAL MEDICINE AND PSYCHIATRY: A DIALOGUE

Joseph Brady

Johns Hopkins University School of Medicine

Baltimore, Maryland

The current lively interest in the relationship between behavioral interactions and virtually all aspects of health and disease reflects the strong empirical and conceptual influence of experimental laboratory science. The historical roots of an extensive and expanding research literature in this behavioral medicine field can be traced to the early studies of Pavlov and Sherrington, before the turn of the century, whose work focused upon the central role of behavioral interactions in the physiological adaptations and adjustments of the internal environment (Pavlov, 1927, 1928; Sherrington, 1906). Of at least as great import, however, was the foundation provided by these early investigations, as well as by those of Beckterev (1932) in the Soviet Union and Thorndike (1898) in the United States, for conceptualizing the behavioral interactions between organism and environment within the framework of an orderly and systematic body of scientific knowledge based upon observation and experiment.

The lessons learned about behavior as a consequence of such controlled laboratory investigations have not always been warmly embraced at the clinical level, however, for reasons which appear to be somewhat unique to the human condition. Unlike other aspects of biology (e.g. anatomy, physiology, biochemistry), where behavior is concerned we bipeds at the "head of the line" harbor strong chauvinistic dispositions which have little to do with gender, race, or other personal characteristics of the species. Acceptance of experimental laboratory science analysis of behavior with "lower organisms" as instructive with respect to exalted human performance repertoires has been hard-won in the face of vigorous "higher order" resistance. It is nonetheless true that traditional appeals to unobserved and unobservable "mental processes" and other explanatory fictions, which have pervaded the fields of both medicine and psychology, must now yield to more operational analyses of clinically relevant behavioral interactions based upon detailed and objective descriptions of observable and quantifiable events.

But what, in fact, have we learned in the laboratory about the nature of those behavioral activities at the interface between individuals and their environments which may be considered relevant to medicine? At the most fundamental level, there appear to be two basic modes which characterize this interactive process. In the first instance, a reactive mode is clearly rooted in the biochemical and physiological adaptations of the organism to the influences of a changing environment (i.e., the environment acts upon the organism and the organism reacts). Since at least the time of Pavlov, this "respondent" paradigm has provided the basis for describing and experimentally analyzing increasingly more complex interactions of direct relevance to clinical medicine in general, and to cardiovascular adaptations in particular. Early "respondent conditioning" studies (e.g. Dykman & Gantt, 1958; Deane & Zeaman, 1985) provided systematic accounts of how neutral environmental stimuli (e.g. tones, lights, etc.), initially producing only minimal changes in circulatory activity, could come to elicit "conditional" cardiovascular responses (e.g. heart rate increases) of substantial magnitude and duration when paired repeatedly with "unconditional" environmental stimulus events (e.g. food or electric shock) which normally elicited such changes. If such conditional tone or light stimuli (i.e. CS) are subsequently presented a number of time without the unconditional food or shock stimuli (i.e. UCS), the magnitude and frequency of the conditional heart rate increase response (i.e. CR) elicited by the CS diminish, and "respondent extinction" occurs. When a period of time intervenes between such extinction and subsequent presentations of the CS, however, "spontaneous recovery" of the CR is observed in the form of temporary reappearance of the response elicited by the CS.

The power to elicit a CR which is developed in one CS by conditioning extends to other stimuli, with the degree of this "stimulus generalization" determined by the similarities and differences between the other stimuli and the CS. Because stimuli other than the CS differ with respect to the magnitude and frequency with which they elicit the CR, "stimulus discrimination" also occurs. Indeed, discrimination can be made increasingly more pronounced by repeated pairings of the UCS only with a specific CS (i.e. respondent conditioning), while insuring that the occurrence of other stimuli is not paired with the UCS.

These basic observations with regard to the reactive or respondent conditioning mode have been elaborated in numerous laboratory and clinical-experimental studies since Russian researchers first introduced this systematic approach to behavior analysis. It has been convincingly demonstrated, for example, that second- or higher-order conditioning can occur when a well-established CS is paired with a neutral stimulus. The neutral stimulus acquires the power to elicit the reactive CR. Although it has not been empirically determined just how far this process can be carried, the development of eliciting properties by CSs two or three steps removed from the original UCS is not uncommon. And the intensive investigative effort, principally Russian in origin, to extend the conceptual framework of such "classical" or "Pavlovian" (i.e. respondent) conditioning to encompass verbal stimuli and semantic responses (Razran, 1961) suggests potentially important directions for development of the theory and the practice of behavioral medicine.

Elicited responses of the type that have provided the primary focus for such basic and important respondent or "reflex" conditioning analyses must nonetheless be seen to represent only a relatively small proportion of the behavioral interactions of higher organisms. The most prominent aspects of such advanced repertoires are represented by the second basic, and generally more _active_ than reactive, _mode_ characterizing behavioral interactions focusing on the operations performed by organisms upon their environment (both internal and external), rather than upon their "reflex" reactions to such environmental influences.

The frequency of such actively "operant" behavior is chiefly determined by the environmental consequences of that behavior. When these environmental consequences increase the likelihood that the behavior will recur, "reinforcement" is defined. When, on the other hand, the consequence of an operant performance decreases the likelihood of that behavior recurring, "punishment" is defined. The important point to be made here is simply that reinforcement and punishment are always defined by the effects of these operations on the frequency or strength of behavioral interactions.

Over the past three decades, a broad range of animal laboratory and human experimental studies have provided important insights into the principles that determine the acquisition, maintenance, and modification of such operant behavior (Honig, 1966; Honig & Staddon, 1976). The basic observation is that the rate of an operant response already in the organism's repertoire can be readily increased by reinforcement ("operant conditioning"). Beyond this, it has been possible to make explicit the process called "shaping", whereby operant conditioning can extend existing simple responses into new and more complex performances. Of critical importance for this shaping process is the observation that a reinforcer not only strengthens the particular response that precedes it, but also results in an increase in the frequency of many other similar bits of behavior, and in effect, raises the individual's general activity level.

Thus, the shaping of behavior proceeds as reinforcers are initially presented following a response similar to or approximating the desired one. Since this tends to increase the strength of various other similar behaviors, a response still closer to that desired can be selected from this new array and can be reinforced. Continued narrowing and refinement of the response criteria required for reinforcement leads progressively to new arrays of available behavior. In this way, by successive and progressive approximation, a new and desired performance can be shaped. The importance of this simple but fundamental and powerful shaping process for the management of health-related behaviors cannot be overstated, since the weight of available evidence suggests that a careful and systematic application of such procedures with effective reinforcers is sufficient to establish or alter any operant performance of which the organism is physically capable. This shaping process is obviously of enormous clinical importance in behavioral medicine since many patient performances can only effectively be changed in this way. Without shaping, one might wait for inordinately long periods before a patient performs some critical health related behavior that could be strengthened by reinforcement.

The fact that changes in behavior are not always brought about by deliberate and systematic manipulation of the environment, however, has led to an analysis of "superstitious" behavior. A potentially reinforcing environmental event may, by chance, follow a response, resulting in the adventitious strengthening of that response. If this sequence of events reoccurs even infrequently (i.e. "intermittent reinforcement", as described below), the individual may learn quite elaborate sequences of superstitious behavior which have absolutely nothing to do with production of the event that is influencing the frequency of the behavior (e.g. the exhortations of the gambler do not produce winning dice combinations any more than native dances produce rain; they persist because they are occasionally followed by "7" or "11", in the first instance, and precipitation, in the second).

The powerful effects of reinforcement in establishing and maintaining operant behavior suggest that withholding such reinforcing consequences (i.e. "extinction") will have comparably powerful effects on the strength of previously reinforced responses. Indeed, such extinction procedures do reduce the frequency of response, although the reduction is not usually immediate. Rather, after the onset of extinction, the initial effect is often a brief increase in the frequency as well as the force and variability of the response previously followed by reinforcement. The extent to which operant responding persists in the absence of reinforcing environmental consequences (i.e. "resistance to extinction") depends, of course, on the interaction of many complex influences including motivational factors (e.g. level of deprivation). But both laboratory and clinical experimental evidence now confirm that <u>the single most important variable affecting the course of operant extinction is the "schedule of reinforcement" on which the performance was performance was previously acquired and maintained.</u>

Whenever a reinforcing environmental stimulus follows some but not all occurrences of an operant response, a schedule of intermittent reinforcement is operating. Accordingly then, "intermittent reinforcement" is defined when only selected occurrences of an operant are followed by a reinforcer. Every reinforcer occurs according to some schedule or rule, although some schedules are so complicated that detailed analysis is required to formulate them precisely. Simple schedules of intermittent reinforcement can be classified into two broad categories: ratio and interval schedules. "Ratio schedules" prescribe that a certain number of responses be emitted before one response is reinforced, the term "ratio" referring to the relationship between the required response total (e.g. 50) and the one response followed by the reinforcing event (e.g. "piece-work" schedule requiring 49 discrete labor units before the single 50th performance is followed by "payoff"). "Interval schedules" on the other hand prescribe that a given interval of time elapse before an emitted response can be followed by a reinforcing stimulus. The relevant interval can be measured from any event, but the occurrence of a previous reinforcer is usually used (e.g. "salaried" pay schedules). <u>The recuperative properties of interval schedules under which the mere passage of even long time intervals brings an opportunity for a single response to be followed by a reinforcer contrasts with the "strain" potential of high ratio schedule requirements under which the performance may extinguish before a sufficient number of responses are emitted for one to be followed by reinforcement (Rachlin, 1970).</u>

Even simple ratio and interval schedules can in turn be classified into two general categories based upon whether the required number of responses or lapse of time are "fixed" or "variable," and all known schedules of reinforcement can be reduced to variations of these basic ratio and interval parameters. A single operant performance, for example, may be followed by a reinforcer in accordance with the requirements of two or more schedules at the same time ("compound schedules"). Two or more responses may be followed by a reinforcer according to the requirements of two or more schedules at the same time ("concurrent schedules"). And perhaps the most ubiquitous case of reinforcement schedule complexity is represented by the "multiple schedule" under the requirements of which two or more independent schedules developed and maintained simultaneously are called forth sequentially under discriminably different environmental stimulus conditions. Virtually all operant behavior is followed by reinforcing stimuli according to multiple, compound, and concurrent schedules built out of the same basic elements as the simple ratio and interval schedules. Each schedule, simple or complex, generates and maintains its own characteristic performance, and when reinforcement is discontinued, the course and character of extinction are prominently influenced by the preceding schedule of reinforcement. Significantly, it has also become increasingly clear in the laboratory and the clinic that at least the frequency or rate of a given operant performance can be more effectively controlled by reinforcement schedule manipulation than by any other means.

The detailed experimental analysis of schedules of reinforcement has also served to emphasize another very important set of relationships between operant performances and environmental events encompassed within the general conceptual framework of "stimulus control." The occurrence of a reinforcer following an operant not only increases the likelihood that the response will reoccur but it also contributes to bringing that performance under the control of other environmental stimuli present when the operant is reinforced. After the responses composing the operant have been reinforced in the presence of a particular stimulus a number of times, that stimulus comes to control the operant (i.e. the frequency of those responses is high in the presence of the stimulus and lower in its absence). A "discriminative stimulus" is thus defined by this process as one in whose presence a particular operant performance is highly probable because the behavior has previously been reinforced in its presence. It is important to recognize, however, that discriminative stimuli do not elicit performances as in the respondent or reflex case, but rather set the occasion for operant responses in the sense that they provide the circumstances under which the performance has previously been reinforced. The control over driving behaviors by traffic signals occasioning vehicle braking and accelerating occurs because of systematic relationships between such performances and their consequences (e.g. fines, accidents, etc.), not because of any inherent or conditional eliciting properties of red, green, and yellow lights. This "controlling" power of a discriminative stimulus develops gradually and at least several occurrences of the reinforcer following the response in the presence of the stimulus are required before the stimulus effectively controls the performance.

Such discriminative stimulus control is not an entirely selective process, however, since reinforcement of a performance in the presence of one stimulus increases the tendency to respond not only to that stimulus but also in the presence of other stimuli with similar properties (i.e. "stimulus generalization"). It is not always clear from simple observation, of course, which stimulus or which property of a stimulus is controlling an operant performance, and both laboratory and clinical experiences have documented the hazard of assuming that the similarity casually observed between stimuli provides an adequate explanation of such generalization. There is unfortunately no substitute for experiment in differentiating the many detailed aspects of a stimulus complex which may exercise critical control. Furthermore, related "response generalization" effects have also been observed to occur when following an operant with a reinforcer results not only in an increase in the frequency of the responses composing that operant but also in an increase in the frequency of similar responses.

This very sensitivity to the differential aspects of stimulus and response complexes provides the basis for the other major cornerstone of the stimulus control process identified as "discrimination". A discrimination between two stimuli is said to obtain when an organism behaves differently in the presence of each. Such "stimulus discrimination" is pronounced under conditions which provide "differential reinforcement", and this process is seen to operate in the formation of a discrimination when there is a high probability that a reinforcer will follow a given response in the presence of one stimulus, and a low or zeroprobability that reinforcement will follow the response in the presence of another stimulus. The extent of the generalization between two stimuli will of course influence the rapidity and stability with which a discrimination can be formed, and it is important to recognize that the antecedents of a performance that occurs under one set of stimulus conditions may include events which have occurred under quite different stimulus conditions. The careful application of differential reinforcement procedures can, nonetheless, bring about remarkably precise control of an operant performance by highly selective aspects of a stimulus complex. This "attention" to specific properties of a stimulus can be facilitated and enhanced by the use of "instructional stimuli" which, in essence, tell about features of the environment which are currently relevant to the occasioning of reinforcement (e.g. a treasure map). "Imitation" and "modeling," considered analytically, appear to represent "special case" instances of such instructional control. Furthermore, this precise stimulus control (i.e. "attention") can be transferred from one group of stimuli or stimulus properties to another by simultaneous presentation of the two together followed by the gradual withdrawal (i.e. "fading") of the original stimulus.

The intimate and continuing association between discriminative environmental stimulus events and the occurrence of reinforcement endows at least some originally non-reinforcing stimuli with acquired reinforcing properties. These stimuli have come to be designated as "secondary" or "conditioned reinforcers" (e.g. fraternity pins, stock market quotations, etc.), to distinguish them from innate, primary, or unconditioned reinforcers which require no experience to be effective. Such conditioned reinforcers can, of course, be either "appetitive" (i.e. strengthening prior-occurring responses by their appearance) or "aversive" in which case their removal or postponement is reinforcing. The development or acquisition of conditioned reinforcing properties by a stimulus is usually a gradual process, as is the case with discriminative stimuli in general, and a common interpretive view of the process suggests that conditions reinforcers may owe their effectiveness to the fact that they function as discriminative stimuli for later members of a response chain which are maintained by the occurrence of reinforcers in their presence.

"Response chaining" refers to the observationally and experimentally verified occurrence of a composed series of performances joined together by environmental stimuli that act both as conditioned reinforcers and as discriminative stimuli. A chain (e.g. party-going) usually begins with the occurrence of a discriminative stimulus (e.g. phone invitation) in the presence of which an appropriate response (e.g. acceptance) is followed by a conditioned reinforcer (e.g. "Glad you can make it.") This conditioned reinforcer is also the discriminative stimulus occasion for succeeding responses (e.g., bathing, shaving, dressing, etc.) which in turn, is followed by another conditioned reinforcer (e.g. leaving the house, catching a cab, etc.), which is also a discriminative stimulus for the next response (e.g. joining the party), and so on. While it is doubtless true that the entirety of such chains is most often maintained by the terminal occurrence of potent environmental consequences (e.g. social interactions, food, sex, etc.), laboratory experiments have clearly demonstrated that the overlapping links in the chain (i.e. discriminative stimulus -- operant response -- conditioned reinforcer) are held together primarily by the dual (and demonstrably separable) discriminative and conditioned reinforcing function of environmental stimuli. The significance of this general chaining principle must, of course, be seen to reside in the fact that <u>virtually all behavioral interactions occur as chains of greater or lesser length, and that even performances usually treated as unitary phenomena can be usefully analyzed at various component levels (e.g. golf, bowling, tennis, etc.) for purposes of modification or proficiency enhancement.</u>

Perhaps the most important aspect of this complex analysis of environmental stimulus events in relation to behavioral interactions is the clear implication that some degree of independence can be gained from the factors limiting conditioned reinforcer potency by the formation of conditioned reinforcers based upon two or more primary reinforcers. Such conditioned stimulus events ("<u>generalized reinforcers</u>") gain potency from all the reinforcers on which they are based, and the most prominent operant performances in the human repertoire (e.g. verbal behavior) as well as the most valued stimulus consequences in the social environment (e.g. money) can be seen to share these broadly based discriminative and generalized conditioned reinforcing properties.

This necessarily abbreviated overview of experimentally-derived concepts and principles relevant to medicine in general and health- related behaviors in particular, has thus far maintained the traditionally accepted differentiation between active (operant) and reactive (respondent) behavioral interaction modes based principally upon procedural distinctions identified in the laboratory. The independent and distinctive features of these two coextensive processes are seldom apparent however, in the course of even detailed natural observation. In no investigative aspect of the behavioral universe is the complex interaction between these active and reactive modes more pronounced than in the experimental analysis of aversive control procedures represented (or misrepresented!) by the technical terms "punishment", "escape", "avoidance", and their "emotional" and "motivational" corollaries.

Empirical and theoretical accounts of those aspects of be- havioral medicine concerned with disordered performances frequently assign a central role to historical and contemporary environmental interactions involving aversive circumstances and conditions. Operationally characterized in terms of their behavioral effects, "aversive stimuli" are defined as environmental events which decrease the subsequent frequency of the operant responses they follow, on the one hand, and/or increase the subsequent frequency of operant responses which remove or postpone them. When an aversive stimulus follows an operant, and decreases the likelihood that such performances will recur, a "punishment condition" is defined. Punishment may be made contingent upon the occurrence of an operant which has never before been followed by a reinforcer, an operant currently being maintained by appetitive or aversive reinforcement, or an operant that is undergoing extinction. Under each condition, the short- and long- term effects of punishment will vary as a function of complex operant- respondent interactions, and both discriminative stimulus control and reinforcement schedule factors may operate to further influence the subsequent form and frequency of the performance.

An "escape" condition is defined when a response terminates an aversive stimulus <u>after</u> the stimulus has appeared. The interaction between operants and respondents is especially prominent in escape situations since the aversive stimulus usually elicits reflexive responses which eventually result in or accompany an operant performance followed by withdrawal of the aversive stimulus. Strong generalization effects appear during initial exposures to escape situations, but the gradual development of discriminative properties by the aversive stimulus narrows the performance, and very low intensities of the aversive stimulus may eventually <u>maintain</u> an operant escape performance requiring a much more intense aversive stimulus to <u>establish</u>. Reinforcement schedule effects similar in all essential respects to the appetitive conditions described above are observed when withdrawal of an aversive stimulus is the reinforcer. Extinction of an operant escape response occurs rapidly when presentation of the aversive stimulus is discontinued, or more slowly and erratically if the occurrence of the operant is no longer reinforced by withdrawal of the reoccurring aversive stimulus.

An "avoidance" condition is defined by the occurrence of an operant response which postpones an aversive stimulus. Avoidance performances may be established and maintained either in the presence or absence of an exteroceptive environmental event (i.e. "warning stimulus") which precedes the aversive stimulus. When an exteroceptive warning stimulus precedes the aversive stimulus, respondent conditioning effects operate to endow the warning stimulus with aversive properties, the termination of which following the operant avoidance response probably combines with the continued absence of the aversive stimulus to act as a reinforcer. The complexity of the avoidance process is suggested by the functionally simultaneous properties acquired by the conditioned aversive "warning" stimulus as 1) an eliciting environmental event for respondent behaviors, 2) a conditioned aversive reinforcer, withdrawal of which strengthens the operant avoidance performance effective in removing it, and 3) a discriminative stimulus which provides the occasion for the operant avoidance response to be followed by a reinforcer. In the absence of an exteroceptive warning stimulus, a temporal respondent conditioning process provides discriminitive cues, and the temporal stimulus correlated with the aversive environmental event acquires the same three simultaneous functions as an exteroceptive stimulus.

Such an analysis of aversive control emphasizes the simultaneous operation of active or operant and reactive or respondent conditioning processes in ongoing behavior segments. Whenever the conditioned stimulus in a respondent conditioning procedure is an appetitive or aversive reinforcer, operant conditioning occurs at the same time as respondent conditioning. Similarly, whenever the reinforcer in an operant procedure is an unconditioned stimulus, respondent conditioning proceeds at the same time as operant conditioning. Thus, insofar as the eliciting and reinforcing stimulus classes are composed of the same environmental events, operant and respondent processes are coextensive.

Relevant applications of these basic behavioral principles to clinical medicine in general and to the analysis of health-related behaviors in particular, have emerged in two major forms. In the first instance, procedures have been developed for active (rather than reactive) behavioral control of visceral, somatomotor, and central nervous system processes based upon the arrangement of explicit contingency relationships between specific antecedent physiological events on the one hand, and programmed environmental consequences on the other. It has been convincingly demonstrated, for example, that such behavioral "biofeedback" intervention can produce reliable bidirectional control over both increases and decreases in cardiac rate (DiCara & Miller, 1969; Engel & Gottlieb, 1970) and blood pressure (Pappas, DiCara & Miller, 1970; Benson, Herd, Morse, & Kelleher, 1969). Large magnitude and enduring elevations in heart rate (Harris, Gilliam, & Brady, 1976) and blood pressure (Harris, Findley, & Brady, 1971) have also been described in more chronic operant conditioning studies. Significantly, more recent studies (Harris, Findley, & Brady, 1973) have involved the application of operant "shaping" techniques with both amplitude and duration of blood pressure elevations systematically increased in small progressive steps to diastolic pressure 35 to 40 mmHg above pre-experimental resting levels.

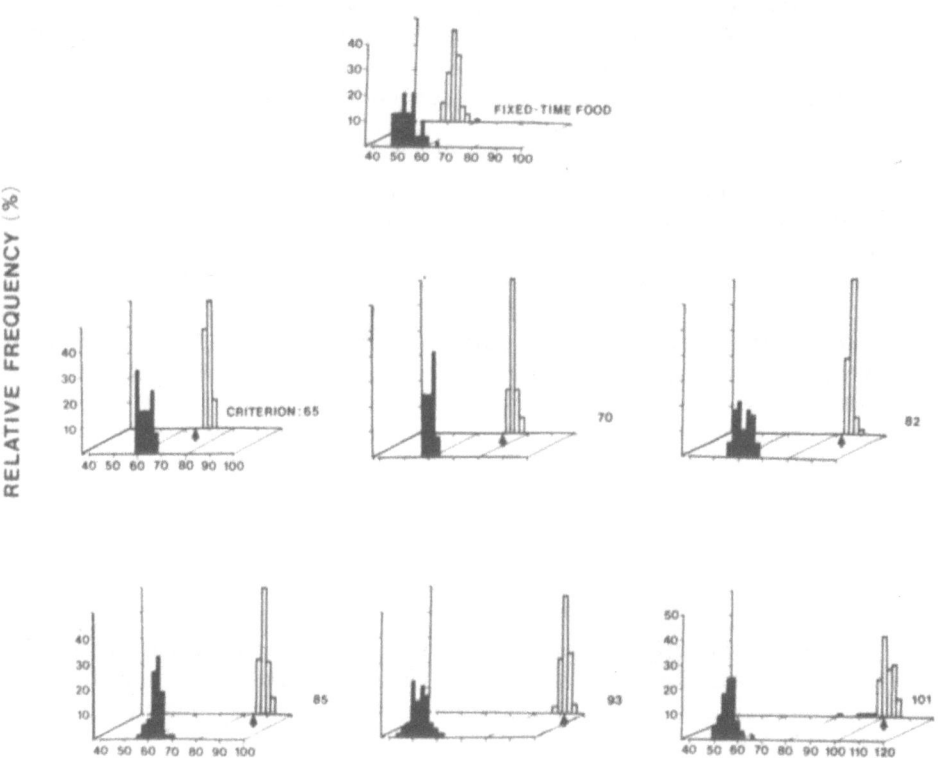

DIASTOLIC BLOOD PRESSURE (mmHg)

Figure 1.   Relative frequency distributions of 40-minute
average diastolic pressures for baboon "#82"
during a baseline condition (Fixed-Time food),
and at successively higher diastolic criteria
(columns, from right to left). Open bars
represents diastolic pressure levels from four
experimental sessions, while filled bars represent data from four associated post-session
periods. Arrows indicate the diastolic
criterion level at each stage of training,
with criterion values shown numerically to the
right of each graph.

Figure 1, for example, shows the relative frequency distributions of diastolic blood pressure from an experiment in which a baboon learned to increase and maintain blood pressure elevations in order to obtain food and avoid shock (Turkkan & Harris, 1981). The shaping procedure involved delivery of food pellets for accumulation of 600 sec of time above the diastolic pressure criterion level, and delivery of a single electric shock to the tail for accumulation of 240 sec of time below that criterion level. When the pressure level was above criteria, white light appeared on the animal's work panel, and when pressure was below criterion, a red light accompanied by 1000 Hg tone was presented. Experimental sessions began at noon each day, and ended at midnight. Criterion levels beginning at 65 mm Hz (i.e. pre-experimental baseline average diastolic pressure level) were progressively elevated at a rate approximating 2-3 mmHg per week. The systematic "shaping" of diastolic pressure elevations over a 10-12 week conditioning period, illustrated in Figure 1, compares the diastolic pressure levels recorded during sessions (open bars) with the levels recorded during the 12-hour intervals between sessions (filled bars) under baseline conditions (top segment) and during successive stages of conditioning. At the highest criterion (lower right segment), diastolic pressures were elevated above 100 mmHg in order to maintain a food-abundant environment throughout the 12-hour experimental session during which less than one shock per hour was delivered. And remarkably, there was absolutely no overlap between the distributions of pressure levels recorded at this highest criterion and those recorded during the baseline period.

While these observations clearly reflect the participation of an active behavioral process in the development and maintenance of cardiovascular functions traditionally considered under more reactive control, there is of course implied no claim to exclusivity. Multiple mechanisms, both behavioral and physiological, must be presumed operative in the mediation of such complex psychophysiological interactions. Certainly, insofar as the environmental stimulus events (both internal and external) involved in these processes have common functional properties (e.g. eliciting, reinforcing, discriminative), both operant and respondent conditioning, at the very least, can be considered coextensive.

The second significant development which has emerged in the context of these behavior analysis principles involves the application of contingency management procedures for the shaping, maintenance, and modification of health-related performances (e.g. food intake, exercise, medication compliance, etc.) in the interest of cardiovascular risk reduction (e.g. Squyres & Coates, 1981). Of particular relevance in this regard would seem to be the experimentally (and clinically) documented

effects of scheduling conditions, stimulus control, and
chaining, which determine under what circumstances and in
accordance with what behavioral requirements a valued commodity
service, or substance, can be obtained. The demonstrably potent
influence of these factors upon the strength and persistence of
behavior is worth emphasizing because these properties of a
performance frequently appear as the most baffling and
recalcitrant aspects of health risk-factor reduction (e.g.
smoking, overeating, etc.). Indeed, it is to the power of the
kind of environmental constraints imposed by scheduling and

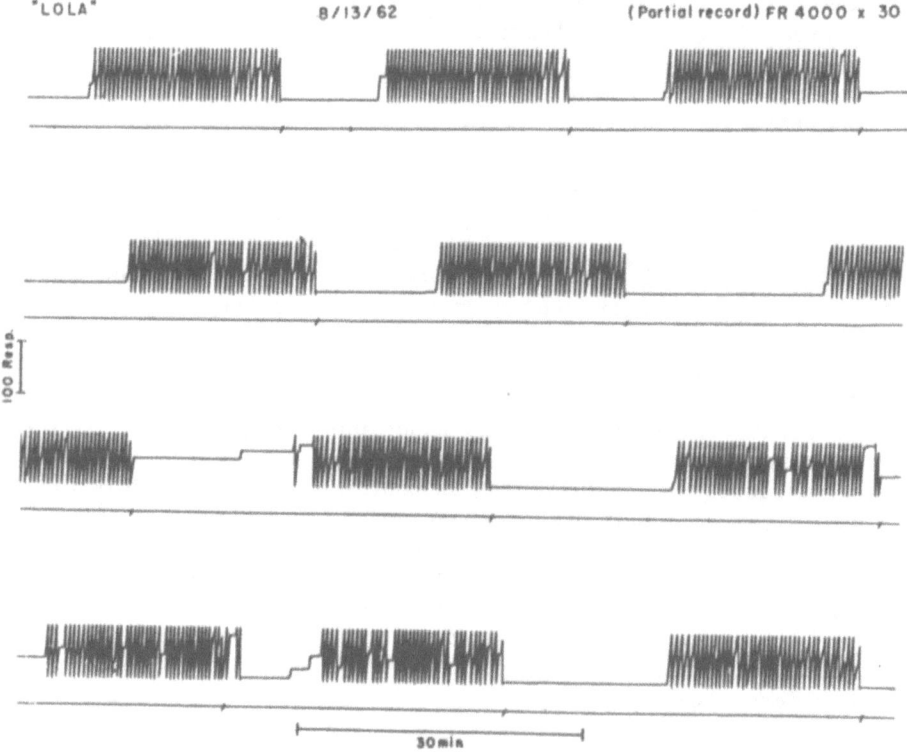

Figure 2. Cumulative record of responses (vertical
excursion of stepping pen -- reset after each 100 responses)
over time (horizontal baseline -- paper speed) for chimpanzee
"Lola" showing pause following each flash of light (conditioned
reinforcer) after a block of 4,000 responses toward the total
requirement of 120,000 for access to food.

stimulus control procedures to entrain performances of remarkable persistence that particular attention would seem appropriately directed.

Figure 2, for example, illustrates a typical segment of a cumulative record from an experiment in which a chimpanzee sustained performance on a ratio-schedule which required 120,000 responses on a heavy push-button manipulandum for access to food (Findley & Brady, 1965). After each 4,000 responses toward the total requirement, a brief flash of light was presented -- the same light that was illuminated continuously during food access once the total ratio was completed. Of particular interest is the pause which follows each flash of light after a block of 4,000 responses, illustrating the control acquired by this conditioned reinforcing stimulus event. Subsequent extension to a 250,000 response ratio and manipulations involving removal and reintroduction of the light flash after each 10,000 responses documented the critical interactions between rule-governance and stimulus control in the establishment and maintenance of such remarkably persistent performance repertoires. It seems important to recognize that while such unusual and extreme examples of schedule and stimulus conditions may appear to push the limits of adaptive functions, they are not tricks or circus acts. They do in fact represent the orderly and lawful operation of general relationships which are common to all behavioral interactions, including health-related performances, and appear to be of particular relevance to the excessive or abusive aspects of such performance.

REFERENCES

Bechterev, V. General Principles of reflexology. Translation: E. Murphy & W. Murphy. London: Hutchinson, 1932.
Benson, H., Herd, J.A., Morse, W.H. & Kelleher, R.T. Behavioral inductions of arterial hypertension and its reversal. American Journal of Physiology, 1969, 217, 30-34.
Deane, G.E. & Zeaman, D. Human heart rate during anxiety. Perception and Motor Skills, 1958, 8, 103-106.
DiCara, L.V. & Miller, N.E. Transfer of instrumentally learned heart rate changes from curarized to noncurarized state: Implications for a mediational hypothesis. Journal of Comparative and Physiological Psychology, 1969, 62(2, P.1), 159-162.
Dykman, R.A. & Gantt, W.H. Cariodvascular conditioning in dogs and in humans. In W.H.Gantt (ed.) Physiological Bases of Psychiatry. Springfield, Ill.: C.C. Thomas, 1958.
Engel, B.T. & Gottlieb, S.H. Differential operant conditioning of heart rate in the restrained monkey. Journal of Comparative and Physiological Psychology, 1970, 73(2), 217-225.

Findley, J.D. & Brady, J.V. Facilitation of large ration
    performance by use of conditioned reinforcement. Journal
    of Experimental Analysis of Behavior, 1965, 8, 125-129.
Harris, A.H., Gilliam, W.J. & Brady, J.V. Operant conditioning
    of large magnitude 12-hour heart rate elevations in the
    baboon. Pavlovian Journal of Biological Sciences, 1976, 11
    (2) 86-92.
Harris, A.H., Gilliam, W.J., Findley, J.D. & Brady, J.V.
    Instrumental conditioning of large magnitude daily 12-hour
    blood pressure elevations in the baboon. Science, 1973,
    183, 175.
Harris, A.H., Findley, J.D. & Brady, J.V. Instrumental
    conditioning of blood pressure elevations in the baboon..
    Conditional Reflex, 1971, 6(4), 215-226.
Honig,   W.D.   Operant   Behavior: Areas   of   Research   and
    Application. New York: Appleton-Century-Crofts, 1966.
Honig, W.K. & Staddon, J.E.R. Handbook of Operant Behavior.
    Englewood Cliffs, N.J.: Prentice Hall, Inc., 1976.
Pappas, B.A., DiCara, L.V. & Miller, N.E. Learning of blood
    pressure responses in the noncurarized rat: Transfer to
    the curarized state. Physiology and Behavior, 1970, 5(9):
    1029-1032.
Pavlov, I.P. Conditioned Reflexes. Translation: G. V. Anrep.
    London: Oxford University Press, 1927.
Pavlov, I.P. Lectures on Conditioned Reflexes. Translation:
    W.H. Gantt. New York: International Press, 1928.
Rachlin, H. Introduction to Modern Behaviorism. San Francisco:
    W. H. Freeman & Co., 1970.
Razran, G. The observable unconscious and the inferable
    conscious in current Soviet psychophysiology: Interoception
    conditioning, semantic conditioning and the orienting
    reflex. Psychological Review, 1961, 68, 81-147.
Sherrington, C.S. The Integrative Action of the Nervous System.
    1947 edn). England: Cambridge University Press, 1906.
Squyres, W.D. & Coates, T.J. A self-management approach to
    cardiovascular risk reduction: Management of the self and
    the environment. Health Education Monographs, 1981 (in
    press).
Thorndike, E.L. Animal intelligence – an experimental study of
    the   associative   processes   in   animals.   Psychology
    Monograph, 1898, 2, 1-106 (Mongr. suppl. whole no. 8).
Turkkan, J.S. & Harris, A.H. Differentiation of blood pressure
    elevations in the baboon using a shaping procedure.
    Behavioral Analysis Letters, 1981, 1, 97-106.

ACKNOWLEDGEMENTS

Research reported in this manuscript supported in part by NIDA
Grant DA00018, NHLBI Grants HL17958, HL17970, and NIMH Grant
MH15330.

BEHAVIORAL MEDICINE AND PSYCHIATRY: A DIALOGUE

Charles Mertens de Wilmars

Universities of Louvain and Harvard

Brussels, Belgium

Since World War II, psychiatry was spurred in many directions by advances in psychology, physiology, ethology and phenomenology, while at the same time, it tried to meet an ever growing consumer demand. Consequences were twofold: a lack of systematization or integration and too much concern for fashion. Both trends made each psychiatrist believe he possessed the key for happiness and made the public at large believe that there are as many "psychiatries" as psychiatrists. Over-involved in the maintenance of their status, psychiatrists lost track of the fact that the "phenomene humain" is a composite. Psychiatrists confounded learning with functioning; they also failed to merge neurological, mental, behavioral and environmental aspects of these processes. Some of them denied the role of mental representations or the link between neurological and behavioral events. Others ignored the retroactive character of behavior; any behavior originates in an object but returns to and modifies that object.

Notwithstanding this confusion, one can find two major movements in the recent development of psychiatry. The first deals with genetically determined or automated neurotransmission; it concerns itself with functioning as the "encoding" of neurological pathways. Pharmacotherapy, electroconvulsive shock therapy or surgery are its therapeutic interventions. The second movement deals with learning. Three major learning processes have so far been described: imprinting, conditioning and motivation. Psychotherapy and behavioral treatments are its interventions.

453

A Nobel prize in medicine (Lorenz, 1973) honored the discovery of imprinting. Imprinting is not restricted to after-birth; it is a life-long, permanent procedure by which we shape our reactions to that of a model without any mental representation. Imprinting coincides (I believe) with einfuhlung or empathy and triggers identification. Identification relays imprinting; it does imply a mental evocation. Identification is a continual sequence of projections and introjections through which we match object and subject, build a neurological print of the outside world (something we can evoke) and shape our (neurological and mental) ways-of-being to a congruent behavior. Identification is probably the conditio-sine-qua-non of any other form of learning.

Conditioning provides for adaptation through predictability; it teaches us to use contingent cues and probability. Conditioning leads to (a) signalling, (b) instrumentation, and (c) cognition. Exteroceptive or interoceptive perceptions become the signals of forthcoming perceptions. Sensori-motor or neuro-vegetative actions become the means to other actions. And stimulations of the verbal-motor-cortex substitute for perceptions or actions. This substitution brings emotions, attitudes, as well as logical, abstract or reflexive thinking into operation. From a therapeutic point of view, conditioning operates through repetitive association and reinforcement; i.e. behavior modification.

Motivation is a third aspect of learning. It provides for better adaptation through evocation, and, henceforth, creative thinking. Motivation spurs the creation of hypotheses, it liberates the self from environmental contingencies and opens the way to self-analysis. Any emotion, attitude, thought, imagination or fantasy may be evoked; and they may be evoked outside their object or the need for it. Therapy on this level is based on the analysis of psychodynamic events, triggering of resistance to remodeling. Psychotherapy looks for conflicting trends or traumatic fantasies which impair functioning. It builds a transference relationship allowing for the revival -- without overwhelming anxiety -- of the traumatic event or fantasy. Psychotherapy has very little to do with consciousness, insight, or symbolism. Insight reinforces already acquired neurological pathways; it does not create new pathways nor does it overcome the inhibitions of certain pathways.

All the psychoneurological processes provided in this taxonomy are closely interdependent. They do not supersede -- but superpose one another. We condition imprintings and motivate conditionings. They are the subsystems of one psychosomatic organization which performs three basic tasks: (a) to match object and subject, (b) to evoke the past and the future, and (c) to seek pleasure or avoid displeasure. Learning integrates all psychoneurological processes in order to keep the organization coherent and congruent.

THE ROLE OF BEHAVIOR MODIFICATION IN PSYCHIATRY - THE CORONARY EXAMPLE

The role of psychological variables in cardiovascular disorders is now widely accepted. Many psychological indices link, however partially, with bioclinical risk factors. The importance of the psychological risk factors is identical to -- although independent from -- that of the bioclinical risk factors. An extended literature supports this view. Moreover, a growing number of researchers advocate the existence of a pathognomonic cluster of psychological risk factors. We know by now that: (a) cardiovascular ailments link with several patterns of overt behaviors (Type A, smoking, etc.); (b) such patterns are never pathognomonic of cardiovascular disorders; they may predict other diseases as well; (c) the factorial components of these psychological patterns correlate with different biological risk factors. However, while it is possible to analyze these patterns in terms of overt observable behavior, they can also be conceptualized in terms of a traditional psychiatric syndrome known as obsessive-compulsive neurosis, a personality structure known as dependent-aggressive type or a psychodynamic character labelled as oral fixation. Most of the patients belonging to these groups fear passivity because they fear their own aggressiveness. This explains why they abandon their medication shortly after its prescription, why they benefit so little from psychotherapy and why relaxation scares them to the utmost.

Based on this more comprehensive perspective, it should be evident that we have to approach the patient on all levels through a therapeutic program where each intervention fulfills a well defined and specific role. Experience has taught us to adopt the following sequence of operations. Any recovery or preventive program starts with a rational and modeling approach of the patient's occupational prospectives. From there on we work backwards. From occupational therapy we move to physical therapy and very gradually to relaxation and body awareness (Jacobson, Schultz). At that level the patient is part of a team and relates more closely to one member of the medical staff (most of the time the physical therapist or the dietician). Relaxation and social integration allow for revival of conflicting trends. At that moment in the program, the patient becomes more anxious and, therefore, more motivated for psychological management of his disease. He utters a demand, not only for help but also for coaching and support in his trial to "take over" from the staff and to care for himself in a realistic way. Maturation and interdependency surface. At this time, we can approach the patient on two levels: biofeedback or any other form of conditioning to modify behavior, and, shortly thereafter psychotherapy (to deal with motivational issues). Medication, diet, smoking and other behaviors are discussed with the patient as therapy progresses. This multiphasic and multifactorial approach calls for a well-trained, multi-disciplinary team where the private coach and tutor play a major role.

CONCLUSION

Behavioral medicine today comprises a large array of techniques and many more indications. In order to improve its efficacy, researchers should concentrate on the understanding of its action and clinicians should locate each of these techniques in a sequence of interventions. Sticking only to a narrow behavioral perspective, which denies the value of pharmaceutical, psychodynamic or sociodynamic interventions, can only serve to retard progress.

REFERENCES

Segers, M. J.& Mertens, C. Preventive behavior and awareness of myocardial infarction: A factorial definition of anxiety. Journal of Psychosomatic Research, 1977, 21, 213-223.

Segers, M.J. & Mertens, C. Personality aspects of CHD related behavior. Journal of Psychosomatic Research, 1977, 21, 79-85.

Van Imschoot, K., Liesse, M., Van Den Abbeele, K.G., Lauwers, P. & Mertens, C. Evolution temporelle des variables neuro-endocrines en reponse au stress. Acta Psychiatry Belgium, 1980, 80, 45-60.

Liesse, M., Van Imschoot, K., Mertens, C., Lauwers, P. & Van Den Abbeele, K.G. Evolution temporelle des lipides en response au stress. Acta Psychiatry Belgium, 1980, 80, 61-78.

Mertens, C. Psychologie Medicale, Bruxelles, De Boeck, 1977.

Rosenman, R.H. & Chesney, M.A. The relationship of Type A behavior pattern to coronary heart disease. Activ. nerv. sup. (Praha), 1980, 22 (1), 1-45.